Wired for Words

Wired for Words

The Neural Architecture of Language

Gregory Hickok

The MIT Press
Cambridge, Massachusetts
London, England

The MIT Press
Massachusetts Institute of Technology
77 Massachusetts Avenue, Cambridge, MA 02139
mitpress.mit.edu

The MIT Press would like to thank the anonymous peer reviewers who provided comments on drafts of this book. The generous work of academic experts is essential for establishing the authority and quality of our publications. We acknowledge with gratitude the contributions of these otherwise uncredited readers.

This book was set in Stone Serif and Stone Sans by Westchester Publishing Services. Printed and bound in the United States of America.

Library of Congress Cataloging-in-Publication Data

Names: Hickok, Gregory author
Title: Wired for words : the neural architecture of language / Gregory Hickok.
Description: Cambridge, Massachusetts : The MIT Press, [2025] |
Includes bibliographical references and index.
Identifiers: LCCN 2025002138 (print) | LCCN 2025002139 (ebook) |
ISBN 9780262553414 paperback | ISBN 9780262384827 epub |
ISBN 9780262384834 pdf
Subjects: LCSH: Neurolinguistics | Language acquisition—Physiological aspects
Classification: LCC QP399 .H53 2025 (print) | LCC QP399 (ebook) |
DDC 612.8/2336—dc23/eng/20250625
LC record available at https://lccn.loc.gov/2025002138
LC ebook record available at https://lccn.loc.gov/2025002139

10 9 8 7 6 5 4 3 2 1

EU Authorised Representative: Easy Access System Europe, Mustamäe tee 50, 10621 Tallinn, Estonia | Email: gpsr.requests@easproject.com

To my parents, Merle and Jean, for a lifetime of love and support

Contents

Preface

This book is twenty-five years in the making. It started as a textbook project based on my "Language and the Brain" course at the University of California, Irvine. I proposed it to the MIT Press, which agreed to publish it. This was around 2000. But this was also the time when David Poeppel and I were starting to develop our dual stream model of speech processing, which made several new predictions that needed testing before I could write what I hoped would be a definitive text. The field was moving too fast to try to take a snapshot summary of the state of the art. After a grant cycle or two of research and theory development, I was ready to start writing. I completed five chapters before I got wind of David Kemmerer's forthcoming textbook *The Cognitive Neuroscience of Language* (Kemmerer, 2015). I was asked to review and comment on a couple of his chapters and found the text to be so good that it made mine redundant. Poeppel and I commented to each other that Kemmerer did a better job than we did at describing our model.

So I turned my attention to a "side project," writing first a journal article (Hickok, 2009) and then a book-length critique of mirror neuron theory (Hickok, 2014b). Realizing that I actually *could* write a book, and feeling guilty that I hadn't delivered on my contract with the MIT Press, I had the idea of writing a monograph rather than a textbook, laying out my own perspective on the neural organization of language. MIT Press editor Philip Laughlin graciously agreed to the shift in focus, so I started again from a blank slate. It took several more years as grant renewals came due, new theoretical ideas needed fleshing out, and general progress in the field needed digesting. This led to several reorganizations, a few complete chapter rewrites, and extensive editing—in short, more delays.

The end product kind of shaped itself, as most of my theoretical papers and grant proposal tend to do. I start with an idea, a general sketch, and begin writing. Then I just follow the logic. Dead ends are where rewrites come in; shifts in focus require reorganizations; and rereads drive edits. I then rinse and repeat until I force myself to stop, usually because of a grant deadline, and click "Send," typos and flubs be damned. I hope that you find the result reasonably understandable, useful, occasionally insightful, or maybe something to build on or at least think about.

As I observed the emerging shape of the volume, I noticed the emphasis on architecture. It dominated so much that I had to put it in the title. Some may consider this slightly old-fashioned, and in some ways, it is. The classical models of the nineteenth century were grounded in architecture. But a lot of old ideas are good ideas, despite claims to the contrary (Tremblay & Dick, 2016), and we shouldn't be too quick to dismiss them entirely because they lack a certain modern panache or are clearly wrong in some respects. Already in the early stages of my career, researchers were dissing functional imaging localizationists, calling them "neophrenologists," arguing that knowing *where* something happens tells you nothing about *how* it works, and so on. I recall even my closest theoretical collaborator, David Poeppel, arguing (reasonably) that what's lacking in cognitive neuroscience is the dimension of "time" in our models.

It's true, of course, that architecture isn't the whole story—dynamics matter, but structure is a critical part of the system and perhaps key to understanding how it works. Consider that the dynamics of neurons throughout the brain are quite similar from one system to the next in the sense that a unified model of single neuron activity can serve as the building blocks of network function (Lin, Chen, Lin, & Huang, 2022). Appealing to single neuron dynamics alone, therefore, isn't enough to explain how a given system works. What matters is the circuitry, the connectivity, and the architecture of the neurons in a system.

Another theme that I noticed as the book took shape is the tendency to frame the issues in a historical context. Some readers don't necessarily want this kind of framing, preferring instead to just get the facts of the modern understanding without mulling over the old ideas and results. I find, however, that the modern understanding is shaped by historical trends, accidents, and even wrong turns in many cases. For me, seeing the historical development of an idea helps me understand our modern biases and proclivities, as well as unnoticed gaps in evidence or missed clues. The result may be a more informed, balanced perspective.

In addition to drawing on a wide swath of varied literature to develop the ideas and perspectives presented in the following pages, I discuss a range of experiments and theories that came out of my own lab. For this, I am indebted to my students, who did most of the hard work. While their contributions are credited in the volume, I would like to thank each of them here as well, because my career would be a shadow of what it is without my former students and postdocs, Nicole Gage, Brad Buchsbaum, Colin Humphries, Kai Okada, Judy Pa, Kevin Smith, Corianne Rogalsky, Stephen Wilson, Kenny Vaden, Paul Fillmore, Diego Almeida, Lisette Isenberg, Jon Venezia, Dale Maddox, Feng Rong, William Matchin, Tepring Piquado, Jack Payne, Grant Walker, Vatche Baboyan, Ryan Stokes, Joselyn Ho, Solena Mednicoff, Daniel Stehr, Meriopi Topalidou, and Alex Teghipco, some of whom are still collaborators of mine today. Special thanks to Corianne Rogalsky, Alex Teghipco, Jon Venezia, and Grant Walker for contributions to this book specifically and for their postgraduate continued collaboration.

I would also like to thank my longtime collaborators at the University of California, Irvine, Kourosh Saberi and Haleh Farahbod, my many non–University of California, Irvine research collaborators over the years (David Poeppel, Tracy Love, Julius Fridriksson, Argye Hillis, and Nitin Tandon deserve special mention), and my mentors, Edgar Zurif, Jane Grimshaw, Steve Pinker, Dave Swinney, Ursula Bellugi, and Ed Klima, as well as many individuals who read parts of the manuscript or participated in related discussions that provided valuable feedback. These include Laurel Buxbaum, Max Coltheart, Gary Dell, Jon Kaas, Alex Marantz, William Matchin, David Poeppel, Oren Poliva, Kourosh Saberi, Jon Venezia, and Jon Winawer.

Besides the academic contributions, I would like to acknowledge the staff at the University of California, Irvine, without whom I would not have had the time to think about or work on any of these projects. Clara Schultheiss deserves special mention for her long-term administrative support.

I also need to acknowledge the source of the magnetic resonance imaging (MRI) graphics in this book, which were generated using MicroGL software (www.nitrc.org) developed by my collaborator, Chris Rorden.

The research from my own lab, as well as the preparation of this book, have been supported over the years by several grants from the National Institutes of Health (NIH) and National Science Foundation (NSF). I have been fortunate to work with many talented co-investigators on these grants, without whom this work could not have been accomplished. These include

Steven Anderson, Roozbeh Behroozmand, Ursula Bellugi, Jeff Binder, Leo Bonilha, Nathan Crone, Hanna Damasio, Dirk den Ouden, Julius Fridriksson, Argye Hillis, Tracy Love, Chris Rorden, Dave Swinney, and Nitin Tandon. Special thanks to my co-investigators, postdocs, students, and staff at the NIH-supported Center for the Study of Aphasia Recovery (CSTAR), which has provided the bulk of my research support over the last decade.

Finally, I am indebted to my family for their continued and unwavering love, support, and anchoring of the important part of my work-life balance. Thank you, Krissy, Taylor, and Ally.

I Preliminaries

1 What Is Language?

> Language! You keep using that word. I do not think it means what you think it means.
>
> —Paraphrased from *The Princess Bride*

This is a book about the biology of language, my primary area of research. Oddly enough, I never much liked the topic during my college years. How I came to study language despite a distinct lack of academic interest is worth a couple of words because I think that it illustrates an important point: language is more than what most people think it is. Let's start with the story of how I discovered the hidden treasures of language science.

Like any school kid, I was compelled by others to study my own language year after year. It was not fun learning how to underline predicates, label parts of speech, fix punctuation, or do any of the other tedious and, to my mind, nonsensical tasks. My dislike for all things language carried on into my college years, where I found myself utterly bored during the language parts of my cognitive and developmental psychology courses. So what if language has a phonology, a morphology, and a syntax? Not surprisingly, I never took a linguistics course as an undergraduate.

What interested me was biology, the brain in particular, and its relation to the mind. I was fascinated with neuropsychological syndromes: amnesia, schizophrenia, prosopagnosia, Tourette's syndrome, split brain cases, and so on. Like many of my generation, Oliver Sacks's book *The Man Who Mistook His Wife for a Hat* made a lasting impression[1] and enticed me to learn more. Still an undergraduate, I started schooling myself by reading some now-classic texts like Kolb and Wishaw's *Fundamentals of Human Neuropsychology* and Heilman and Valenstein's *Clinical Neuropsychology* (both from 1985). I even trekked to San Diego to meet with a mogul in the field, Larry Squire, to soak up some sage advice on one of my then favorite topics, the neuropsychology of memory. Perhaps predictably, though, aphasia, an acquired disorder of language, was not among the neuropsychological disorders that attracted my attention. I recall skimming a chapter describing the various aphasia types—Broca's, Wernicke's, conduction, and so on—and begrudging the task of having to learn about them as part of my future training as a neuropsychologist. Nonetheless, through the usual twists of fate that carve out academic career paths, I ended up a grad student in the lab of a famous aphasiologist, Edgar Zurif. I was to become a language scientist.

This meant taking some linguistics courses. My first was a syntax course taught by linguist Jane Grimshaw. What I learned really surprised me. We did not spend time cataloging nouns and verbs and prepositions in this or that sentence; instead, we learned about the hidden machinery that generated all those objects in the first place, and about the methods for studying it. Language, it turned out, was more interesting than I thought. It was not, I learned, a large and nearly arbitrary collection of rules for combining sounds to make words and words to make sentences, but rather a relatively small set of deep, interacting principles that give rise to all the possibilities for combining the elements of language. Even more profound, variants of those principles seemed to explain much of the variation that is found from one language to the next. Studying language from this scientific perspective is not about cataloguing the endless variation. Instead, it is the search for a core set of deeper principles that explains that variation in a simple way, much as physics aims to do for the physical world or evolutionary biology does for the biological world. The implication is profound. If the core of language is not what we see on the surface, but some underlying principles shared by all humans, then we are dealing with a biological system—a human trait.

Another thing I learned was that theoretical linguistics is an empirical science, just like biology, cognitive psychology, and physics. What little I knew about linguistics led me to believe that it was some form of logic or philosophy in which theoreticians just think really hard about the formal systems that might describe the patterns that one finds in language (not that there's anything wrong with that!) and never do any data collection. But in fact I learned how practicing linguists examine data to develop theories that make predictions, which are then tested with more data. Yes, you end up with formal descriptions of linguistic patterns—just as you end up with formal mathematical descriptions in physics—but you get there through the normal theory development and experimental testing that drive any science.

The only difference between linguistics and the fields that I was more familiar with (say, cognitive psychology) was that most linguistic data is really easy to collect and the results are easy to replicate: you ask yourself and a couple of friends whether you can say this or that form of a word or sentence and your experiment is done! A single linguistics paper might have a dozen or more experiments testing different predictions of a new hypothesis. Then, every time someone reads the manuscript, each of those experiments is effectively replicated as the reader evaluates his or her own judgment about the data. In light of the recent replicability crisis in psychology and some other fields, where there is much discussion of the reliability of data, it is noteworthy that in linguistics, there has traditionally been very little debate over the data. This is no surprise, given the nature of the experiments and their repeated replication. That is not to say that there aren't debates over data, and indeed nowadays these debates include some people arguing that linguists need to do traditional psychological experiments with quantitative analyses (Gibson & Fedorenko, 2013; Sprouse & Almeida, 2013). I agree that as the theories develop in sophistication and the effects grow more subtle, quantitative methods have an important place in linguistics. But this just reinforces my main point here: linguistics is an empirical science.

By the end of my first year of grad school, my understanding of language science was completely reframed. It is not linguistic stamp collecting, as I thought, but a rigorous experimental science aimed at elucidating a deep biological human trait. Moreover, I had started to appreciate that language is a trait that already had a long history as a critical testing ground for fundamental questions in human neural and psychological sciences, and during my graduate career in the late 1980s and early1990s, it was showing hints of continuing to do so.

Language as a Testing Ground

One of the big neurological questions in the nineteenth century was whether the brain's cerebral cortex is functionally subdivided into separate mental organs or whether it is equipotential. Franz Joseph Gall (1758–1828) championed the functional neuroarchipelago view with his theory of "organology," as he called it (van Wyth, 2002), which later came to be known as *phrenology*. One of the mental organs localized in Gall's scheme was language. In fact, he later recalled that the inspiration for his organology theory came from a childhood observation that people with exceptional verbal memory had particularly bulgy eyes: "If [verbal] memory manifests itself by an external character," Gall (1835, p. 8) wrote, "why should not the other faculties have their characters outwardly visible? It was this which gave the first impulse to my researches, and which was the occasion of all my discoveries."

Language was an ideal test case for this theory. It was already known in Gall's time that language deficits could occur following brain injury (Benton, 1991), and Gall himself described such cases in support of his mental organ claims. He even distinguished two types of language-related mental organs, both in the frontal lobes: one for the memory of words and the other for the ability to articulate speech (Gall, 1835). Given that aphasias, as we call them today, occurred relatively frequently, and given that many of Gall's other mental faculties were rather intractable as topics for neurological investigation—*constructiveness, comparative sagacity, metaphysical depth of thought, wit,* to name just a few—it is no surprise that language emerged as a centerpiece of the debate. Although many scientists of the day reported, debated, and even wagered on cases of language disturbance and their neuroanatomical source (Hunt, 1869), it was Paul Broca (1824–1880) who gets credit for settling it in favor of localization of function with his famous cases, Leborgne and Lelong, about whom he reported loss of the ability to speak, with corresponding damage to the inferior posterior portion of the left frontal lobe (Dronkers et al., 2007). Language, then, played a key role in settling the big neurological question of the nineteenth century.

Language was also a major catalyst in the cognitive revolution of the 1950s and 1960s, which opened the mind to detailed investigation after decades of behaviorism's obsession with stimulus-response observables. Among the earliest volleys of the revolution was Karl Lashley's (1890–1958) paper *The Problem of Serial Order in Behavior,* delivered at a meeting at Cal Tech in 1948 and published later (Lashley, 1951). Lashley leaned heavily on language as a test case, proving that human behavior cannot be explained simply in terms of simple stimulus-response associative chains. To do so, he presented a range of examples, including Spoonerisms such as "Let us always remember that waste makes haste" and "Our queer old dean" for "our dear old queen." These sorts of errors demonstrate unequivocally that speech is planned internally prior to its overt expression; otherwise, an element from later in the utterance couldn't appear prior to its proper position. Lashley's conclusion from his discussion of language is worth an extended quote:

> I have devoted so much time to discussion of the problem of syntax, not only because language is one of the most important products of human cerebral action, but also because the problems raised by the organization of language seem to me to be characteristic of almost all other cerebral activity. There is a series of hierarchies of organization; the order of vocal movements in pronouncing the word, the order of words in the sentence, the order of sentences in the paragraph, the rational order of paragraphs in a discourse. Not only speech, but all skilled acts seem to involve the same problems of serial ordering, even down to the temporal coordination of muscular contractions in such a movement as reaching and grasping. Analysis of the nervous mechanisms underlying order in the more primitive acts may contribute ultimately to the solution even of the physiology of logic. (Lashley, 1951, pp. 121–122)

To boil this argument down, coordination of sequences of behavior, which underlies much of human cognition, requires hierarchical plans organized by the brain. Language is the prototypical example of this.

Lashley's arguments provided a strong linguistic case against behaviorism, but few details regarding what might replace it. A few years later, a young linguist filled the gap. Noam Chomsky showed that the particular sequences found in a given language could be explained quite well using a hierarchical phrase structure and a set of transformations that apply to them. He also provided more detailed and rigorous arguments against associative chain accounts and specific behaviorist attempts to explain language (Chomsky, 1956, 1959). His work showed that cognitivist approaches can prevail where behaviorist approaches fail, and together with discoveries in other domains of psychology and computer science, resulted in the cognitive revolution. Language, then, played a key role in one of the biggest psychological debates of the twentieth century.

Language's track record as a theoretical test bed continued into the late twentieth and early twenty-first centuries. In the 1980s, cognitive scientists debated whether mental operations were powered by symbolic rule systems or parallel distributed processing associative machines (Rumelhart & McClelland, 1986b). A subsystem of language, the past tense, played a major role in the theoretical and empirical ping-pong match (Pinker, 1991; Pinker & Prince, 1988; Pinker & Ullman, 2002; Rumelhart & McClelland, 1986a), and it is an issue that I dabbled in (Ullman et al., 1997).

More recently, another idea has taken hold in psychology and neuroscience—namely, that so-called mirror neurons, which were discovered in macaque monkeys in the 1990s (di Pellegrino et al., 1992; Gallese et al., 1996), explain a range of human cognitive capacities ranging from understanding actions to language, empathy, theory of mind, and beyond (Rizzolatti & Craighero, 2004). This hypothesis has been tested in a variety of domains, but some of the strongest and clearest results have come from language science, which makes a strong case against the mirror neuron claim (Hickok, 2014b).

I think there are a couple of reasons why language turns up as a central test case in debate after debate regarding the nature of the human mind. One is obvious: we are the only species that uses spoken language, and it's complex enough that it sets a lower bound on what kind of computations the mind must be able to carry out. For example, as Lashley first pointed out, association doesn't cut it. Another reason is that language contains bits and pieces that represent many of the cognitive and neural processes that we seek to understand generally. Language has perception, motor control, working memory, long-term memory, attention, categorization, concept representation, multisensory integration, sequencing, combinatorial processes, hierarchical organization, and more. If we can understand how language is created by the brain, we'll have all but solved the "How does the brain work to create the mind?" problem.

Language: Not So Easy to Do

Every typically developing human child will acquire any human language they are exposed to and do it automatically and effortlessly. Given the same opportunity, other intelligent species—chimps, bonobos, or gorillas, for instance—won't, and indeed can't. A survey of forty years of research on language capacities in nonhuman apes sums it up nicely:

> Human and ape language users are clearly distinct. Although apes are able to utilize gestures, plastic chips, and keyboards as symbols, the most generous description of the productive vocabularies tops out at 500 or so. By the time humans graduate high school, vocabulary estimates are usually about 50,000. Although apes may be able to understand some word-ordering rules at the level of a two-and-a-half-year-old child, by the time a child is four, they can understand much more complex sentences than can be tested with a chimpanzee. Chimpanzees seem to

> order their own sequences according to semantic categories, but again, four-year-old children use complex syntax that no ape has yet been shown to master. (Lyn, 2012, pp. 373–374)

Similar, if less formal, experiments are carried out every day on an even more intelligent critter, adult *Homo sapiens*, who have substantially more difficulty learning a new language than does a demonstrably less intelligent two-to-four-year-old child (Hartshorne, Tenenbaum, & Pinker, 2018; Johnson & Newport, 1989). This tells us two things. One is that raw general intelligence isn't the basis of our linguistic ability. There must be something in the neural wiring of human children that enables language. The other is that learning a language is a complicated thing; it must be, if the most intelligent adult animals on the planet have a hard time doing it, even after having done it as children.

As a postdoctoral student at the Massachusetts Institute of Technology (MIT), I had the opportunity to work with Susan Curtiss at the University of California, Los Angeles (UCLA) to collect some language data from a unique individual called Chelsea (Curtiss, 2014). Chelsea was born with a severe hearing impairment and raised in a warm and loving family, but she never had access to sign language and never developed any kind of basic "home sign" system. She was therefore raised without any appreciable language exposure. It wasn't until she was thirty-two years old that it was discovered that her hearing was correctable to near-normal levels. Her nonverbal IQ was measured on formal testing to be between 77 and 89, and her mental age equivalent between nine and ten years—not bad from a language acquisition perspective. This was in 1980, and Chelsea received speech and language training, as well as everyday language immersion, for about thirty years. We know the range of proficiencies in learning a *second* language that a typical adult can achieve, but what about acquiring a *first* language as an adult?

Decades of study by Curtiss and others revealed that Chelsea was able to learn a reasonable vocabulary, estimated in 1992 at approximately 500 words, but she never acquired even the rudiments of syntax. Her sentences were little more than lists of meaningful words. Here's some examples:

Penny dollar man girl

The they are is hugging the dinner

The girl is with the girl get wins

Morning up outside the mushroom

Cindy baby stomach hospital

Ride tree bush splash

In contrast, Chelsea has good number sense. She understands number quantity, can count as high as she's asked to, keeps and balances a checkbook, and can add, subtract, multiply, and divide.

Chelsea's case reinforces the two points made earlier about language—more is needed beyond some basic level of intelligence, and it is more complicated than it seems. The case also bolsters the argument that there is a critical or sensitive period for language acquisition, a maturational time window during which language can optimally be acquired (Curtiss, 2014; Hartshorne et al., 2018; Johnson & Newport, 1989; Senghas et al., 2004). Without the benefit of having learned a language during the normal time window, Chelsea was even more limited in her abilities than a typical adult trying to learn a second language.

Once acquired, the task of processing language remains a nontrivial computation, as you are surely aware of from trying to interact with automated speech recognition systems. Speech recognition would be easy if each phoneme or speech sound mapped uniquely onto a particular acoustic pattern. But it was discovered decades ago that it's way more complicated than that (Liberman, 1957). A given sound (say, /d/ as in "dog") is acoustically quite different depending on the sound that follows it and varies quite a bit depending on who is saying it, how fast the person says it, whether it is enunciated clearly, in conversational speech, with food in the mouth, or with a cold. Even in a single word, you can't cut up the sounds to isolate the "d," the "o," and the "g" in "dog." The sounds blend together at the boundaries due to a process called *coarticulation,* in which we preshape our vocal tract for the next sound while articulating the present one. You can hear the consequences of this by saying the words "key" and "koala" slowly and listening to the first /k/ sound; they are different, reflecting the different lip positions necessary for forming the following vowel.

There are other difficult problems our language system encounters when trying to extract meaningful linguistic bits from a speech stream. One is identifying which parts of sound go with which linguistic unit, a process called *segmentation*. The problem is illustrated by *oronyms,* utterances that have ambiguous segmentations. Say the following out loud, moderately fast:

The stuffy nose—The stuff he knows
Some others I know—Some mothers I know
S'cuze me while I kiss the sky—S'cuze me while I kiss this guy

Our brains have solved the segmentation problem so well that it leads to an illusion: we hear speech in our native tongue as presegmented into separate words as if there are little acoustic gaps between each one, similar to the gaps between words on this page. It's not the case. Acoustic gaps occasionally occur in streams of connected speech, but they are not necessarily between words. In connected speech, energy gaps occur mostly around unvoiced stop consonants like /p/, /t/, and /k/, which involve a complete interruption of airflow by closing off the vocal tract at different points while at the same time relaxing the larynx for a couple dozen milliseconds so that it doesn't vibrate and make a sound. There are plenty of modulations of energy that produce relative peaks and valleys in the acoustic signal, but these mostly track syllables, not words. This is why foreign languages sound like the words all run together. They do! And that is true for your language too.

Language remains difficult to compute even after speech sounds and word boundaries are identified. Words are often ambiguous ("plant," "arm," "watch"), used metaphorically ("the brass killed the project"), or idiomatically ("she gave him the cold shoulder"), and sentences often add to the interpretive possibilities with structural ambiguities ("enraged cow injures farmer with ax"). Layer on stress patterns and prosody, and the same sentence can be a statement, an imperative, a question, or a sarcastic negation of the literal meaning of an utterance.

All these problems occur under perfectly articulated laboratory conditions. But consider wild-type language, where speech is littered with false starts, interruptions, and truncations; masked by extraneous noise; carried by a cornucopia of different voices, voice qualities, and accents; and spewed out at various rates. Most of the time, we easily manage this chaos using context and our general world knowledge to fill in gaps, resolve ambiguity, and arrive at a meaningful interpretation of what was said.

And that is just the receptive side of language.

Producing speech entails the exquisite coordination of more than 100 muscles controling several independent articulators from lips to larynx and even down to the diaphragm, which transforms normal breathing into a regulated engine of airflow that is filtered and massaged to yield all the chirps, swooshes, clicks, and hums that combine to form speech. All of this is accomplished at an astounding rate of about three to four syllables per second, with an error rate of only about 1 percent.

How do we choose which words to articulate? This requires higher-level computational gymnastics. Imagine a scenario in which a baby is sitting on a couch screaming his head off and a toddler is sitting next to him, covering her ears and grimacing. Ask a bunch of people to describe that scene, and you'll get a bunch of different responses. Here are some that I got when I ran this informal experiment in my classroom:

The baby is crying.

Both kids are upset, but for different reasons.

The girl covered her ears as her baby brother cried.

The little girl is covering her ears at her little brother's crying.

Transmitting an idea from thought to speech involves a host of largely unconscious decisions even prior to muscle coordination. Here's an informal sketch of the to-do list:

- Know your audience. Does the person you are telling know these kids (use proper names) or not (use common nouns)? Have you been talking about them already (pronouns are an option) or not (don't use pronouns)?
- Pick a focus: Is your utterance going to be about the baby's crying, the toddler's reaction, their joint angst, or something else?
- Pick one of many possible structures that matches your focus:
  - The baby's cry . . .
  - The cry of the baby
  - The girl covered her ears . . .
- Pick the words that you want to use to describe the objects and actions:
  - Baby, brother, sister, little boy, toddler . . .
  - Cry, scream, yell, screech, angst . . .
- Add some morphology:
  - crying, cries, cried
- Sequence the words, syllables, sounds.
- Integrate intonation:
  - The KIDS are screaming.
  - The kids are SCREAMING.
  - The kids ARE screaming.

You can do all this and still get syllables out of your mouth at a clip of about three or four per *second*. Humans are indeed elite linguistic and vocal motor control athletes.

Defining Language

The concept of "language" means different things to different people, so if we are going to describe its neural organization, we need to define the sense of *language* that matters here.

First, we need to distinguish between two senses of language. The first is what I will call "language phenotypes," exemplified by world languages such as Hindi, Mandarin, Arabic, Italian, and the many subdialects of each. These are the outward forms that language takes among groups of people. When looking at the phenotypic level, what we mostly notice are the differences among them.

The second sense of language, and the focus of this book, refers to the underlying biological and computational machinery—neural circuits, anatomy, and algorithms—that, when combined with a particular linguistic environment, gives rise to one surface form or another. Here, we are not so much interested in the differences among language phenotypes, but rather their commonalities. Unless otherwise stated, I will use the word "language" in this second, biological sense.

How do we know that there *is* an underlying biology that constrains and enables the shape of language phenotypes—some kind of *language organ*, in Chomskian terms? That is a question for another book, but here is a thumbnail summary of the evidence that is typically used to argue the case.

A uniquely human trait. As discussed previously, no other species has been able to express a language phenotype. No matter how hard you try to teach an otherwise highly intelligent animal to use language, it never comes close to what humans are capable of. Something about our biology has to be responsible for our linguistic capacity. What are the relevant bits of biology? That is a topic of much debate and part of what we will discuss in this book.

Universality of language among humans. There is no human society that does not have sophisticated language ability. If language phenotypes were invented as a kind of cultural technology rather than underwritten by biology, we would expect substantial variations across the world's societies.

Specifically, we would expect to find some cultures where language either doesn't exist at all or takes a wide range of forms, from primitive to complex. This is not what is found. In contrast, written language *is* an example of cultural technology. It was invented only relatively recently in human history, took different forms in different societies where it was separately invented, and was/is completely absent in plenty of human groups.

Core similarity of linguistic structure. Languages at the phenotypic level appear to be very different; indeed, they are often mutually incomprehensible. But linguistic research on a range of language phenotypes has uncovered a common blueprint. Broadly, all languages are hierarchical, productive (i.e., they combine discrete elements to yield a potentially infinite number of utterances), and combinatorial (i.e., meaning is derived by combining the meanings of the individual words with their structural relationships).

Learnability. There are some things that you just can't learn no matter how hard you try, such as ear-wiggling for many humans or language for nonhuman apes. This is one kind of learnability argument for the existence of a biological contribution to language phenotypes, as pointed out in this discussion. Consider a more specific example: nonhuman apes can't produce spoken words despite their ability to learn and produce signed words. For decades, it was believed that the limitation was anatomical; their vocal tracts simply can't make the sounds. But recent research has proved this false; even macaque monkey vocal tracts are "speech ready" (Fitch et al., 2016). The limitation is due to the lack of neural control of the speech articulators. Another kind of learnability argument arises when an ability emerges in a form that couldn't have been learned from environmental input alone. This is obvious in cases of prewired simple instincts such as egg retrieval in geese, temperature-triggered movement rates in woodlice, and the human gag reflex. More subtle examples are found in language, where kids end up learning systematic grammatical patterns that, it is argued, they could not have gleaned from the sample of utterances that they are exposed to (Lasnik & Lidz, 2017).

In short, language phenotypes are argued to exhibit features that are not present in the linguistic environment of the learner, which means that there must be something beyond the environment that shapes them. This is the *poverty of the stimulus* argument. A now-classic and rather dramatic example of this comes from documentation of the emergence of a new language in Nicaragua, where young deaf children were exposed to a cobbled-together, signlike, gestural system (a pidgin) and transformed it into a full-fledged, rule-governed sign language (a creole) (Senghas et al., 2004).[2]

Critical periods. Children learn language more readily than do adults. This holds for the common scenario of second-language learning and the rare situation of first-language learning in adulthood, as described earlier. This fact has led to the hypothesis that there is a critical or sensitive period for language acquisition (Johnson & Newport, 1989; Newport, Bavelier, & Neville, 2001). Critical periods have been documented for a range of behaviors and abilities in animals (Hensch, 2004; Sengpiel, 2007), and in general they have been interpreted as evidence for a biological bias toward learning at an early stage of development. For our purposes, the relevant fact is that the existence of a critical period for language acquisition rules out a simple explanation relying on general intelligence as the main driver of linguistic sophistication in *Homo sapiens*. If it were just about intelligence, four-year-olds should never outperform adults at language learning, and yet they do.

Genetics. Perhaps the ultimate demonstration that language phenotypes are shaped in part by biology is the demonstration that an individual's genotype has measurable effects on his or her language phenotype. Such a linkage has been found between at least some variants of developmental speech and language disorder associated with genetic variation in and around the FOXP2 gene (Vargha-Khadem, Gadian, Copp, & Mishkin, 2005). There has been significant debate about whether such genetically associated language disorders are language specific or more general, but as we will see, a relation between language and homologous nonlanguage abilities is to be expected.

Homologies

Taking a biological perspective on language changes how we approach the scientific problem of understanding it. We've already seen how it takes the spotlight off the differences among language phenotypes and focuses our attention instead on their underlying similarities. Another way that a biological perspective changes our approach is that it opens the door to exploring the neural basis of language using homologies with other nonlinguistic systems. Let me explain.

The concept of a biological homology is relevant to understanding why different species can exhibit similarities in a particular trait. The mammalian forelimb is the classic example. Primates, dogs, bats, and whales have similar sets of bones in their forelimbs, even though the size, shape, and use of the whole structure are very different. The similarity is said to be *homologous* if it derived from the evolution of a shared ancestral trait. So the reason why vertebrate forelimbs are similar is that they evolved from a common ancestor that had that structure. The differences are a consequence of subsequent evolution or descent with modification. Homology can be distinguished from *analogy*, where similarities of function evolved in different species but with independent evolutionary histories, such as bird, bat, and insect wings.

Homologies exist in brain organization too. In fact, the existence of neural homologies is the biological justification for using animal models as a way of understanding human neural function. We can learn something about human vision, say, by studying cats and monkeys precisely because our visual system has a common ancestor as that of cats and monkeys. Of course, the more closely related two species are, the more similar the two systems are likely to be because there has been less time for evolutionary tinkering.

It is a common view that language scientists can't use homologies because there is no animal model for language. This view entails that comparative research is limited to the case of analogy, such as in birdsong, which can provide clues for how one might design a vocal communication system in some classes of brain. But homologies to language systems *do* exist.

Consider the fact that language evolved after our split with Pan (chimps and bonobos). These species are among the most intelligent in the world, with highly developed sensory, motor, and higher cognitive systems, including abilities in learning and memory, attention, decision making, categorizing, social communication, sequencing processing, and others. It is therefore safe to assume that language evolved in something like this neurocomputational context. When we think broadly about how language might have evolved, we have two options. One is that something entirely new evolved from scratch and was wedged into the preexisting neural and computational contexts. The other is that it evolved by tinkering with what was lying around the house. Certainly, our brains are bigger than those of our Pan cousins, so there is room for new machinery, but most evolutionary biologists and language scientists would favor the tinkering hypothesis because this is how evolution tends to work.

If you buy the idea that language evolved by tinkering with existing neural machinery, then it stands to reason that homologies between linguistic and nonlinguistic systems should exist *in our own brains* because the linguistic systems will have evolved from a common neurocomputational "ancestor" with other systems. And if we can find such homologies, we can use animal models of these nonlinguistic systems to build a neurocomputational bridge to traditional animal models. For example, if aspects of our linguistic system evolved out of nonlinguistic sensorimotor control systems (as I'll argue later in this book), we can learn something about the neurocomputational architecture of language by studying visuomotor control in humans, or even monkeys.

There are important caveats to this broad logic. Language, even as I've framed it here, is not monolithic. Chomsky and colleagues have distinguished between a broad and narrow sense of language (Fitch, Hauser, & Chomsky, 2005). The broad sense includes all the interfaces: linkages between language and perceptual-motor systems on the one hand and to conceptual systems on the other. In this broad sense, it is uncontroversial that we should find at least some homologies to nonlinguistic systems in auditory or motor processing, for example. The narrow sense of language refers to the core combinatorial system. Exactly what this comprises is up for discussion, but it could refer to syntax generally, the recursion property of syntax, or simply the ability to take two bits of information and merge them into one, an operation called "Merge" (Chomsky, 2010).

For our purposes in this book, I will adopt a broad conceptualization of language and consider the neural organization of this broader range of parts. I also adopt the view that these different parts—including vocal motor control, speech perception, conceptual combination, and syntax—likely evolved along different trajectories and for different purposes, coming together only recently in our evolutionary history to form the collective language trait that we know today (Fitch, 2017). At every step of the way, I will approach the topic from this biological, evolutionary perspective, using possible homologies to constrain and inform the models. As we will see soon enough, a number of homologies have been discovered, which suggests that this approach is fruitful.

A Brief Preview

I've organized this book as a kind of journey, roughly following the receptive language path (comprehension) from speech perception and word recognition up to semantics and syntax. We then start our descent along the expressive language path (speaking) from the highest levels of planning down to speech-motor coordination. There will be a few scenic side trips along the way. For example, in chapter 4, we'll loop around the oft-trodden and, in my view, labyrinthine trail of hemispheric asymmetries; and in chapter 6, we'll climb the hillock of questions concerning how speech is coded by the brain and the role of neural oscillations. We'll take several smaller detours within various other chapters. Once we've traversed the full landscape, the final chapter represents my attempt to craft a broader map of the terrain. Feel free to skip ahead to that overview in chapter 12 if you'd like a bird's-eye view before we start. I've also included an appendix that was originally part of chapter 2 but deviated a bit too far off the path to be included on the main journey.

While our survey of the neurolinguistic landscape is broad, it is not exhaustive. There are many important topics that we will not cover at all, or only in passing. These include reading and writing, signed language, audiovisual speech, pragmatics, discourse, and other subjects. Our focus instead is on the neural systems supporting language abilities that have been central to the field of aphasiology and neurolinguistics since the 1860s: speech perception, word and sentence comprehension, conceptual knowledge representation, sentence production, word production, and motor-speech coordination.

We start our journey with some necessary background on a major principle of functional brain organization that is both extremely useful as a guide to understanding the neural architecture of language and potentially confusing. I'm referring to dual stream models of cortical function generally (Milner & Goodale, 1995) and to the dual stream model of speech processing more specifically (Hickok, 2022; Hickok & Poeppel, 2000, 2007; Poeppel & Hickok, 2004), which, for better *and* worse, has become the dominant model in the field. To see why this approach is both highly useful and potentially obstructive to further progress, we need to situate it in the broader context of brain organization. To preview, the existence of one processing stream linking perception and action and a separate (but interacting) processing stream linking perception and understanding is very strongly supported by a range of evidence from multiple sources in both language and nonlinguistic research. However, the dual stream architecture makes sense only from the sensory perspective. When we consider the neural architecture of language more broadly—for example, from the perspective of expressing thoughts through speech—the dual stream framework leads to confusion. Chapter 2 unpacks these ideas.

2 The Dual Stream Brain

> Our motor apparatus does not wait for our conscious awareness, but does restlessly its duty, and our consciousness watches it and has no right to give it orders.
> —Hugo Munsterberg, 1889

> I know kung fu.
> —Keanu Reaves in *The Matrix*, 1999

The Will of the Ball

In the movie *Awakenings*, a young, fictionalized Oliver Sacks, played by Robin Williams, examines a woman who appears to be awake but is completely unresponsive. She is a longtime sufferer of "sleepy sickness," a rare form of encephalitis, *encephalitis lethargica*. The disease swept the world in the years between 1917 and 1926, affecting half a million people, one-third of which were left with postencephalitic Parkinsonism. In his book *Awakenings*, Sacks (1973, p. 9) describes their condition: "They would be conscious and aware—yet not fully awake; they would sit motionless and speechless all day in their chairs, totally lacking energy, impetus, initiative, motive, appetite, affect or desire; they registered what went on about them without active attention, and with profound indifference. They neither conveyed nor felt the feeling of life; they were as insubstantial as ghosts, and as passive as zombies."

In a critical scene from the movie, Williams holds a pair of reading glasses just in front of the woman's face. She yields not a twitch of a response—until, that is, he releases the glasses. The falling object compels the woman's hand to swoop up and snatch them in midair. In the next scene, Williams is presenting the patient to a small group of more senior doctors. He demonstrates her chronic catatonic state as he waves his hand in front of her statuesque face. Then he steps back and tosses a ball toward the woman, which she adeptly catches. The skeptical doctors dismiss the act as a mere reflex. Williams counters that if she had batted it away, it might be a reflex, but catching it indicates something more. He goes on to explain, "It's as if having lost all will of her own on which to act, she borrows the will of the ball." The doctors scoff and excuse themselves.

Both Williams (Sacks) and the incredulous doctors have valid points. Williams (Sacks): It is not a simple, touch-induced grasp reflex like that found in newborns, but a complex, coordinated act that aligns movements of the body with the flight dynamics of an external object of a particular size and shape. Incredulous doctors: Well, neither is it a fully willful act, in that it is *triggered* by an external stimulus. Williams's line, "She borrows the will of the ball," captures this tension.

A more controlled ball-catching experiment was carried out in the 1990s in a group of people with Parkinson's disease (Majsak, Kaminski, Gentile, & Flanagan, 1998). The study's participants were seated in a chair facing a small ramp that tilted downward from left to right in front of them. Their task was to grasp a ball on the ramp in one of two conditions. In the *stationary condition*, the ball was fixed to the ramp at a reachable target position, and participants were asked to reach for and grasp it as quickly as possible when cued to do so. In the *moving condition*, the ball was released from behind a screen at the top of the ramp, and the participants were asked to reach for and grasp it as it rolled past the same target zone. People with Parkinson's disease were substantially slower to grasp the ball in the stationary condition than in the moving condition. In fact, like the woman in *Awakenings*, the speed of the grasp in the moving condition was impressive and on par with healthy control participants.

Similar will-of-the-ball phenomena have been documented in various neurological conditions involving different sorts of behaviors, including speech. In 1885, a young physician, Gilles de la Tourette (1857–1904), was among the first to describe a neurological disorder in which involuntary movements and vocalizations are the primary symptom, which later became known as *Tourette's syndrome* (de la Tourette, 1998). Gestural and verbal tics, often profane, dominate the popular neurological view of the syndrome, but Tourette himself noticed the cooccurrence of another characteristic symptom: the tendency to mimic both speech and gestures. In his classic report, Tourette described a patient, "S . . . ," whom he personally examined at Salpêtrière, the hospital in Paris where he worked under the prominent neurologist, Jean-Martin Charcot (1825–1893):

> Shortly after his admission to the hospital and thanks to a closer and more detailed examination, an utterly characteristic phenomenon was noted. In some circumstances, S . . .'s cries acquired a new, quite special character;
>
> . . . the patient now echoed faithfully single words and even short phrases which he heard pronounced by others. Thus, 'Here's Monsieur Charcot' would be succeeded immediately by a repetition of . . . 'Ah, here's Monsieur Charcot, Monsieur Charcot, Monsieur Charcot'. . . . (1885, pp. 107–108)

Tourette describes a second tendency:

> S . . . showed a compulsion to imitate gestures just as he imitated by repetition words that he had just heard spoken, albeit in a rather less extreme way. S . . . was standing in the courtyard of the Salpetriere when X . . . approached him, 'Ah! ah! X . . . , X . . . shit, shit', said S . . . alternately raising his arms in the air and then lowering them to the ground whilst lifting up his right leg. The movements then stopped or at least subsided. Making fun of S . . . , X . . . and numerous others repeated a series of everyday gestures and words which he could not help but imitate. (1885, p. 108)

Today, these symptoms are collectively called *echophenomena* and are subdivided into two major types, one involving speech (*echolalia*) and the other gestures (*echopraxia*). Modern work has confirmed Tourette's nineteenth-century observation that echophenomena are core features of his namesake syndrome (Cavanna et al., 2011).

Echophenomena are found in a wide range of neurological conditions beyond Tourette's syndrome, including autism spectrum disorder, catatonic schizophrenia, dementia, epilepsy, and aphasia (Ganos, Ogrzal, Schnitzler, & Munchau, 2012). The symptom can also appear as a result of tumors in certain areas or surgical attempts to remove them (Suzuki et al., 2012). In short, echophenomena are common neurological symptoms.

A particularly dramatic example of echolalia was described in a 1968 report by one of the most influential aphasiologists of the twentieth century, Norman Geschwind (1926–1984) and two of his colleagues, Fred Quadfasel and Jose Segarra. The patient was a twenty-two-year-old woman who was found unconscious and apneic (i.e., not breathing) on her kitchen floor. She had succumbed to carbon monoxide exposure from an unlit gas jet. The woman was revived on the spot and her breathing restored, but she spent the next seventeen days in a coma. When she finally regained some degree of animation, the staff noticed that she sang songs and repeated questions despite her otherwise unresponsive and near-vegetative state. Even more bizarrely, they reported a kind of mindless, sentence-completion behavior. For example, when prompted with "Ask me no questions," she would respond, "I'll tell you no lies." She was subsequently transferred to a Veterans Administration hospital, where she remained in the care of Dr. Quadfasel until her death nine years later.

On initial examination, the patient was found to exhibit few spontaneous movements despite having normal strength in her arms. When presented with objects or the examiner's hand, she would grasp them strongly and not release them, a behavior reminiscent of the catatonic patient in *Awakenings*. She would also follow the examiner's movements with her eyes. Her spontaneous speech was similarly sparse, limited to a few stereotyped phrases, such as "Hi, daddy," "So can daddy," "Mother," "Dirty bastard," and some unintelligible words. She failed to follow commands, such as "Point to the ceiling" or "Touch your nose," and otherwise gave no indication of comprehension. She responded to pain (pinprick testing) and other forms of stimulation with agitation or displays of fear. She was also incontinent.

The overall neurological picture was that of a woman with no conscious awareness or will of her own. What movements she did exhibit were nearly entirely driven by external stimulation. Indeed, the authors reported that in nine years of observation, she never used her arms for anything except stroking her hair repetitively if disturbed, grasping for objects, or striking at hospital staff.

Remarkably, this pattern extended to language. She never exhibited a single instance of propositional speech, never asked for anything, and never responded appropriately to a question. Instead, when spoken to, she would repeat the utterance again and again. Occasionally, instead of repeating, she would exhibit the completion behavior noted shortly after her accident. Her completion of "Ask me no questions" with "Tell me no lies" persisted. Additional examples noted during her chronic institutionalization included the completion of "Close your eyes" with "Go to sleep" and the completion of various proverbs or song titles. Longer completions were also recorded. Prompted with "Is this a rose?" she might say, "Roses are red, violets are blue, sugar is sweet, and so are you." When hearing "coffee," she might respond, "I love coffee, I love tea, I love the girls and the girls love me" (a riff on the popular 1940s song "Java Jive" by the Ink Spots). Although particularly dramatic in this case, the tendency to complete phrases was not a novel observation. Austrian psychiatrist Erwin Stengel (1947) reported similar "completion phenomena," as he called it, coexisting with echolalia in cases of transcortical aphasia.

The young patient's tendency to sing along with familiar songs on the radio also persisted. If the radio was turned off by the staff, the patient would continue singing correctly for a few lines and then stop. But if a staff member continued the song by humming the melody, she would continue singing the words to the end of the song. This behavior even extended to new advertisement jingles that did not exist before her injury. To test this apparent ability to learn, new songs were introduced and played repeatedly. Before long, the patient was singing along with these too.

What this case shows us is that fairly complex, well-formed actions—grasping at objects and recapitulating long strings of speech—does not require an understanding of the stimulus that triggered the action. Nor does it even appear to require conscious awareness.

Another fascinating case of echolalia, reported by the neurolinguist Haiganoosh Whitaker, illustrates the degree of complexity that this automatic, reflexlike action system can achieve, at least for language (Whitaker, 1976). The patient was a fifty-nine-year-old woman, "HCEM," who was suffering from early-onset dementia and rendered incapable of caring for herself. She needed assistance with dressing, getting into and out of bed, going to the bathroom, and other daily tasks. Her days were spent silently watching television or staring out the window. She exhibited no spontaneous speech. When prompted to name objects, her typical response was no response. When she did attempt an answer, she would sometimes produce a word in the semantic or visual ballpark: "library" for a book, "razor" for a toothbrush, "clock" for a telephone, or "ring" for a quarter. She correctly named only two items, a pencil and a cup. She could not follow simple commands and showed little evidence of comprehension, even of single words. Dr. Whitaker summarized HCEM's behavioral state as indolent, apathetic, and totally noncommunicative.

Despite her cognitive and volitional linguistic ineptitude, HCEM was an echolalic savant of sorts. Similar to the case seen by Geschwind and colleagues, HCEM showed completion tendencies and could sing along with familiar melodies. Here's an exchange between Dr. Whitaker and HCEM:

HW: Can you sing?

HCEM: Can you sing?

HW: Sing a Christmas carol for me.

HCEM: Sing a Christmas carol for me.

HW: Jingle bells . . . (*sung*)

HCEM: Jingle bell, jingle bells, jingle all the way . . . (*sung the whole stanza with correct tune*)

HW: Silent night . . .

HCEM: Holy night, all is calm, all is great (*sung the whole stanza but at the end only hummed*)

HW: What other Christmas carols can you remember?

HCEM: What Christmas carols can you remember?

HW: Well, sing something else for me.

HCEM: Sing something else for me.

HW: I wish you a merry . . .

HCEM: Christmas.

A more striking ability was uncovered when Dr. Whitaker presented HCEM with grammatically incorrect sentences. HCEM repeated them back, but in correct grammatical form. Here are several examples:

HW: *Can you told me your name?

HCEM: Can you tell me your name?

HW: *Have you dinner?

HCEM: Have you had dinner?

HW: *Have you dinner?

HCEM: Have you eaten your dinner?

She also corrected wonky stress patterns ("strawberry SHORT CAKE" → "strawberry SHORT cake") and corrected phonemic errors if the item being named was visible ("this is a dolly-pop" → "lolly pop"); otherwise, she repeated the nonsense word. However, she did *not* correct semantically anomalous sentences, such as, "I had a building for breakfast" or "the door opened the woman."

Related but milder examples of will-of-the-ball phenomena have been reported in the neurological literature. These include the following:

- *Ambient echolalia*—the tendency to speak words heard in the background of a conversation, often in midsentence (Fisher, 1988).

- *Alien hand syndrome*—the patient's limb performs purposeful and often mischievous acts, Dr. Strangelove–style, outside the person's will (Biran & Chatterjee, 2004).
- *Utilization* behavior—the patient is compelled to use an object that is placed in front of her or him, even if it is inappropriate for the time, such as brushing one's teeth immediately upon being presented with a toothbrush (Iaccarino, Chieffi, & Iavarone, 2014; Lhermitte, 1983).
- *Environmental dependency syndrome*—a more florid version of utilization behavior that applies to settings (things you do in a grocery store, museum, or bedroom) rather than objects. For example, one patient proceeded to undress and get in bed upon entering a room with a bed in it (Archibald, Mateer, & Kerns, 2001; Lhermitte, 1983).

I've given you a brief tour through this neurological landscape to drive home a point: the tendency to borrow an external will—whether that of a ball, someone else's speech or gestures, or even a complex setting—appears to be a proclivity that lies just below the surface of our own consciously willed behavior. Any number of neurological insults can release the urge. Indeed, even healthy adults can exhibit echophenomena in states of severe fatigue or semiconsciousness and it is found regularly in typically developing infants and toddlers where it is proposed to serve critical functions in language learning and the development of social cognition (Meltzoff, 1988; Meltzoff & Decety, 2003; Meltzoff, Kuhl, Movellan, & Sejnowski, 2009).

I'm willing to bet that you've experienced something similar. Consider the state of having driven several miles along a familiar route with no memory of how you arrived at your current location. Your conscious mind was wandering through a more interesting landscape while your limbs borrowed the will of the road to do their job. Ditto for that drowsy morning shower when you find yourself turning the water off and have no recollection of ever washing your hair (but it smells and feels clean). The will of the shampoo? Or perhaps you've been struggling to complete a task at work when you suddenly find yourself helplessly distracted by the sight of a book or magazine, and you compulsively pick it up and start thumbing through it—utilization behavior. And no doubt you've spent hours of your life pen twirling, hair twisting, beard twiddling, or fidget spinning; yes, the compulsive performance of repetitive, nonpurposeful movements is also commonly seen in excess in the clinic, where it is referred to as *manual groping* (Archibald et al., 2001).

Take one more step with me. As I'm typing these words, I have no conscious sense of guiding each finger toward the appropriate key; each one appears to borrow the will of its key. In fact, I have no conscious awareness of the vast majority of movements that I generate, even those that are complex: preshaping my hand as I reach for an object; the grasps, flips, twists, and pulls involved in tying my shoes; the increasing strength of my grip on a cup as I fill it with liquid; the coordinated dance of my lungs, larynx, tongue, and lips as I speak; and on and on. This tells us that will-of-the-ball phenomena are not just neurological curiosities or rare lapses of mental function in healthy folk. They are a constant undercurrent of everyday behaviors. Such behaviors seem remarkable in the clinic simply because they have been laid bare by diseases that fog consciousness or the ability to suppress contextually inappropriate, stimulus-driven actions.

The Conscious and the Automatic Self

In psychology, of course, we have other words for everyday will-of-the-ball-like behaviors; we call them *automatic behaviors, motor chunking,* or in the case of oft-repeated actions, *habits*. They are composed of highly practiced actions, often triggered by external stimuli and capable of being executed with minimal conscious will. Automatic behaviors are often contrasted with *goal-directed* behaviors, which require conscious awareness of the situation at hand (there's a lot of traffic on my normal route) and a conscious decision on a course of action (I'm going to try a new route). This is not to say that you *can't* be conscious of an automatic behavior—only that such awareness isn't required for successful execution.[1]

The distinction between automatic, largely unconscious actions and more conscious, goal-directed actions has a long history in psychology. Based on a variety of reaction time experiments, the nineteenth-century German psychologist Wilhelm Wundt (1832–1920) and his students concluded that a sensory stimulus can trigger a motor response in two ways (Neumann, 1990). It could either be *apperceived*—an old term for "consciously perceived"—and then a conscious decision is made to react, or, for well-practiced actions, the stimulus can shortcut awareness and trigger a motor response directly. The latter led to faster reaction times. One of Wundt's students, Hugo Munsterberg (1863–1916), stated the distinction even more strongly as follows (Munsterberg, 1889, cited and translated by Neumann, 1990, p. 173): "When we apperceive the stimulus, we have as a rule already started responding to it. Our motor apparatus does not wait for our conscious awareness, but does restlessly its duty, and our consciousness watches it and has no right to give it orders."

In modern times, David Milner and Melvyn Goodale documented Munsterberg's proposed distinction in a case report about a thirty-five-year-old women, D. F., who suffered *visual form agnosia* (the inability to recognize objects or object forms by sight) as a result of carbon monoxide poisoning (Goodale, Milner, Jakobson, & Carey, 1991). The injury damaged her ventral visual cortex bilaterally but spared most of her primary visual cortex. The visual deficit was profound: on a simple task of telling the difference between square and rectangle plaques, she performed at chance level, even when the distinction was obvious to control subjects. Nor could she indicate the width of the various-sized plaques by adjusting the gap between her index finger and thumb. In stark contrast, when asked to reach out and pick up a plaque, she preshaped her grip size precisely during her reach, just as neurologically typical people do. Goodale et al. (1991, p. 155) echoed Munsterberg, if less poetically, in their conclusion.

> Our data show that a person with brain damage may retain the ability to calibrate normal aiming and prehension movements with respect to orientation and dimensions of objects, despite a profound inability to report, either verbally or manually, these same visual properties. This dissociation suggests that at some level in normal brains the visual processing underlying 'conscious' perceptual judgments must operate separately from that underlying the 'automatic' visuomotor guidance of skilled actions of the hand and limb.

In support of this claim, Goodale and Milner (1992) also pointed to the existence of a syndrome in which the reverse dissociation is found, *optic ataxia*. Patients with this disorder have more dorsal lesions in the parietal lobe and have difficulty using visual information about the shape or location of an object to guide grasping actions; they grope around for objects as if reaching in the dark. But the problem is not a global cortical blindness. When presented with an object and asked to name it, such patients do so without difficulty.

On the basis of these and other observations, Goodale and Milner proposed that the cortical visual system is partitioned into distinct "what" versus "how" processing streams: the modern visual processing *dual stream model* (Goodale & Milner, 1992; Milner & Goodale, 1995). The "what"

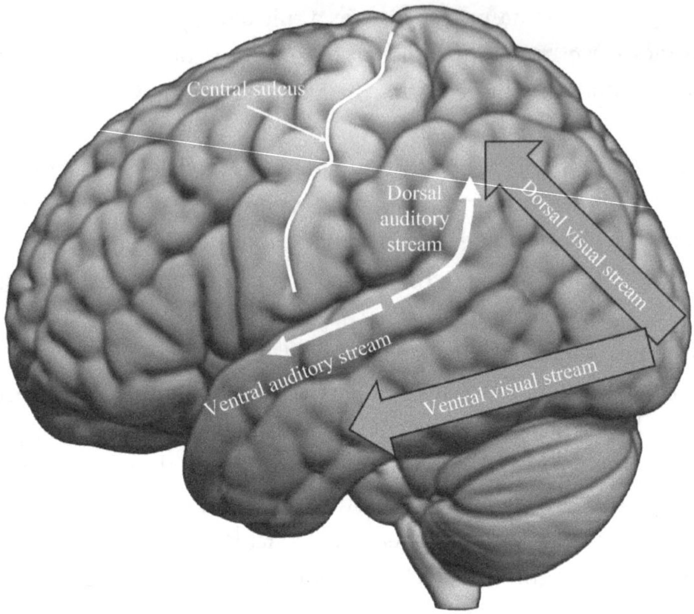

Figure 2.1
Schematic depiction of the dual stream architecture of visual and auditory processing.

stream courses from the primary visual cortex ventrally into the temporal lobe and is responsible for identifying *what* an object is (figure 2.1). The "how" stream courses dorsally from the primary visual cortex into the posterior parietal lobe and is responsible for converting visual information, such as shape, orientation, and location, into action plans appropriate for the perceived object; that is, *how* to interact with the environment motorically.

The proposal built on a previous theory by Ungerleider and Mishkin (1982), the "what" versus "where" model, in which "where" refers to computing an object's location in space. The models seem quite similar on their face, but there is a fundamental difference. Whereas the "what/where" model differentiates the streams in terms of processing distinct sensory features—shape features used to recognize objects, retinal position features used to compute location—the "what/how" model differentiates the streams in terms of how visual information is used for different *functions*—shape information for object recognition versus grasping. Goodale and Milner (1992, p. 20) put it this way: "Our alternative perspective on modularity in the cortical visual system is to place less emphasis on input distinctions (e.g. object location versus object qualities) and to take more account of output requirements. It seems plausible from a functional standpoint that separate processing modules would have evolved to mediate the different uses to which vision can be put."

We now have a functional-anatomic foundation for will-of-the-ball phenomena and the distinction between conscious apperception (ventral stream) and unconscious, stimulus-driven action (dorsal stream). Milner and Goodale spelled it out for visuomotor functions in the 1990s, but, as I detailed earlier, the functional distinction between unconscious will-of-the-ball-like phenomena and conscious understanding exists for speech and language as well. The preceding discussion focused on cases where speech/language ability was preserved without apparent conscious awareness, but the reverse dissociation also holds. In a similar way to how people with optic ataxia can recognize objects but fail to grasp them normally, people with *conduction aphasia* can recognize and understand words but fail to reproduce them normally (Buchsbaum et al., 2011; Goodglass, 1992; Hickok et al., 2000; Wernicke, 1874/1969). This suggests that the brain instantiates a *homologous* functional-anatomic division of labor in linguistic and visual domains.

The Dual Streams of Speech

I proposed and developed this precise hypothesis with my collaborator, David Poeppel, in a series of papers starting in 2000 (Hickok & Poeppel, 2000, 2004, 2007). We proposed a function-based dual stream model (figure 2.2) that is homologous to that put forward by Milner and Goodale. We argued that the ventral stream supports speech comprehension, a *what*-does-this-sound-mean function, whereas the dorsal stream supports speech production, a *how*-can-I-reproduce-this-sound function. Before our proposal, Josef Rauschecker and colleagues had followed Ungerleider and Miskin's lead in proposing a feature-based dual stream model for the auditory system that comprised a ventral-anterior "what" stream and a dorso-posterior-frontal "where" stream (figure 2.1) (Rauschecker, 1998; Rauschecker & Tian, 2004; Rauschecker, Tian, & Hauser, 1995). Thus, there were parallels between vision and speech/hearing, even in how the models evolved theoretically from what/where to what/how. The parallels between very different domains, vision and speech, seemed promising to us because it seemed that we had hit on some general

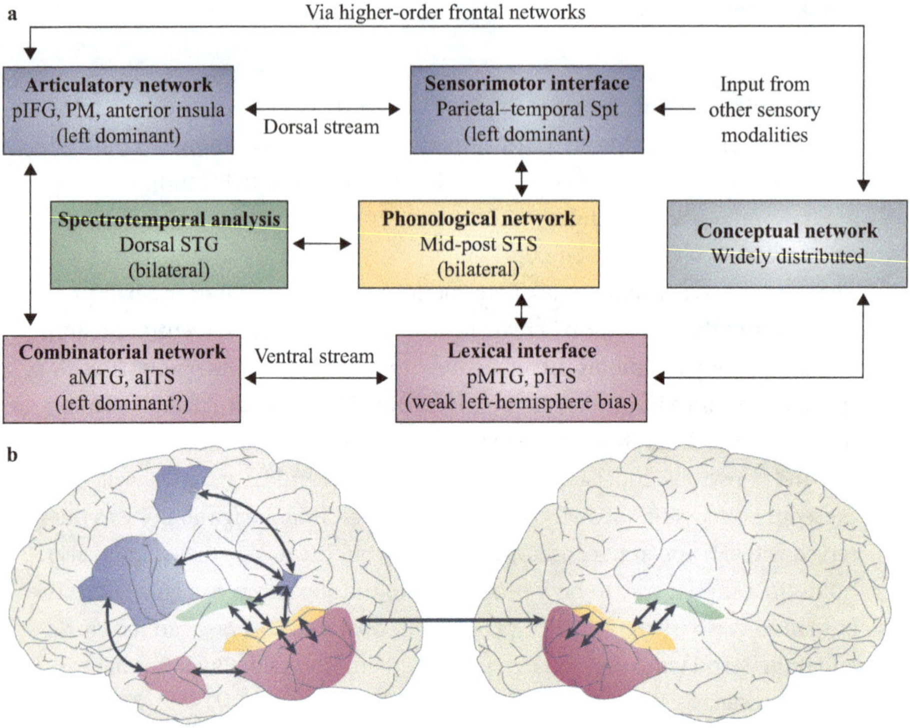

Figure 2.2

The Hickok and Poeppel (2007) dual stream model of the functional anatomy of language. (a) Schematic diagram of the dual stream model. The earliest stage of cortical speech processing involves some form of spectrotemporal analysis, which is carried out in auditory cortices bilaterally in the supratemporal plane. These spectrotemporal computations appear to differ between the two hemispheres. Phonological-level processing and representation involve the mid-to-posterior portions of the superior temporal sulcus (STS) bilaterally, although there may be a weak left-hemisphere bias at this level of processing. Subsequently, the system diverges into two broad streams: a dorsal pathway (blue) that maps sensory or phonological representations onto articulatory motor representations, and a ventral pathway (pink) that maps sensory or phonological representations onto lexical conceptual representations. (b) Approximate anatomical locations of the dual stream model components, specified as precisely as available evidence allows [circa 2007]. The green-shaded regions depict areas on the dorsal surface of the superior temporal gyrus (STG) that are proposed to be involved in spectrotemporal analysis. Regions shaded yellow in the posterior half of the STS are implicated in phonological-level processes. Regions shaded pink represent the ventral stream, which is bilaterally organized with a weak left-hemisphere bias. The more posterior regions of the ventral stream, posterior middle, and inferior portions of the temporal lobes correspond to the lexical interface, which links phonological and semantic information, whereas the more anterior locations correspond to the proposed combinatorial network. Regions shaded blue represent the dorsal stream, which is strongly left dominant. The posterior region of the dorsal stream corresponds to an area in the Sylvian fissure at the parietotemporal boundary (area Spt), which is proposed to be a sensorimotor interface, whereas the more principles: "This similarity in organization across sensory systems suggests that the present dual-stream proposal for cortical speech processing might be a reflection of a more general principle of sensory system organization" (Hickok & Poeppel, 2007, p. 401). For evidence of a dual stream organization in somatosensation, see the appendix of this book.

While we were excited about our "new" model, it was far from new.[2] Carl Wernicke (1848–1905) owned the claim to the first dual stream model of speech processing—indeed, the first cortical dual stream model, period. Wernicke's scheme is familiar to modern students of language science, but not all of them explicitly recognize that it has a dual stream architecture. He proposed that auditory word images, stored in auditory cortex, were connected to two systems: a widely distributed conceptual system for comprehension and the more focal motor speech system (Broca's area) for speech production (Wernicke, 1874/1969). The model that I proposed with Poeppel is a modern update of Wernicke's original. We will dig into the details of the models in later chapters. Here, we are concerned only with the issue of dual stream architectures in general.

Why Two Streams?

If we have hit on a general organizational blueprint, it's fair to ask why this should be the case. Specifically, why does the brain bother divvying up its resources to compute "what" information separately from "how" information? The answer is straightforward. These are different computational problems that require different computational solutions.

To recognize a cup as a cup, it doesn't matter whether it is near or far, upright or sideways, with or without a handle, toy-sized or fit for a giant, partially occluded from view, or indeed shaped like an owl. What matters are the *invariant* features of cupness, whatever those are (an interesting and complicated problem itself that we will not pursue here). To perform the "what" task, the brain has to abstract away from all the idiosyncratic details

Figure 2.2 (continued)

anterior locations in the frontal lobe, probably involving Broca's region and a more dorsal premotor site, correspond to portions of the articulatory network. (Figure description quoted from Hickok & Poeppel, p. 395, under copyright retained by the authors.) *Abbreviations:* aITS, anterior inferior temporal sulcus; aMTG, anterior middle temporal gyrus; pIFG, posterior inferior frontal gyrus; PM, premotor cortex.

of the *particular* cup that you are looking at and compute a representation that can be compared somehow with your abstract concept of a cup.

Now, if your goal is to pick the thing up, then all those particulars that didn't matter for determining cupness suddenly matter a great deal. You are going to have to tweak your action plans differently depending on whether it is near or far, upright or lying down, with or without a handle, small or big, positioned behind an obstacle, or oddly shaped. Moreover, suppose that there were two owl-shaped objects of identical shape and size, but one had an opening at the top suitable for pouring in liquid and the other was just a solid owl. One is a cup and one is not, but the "how" of reaching for it constitutes an identical computational task (assuming the same location, orientation, etc.). We have a classic double-dissociation in these computational scenarios: we can have the same "what," such as a cup on the left versus the right side of your body, which requires distinct "how" computations; and we can have different "whats," a cup or an owl, each of which requires the same "how" computation.[3]

We might ask another question: Why do we need a "what" system at all? Why can't we just reflexively respond to our environment without bothering to compute an abstract interpretation of the object? The broad answer is that it is highly adaptive to use "what" information to inform our decisions about whether and how to interact with an object motorically. It makes a difference to our survival whether the long, tan, cylindrical object on the ground is a snake or a stick. Less life threatening, but perhaps equally important to our social life, it is important to reach for one's own beverage glass at a dinner party rather than any glass within arm's length. Computing "what" information allows us to evaluate a range of opportunities for action and choose the course that is most appropriate to achieving our current goals. If we lacked a "what" system, we would be reduced to often-maladaptive reactionism, as we saw in the neurological patients discussed in the beginning of this chapter.

Given this computational necessity, it is no surprise that two broad streams of sensory processing have been identified in different sensory systems and across a range of diverse species, including frogs (Ingle, 1973), hamsters (Schneider, 1969), monkeys (Rauschecker et al., 1995; Ungerleider & Mishkin, 1982), and humans (Hickok & Poeppel, 2007; Milner & Goodale, 1995; Wernicke, 1874/1969).

We might also ask, Why does the "how" system set up shop in the dorsal fronto-parietal portions of the primate brain, while the "what" system occupies the ventral occipital-temporal region? I think it boils down to the location of systems with which the two processing streams primarily interact—the motor system, which is in dorsal cortical regions, and the conceptual memory systems, which tend to be more ventral in the temporal lobe. A more detailed discussion of this issue would take us too far afield, so I'll leave it there. For the interested reader, however, I've included an appendix in this book that delves into the anatomical questions in more depth.

Challenges

We have a tidy functional anatomical picture that explains the pervasiveness of will-of-the-ball phenomena, the dissociability of "what" versus "how," and the general organization of sensory systems in the cortex. Now it's time to muddy the waters.

Consider the following simple request that you might receive at the dinner table: "Please pass the salt." Does your response to this request draw on the "what" or the "how" system—or both? Of course, it's both a "what" and a "how" problem, in that you need to distinguish the saltshaker from the pepper shaker, the butter tub, and the flower vase (a "what" problem), and then you need to execute the action targeting the relevant object's position, shape, and orientation (a "how" problem). This isn't news so far, as we already discussed the fact that knowing what an object is helps us decide which actions to take. But now suppose that the salt and pepper shakers are the same size, shape, and color. How do you know which shaker to pass? One option is to grab one and look at the top, where you might find an "S" or "P," or perhaps you need to shake some of the contents out to determine which is which if the markings are ambiguous. In either case, your ability to determine "what" the object is depends on your ability to use action to further explore it. If "how" can be dependent on "what" (the first scenario) and "what" can be dependent on "how" (the second scenario), maybe the two streams are not so neatly segregated, either functionally or anatomically.

Some researchers have argued precisely this point. Indeed, Thomas Schenk and Robert McIntosh (2010, p. 52) go so far as to suggest that "we should reject the idea that the ventral and dorsal streams are functionally independent processing pathways." Their argument is based on a range of observations. One is that long-term memories about objects affect how we grasp them (McIntosh & Lashley, 2008; Schenk & McIntosh, 2010). For example, the grip strength that you apply when grasping a carton of milk takes account of your past experience with their typical weight, such that you may awkwardly overapply grasping and lifting forces if the carton turns out to be empty. Long-term memories about objects are part and parcel of "what" stream computation, indicating that online guidance of actions is influenced by this information. Schenk and McIntosh also point out that Milner and Goodale's famous case, DF, is not entirely normal in her visuomotor ability. She can grasp familiar objects, but she makes mistakes in selecting the functionally appropriate place to grasp them and has difficulty using shape information to preorient her hand when the object does not have a clear principal axis, such as a symmetrical cross shape (Carey, Harvey, & Milner, 1996). There are other similar examples as well.

Goodale and Milner (2010) respond by agreeing wholeheartedly: of course the two streams interact! In fact, they acknowledged the fact of the interaction between the streams in their earliest formulations, citing evidence from anatomy and physiology. I would go a step further and point out that a "what" stream probably could not have evolved in the first place if there were no way for it to directly affect the "how" stream in real time. What adaptive benefit is there to recognizing an object as a snake instead of a stick if that information can't be put to immediate use in guiding actions? That said, interaction between the streams does not imply that they are computing the same function, as Goodale and Milner argue.

To take an extreme example from speech, the visual information coming from the lips of a talker interact with the acoustic speech information sufficiently to change what is heard, the McGurk-MacDonald effect (McGurk & MacDonald, 1976)). By Schenk and McIntosh's logic, we should (incorrectly) reject the idea that the auditory and visual systems are functionally independent processing pathways. Even though the two sources of information end up interacting impactfully, the functions computed by the visual and auditory systems to get to the same representational point are quite different. In the same way, even though information about the abstract object category can influence your reaching and grasping actions, the computational functions that take you from a retinal image to the category versus hand trajectory, orientation, and grip type are quite different, dissociate following brain injury, and so on, as detailed extensively in this chapter.

Milner and Goodale (2008) make a distinction between ventral stream action *planning* (do I reach for the snake or the stick?) and dorsal stream action *programming* (how do I reach for whatever I decided to reach for?). I made a similar proposal in *The Myth of Mirror Neurons,* where I argued that the dorsal and ventral streams can be thought of as different hierarchical levels of motor control, with the dorsal stream managing the reflexlike, moment-to-moment details and the ventral stream contributing to the higher-level decisions regarding the immediate targets for action (Hickok, 2014b, pp. 162–165). The distinction between will-of-the-ball phenomena (dorsal stream) and higher-level planning (ventral stream) falls out naturally from this kind of division of labor.[4]

Schenk (2010) has countered that the distinction between planning and programming is not so clear. How do you decide what counts as one or the other? If DF grasps an object in the wrong place, how do we know whether she made a lower-level programming error or a higher-level planning error? This is a legitimate question, but not one that fundamentally undermines the existence of a distinction. Specifically, ambiguity in the source of an error does not change the computational demands involved in extracting "what" versus "how" information from the visual input, nor does it erase the clear-cut neurological dissociations. Rather, it simply suggests that there is more work to do in mapping out the details of the system.

Importantly, a very similar question arises in speech production research. We know from much evidence that there is a distinction between a higher-level, abstract representation of a word (its so-called *lemma*) and its lower-level phonological form (see chapter 7). Often, speech errors can be ambiguous with respect to the source of the glitch in the system. For example, if someone says "rat" when she or he meant "cat," is this because the wrong word or the wrong initial phoneme was selected? We can and should pose this ambiguity as a scientific problem, but it doesn't imply that the distinction between word and phoneme is not real. In fact, with the right kind of modeling, we can sort out the sources of these errors as my colleagues and I have shown (Walker & Hickok, 2016; Walker, Hickok, & Fridriksson, 2018). See chapter 10 for details on how this is done.

The Problem with Dual Stream Models

Dual stream models tend to be proposed by perceptual scientists. Milner and Goodale are vision scientists, Rauschecker is an auditory scientist, and Poeppel and I were mostly interested in the neural basis of speech perception when we started to work on what morphed into our dual stream model. From the perspective of sensory systems, dual stream models make perfect sense. Sensory information entering the cortex through the primary visual or auditory cortex can be mapped onto two very different systems—the motor system for action or the conceptual system for understanding. However, if one starts an action with a thought instead of a percept and intends to act on that thought, it doesn't make sense to frame the process as involving two separate streams.

We can see this very clearly in the language domain. If I have a thought that I'd like to express verbally, several steps are involved. Minimally, starting with the conceptual message or concept, a set of words must be selected, their phonological forms accessed, and the sequence of words, syllables, and phonemes must be articulated. This is the basic architecture of psycholinguistic models of speaking, as we'll see in chapter 10, and there's nothing dual stream about it. Moreover, some parts of the process involve ventral, temporal lobe structures, while other parts involve dorsal fronto-parietal structures; that is, both ventral and dorsal stream systems. Are we justified at all, then, to claim the existence of separate ventral and dorsal streams for language processing?

The answer is yes, when looking at the system from the perspective of the auditory system. From there, we can see two diverging pathways—one heading toward the motor system in the dorsal brain and the other heading toward the conceptual system in the ventral brain. But looking at the same architecture from the perspective of the conceptual system, we see mostly one pathway, mapping from thought to speech through a set of hierarchical stages of processing that involve both ventral and dorsal brain areas.

The same holds for the dorsal and ventral streams in visuo-manual control. If we start with a thought ("I need more coffee"), we can activate a semantic representation of the goal ("coffee cup"); identify a corresponding object in the real world ("my coffee cup"); use its particular shape, size, and location features to plan a grasp; and then execute the grasp. This involves the whole system, both ventral and dorsal. This doesn't erase the distinction between the dorsal and ventral systems; rather, it is just looking at it from a broader, nonsensory-centric perspective, where the ventral and dorsal networks are involved in different hierarchical levels of the process, just as they are in speech.

I think this failure to appreciate the sensory-centric perspective of dual stream models has led to much confusion and unnecessary debate. Dual stream models have their place, but one should not try to pigeonhole all linguistic or visuomotor function into such a framework. Similarly, one should not contest the validity of dual stream models because they assume too sharp a distinction for some tasks or scenarios.

The Big Picture

Take a step back from this discussion and notice, again, a kind of homology between research on language and nonlanguage processes. Not only do we see homologies in the broad, dual stream organization, but we see similar arguments for hierarchical levels of planning/programming, and similar kinds of problems cropping up in the trenches of these prima facie very distinct lines of neuroscience inquiry. A major thesis in my approach to understanding how the brain creates language is that we can take advantage of the existence of homologies between systems and make some theoretical advances by employing a comparative research approach (see also Fitch, 2010). Just as vision scientists have learned heaps of information about the human visual system by studying vision in other species, so can we learn about the neural organization of the human language system by studying homologous networks in both humans and nonhumans. For example, if aspects of our language system are homologous to our visuomotor system, as appears to be the case, and if the macaque visuomotor system is homologous to our own, as is widely acknowledged, we can actually learn something about the neural organization of language by studying macaque vision.[5] In other words, and in contrast to what we language scientists have thought for decades, we *do* have animal models for language. Indeed, we may have a lot of them in the form of animal research on systems that hold a homologous relation to language: memory, categorization, decision making, sensation, motor control, and others. Moreover, we have even closer neurophylogenetic relatives to language systems in our own brain where such homologies exist.

The benefit of comparative research goes the other way too. Given that aspects of language and motor control are homologous, not only can language theories be advanced by looking at research on motor control, but research on motor control can be advanced by looking at research on language. I hinted at one example a few paragraphs back, where linguistic theories of levels of representation are much more advanced that those in motor control. Perhaps linguistics can inform debates in motor control by sharpening the notions of "planning" versus "programming." To give another example, theories of the two visual streams lean heavily on a single case study, DF. Much can and has been learned from such rare cases, but one can always question how generalizable the data are. The reason for this reliance on DF is that ventral stream deficits are rare in vision. But they are fairly common in language, where some of the fundamental questions in the vision world have been clearly answered in the language world. Can you have normal repetition (a "how" function) without comprehension (a "what" function)? Yes, as shown by the syndrome of transcortical sensory aphasia (Berthier, 1999; Boatman et al., 2000) and the two cases described earlier in this chapter. Using the logic of homology, this fact bolsters claims for the neurocomputational distinction between the *visual* streams.

Summary

Dual stream architectures have been proposed repeatedly and independently in multiple sensory-processing domains, including speech and language. The evidence for their validity is strong and widespread, as we saw from the many examples of how sensory-driven action can function at very high levels without apparent conceptual understanding or conscious awareness. Challenges to dual stream models typically come from misunderstandings (or overextensions) of what they were developed to explain. When considering these two streams from the perspective of volitional, concept-driven action, it is helpful to think of ventral stream systems as higher levels of the action-planning hierarchy and dorsal stream systems as lower levels of the hierarchy.

Enough with the preliminaries. We are now ready to start our journey through the architecture of language. We'll start with a close look at the seemingly simple process of perceiving and recognizing speech.

II The Clangs of Language

3 The Hidden Symmetry of Speech Perception

> Through the destruction of Wernicke's area, one looses the perception of words.
> —Salomon Henschen, 1918

At fifty-five years old, Mr. L, a teacher and journalist, suffered a stroke. Except for a slight weakness on the left side of his face, his symptoms were relatively mild and almost entirely restricted to language: he had trouble coming up with the right words and sounds when talking. This occurred during normal conversational exchanges, as well as clinical assessments when he was asked to repeat phrases verbatim or read aloud. His writing ability appeared to be in a similar state. The content of his speech was otherwise coherent, and he seemed able to understand what was said to him. Happily, Mr. L gradually improved to near-normal levels, save for the occasional *paraphasia*, as speech errors are called.

The year of Mr. L's stroke was 1877, and his pattern of deficits and preserved abilities would have been recognized by some neurologists of the day as a case of conduction aphasia, a condition first identified a few years earlier by Carl Wernicke (Wernicke, 1874/1969). But in June 1882, the plot thickened when Mr. L suffered another attack. Again, it seemed relatively mild in some respects. His facial paralysis had increased and, according to his wife, his problems with speech errors were accentuated, although this was not confirmed in formal neurological examinations. He could not read or write initially after the attack, but these symptoms later subsided completely. Mr. L would have escaped serious disability once again, except for one glaring additional change in his neurological status: while he could hear sound, when it came to speech, he behaved as if he were completely deaf. His doctor was Ludwig Lichtheim (1845–1928), who later described Mr. L's case in his famous 1885 paper:

> Patient gives the impression of an absolutely deaf man. . . . One cannot have the least communication with him except through writing; even his wife cannot make herself understood otherwise. . . . I should probably have taken him for really deaf had I not been assured that he was acute of hearing, and could perceive all noises, which he said hurt him; and had I not tested the fact myself (Lichtheim, 1885, p. 462).

Mr. L's case differed from previous cases of receptive or "sensory" (now Wernicke's) aphasia, where there remained some residual ability to comprehend at least bits and pieces of spoken language. He also deviated from the typical pattern in terms of his ability to produce coherent, nearly error-free speech, which remained unchanged after his second stroke, except perhaps initially as his wife had reported. Wernicke's aphasia, in contrast, typically includes not only comprehension deficits, but significantly disordered speech output as well. Remarkably, Lichtheim reports that Mr. L was able to write skillfully and continued to publish his work in the local newspaper.

For these reasons, Lichtheim diagnosed Mr. L with "isolated speech-deafness" rather than Wernicke's sensory aphasia. He attributed it to an interruption of the nerve fiber pathway *a-A*, illustrated in his diagram (figure 3.1), which connects incoming auditory impulses with the cortical center for auditory images (*A* in the diagram), where Lichtheim, following Wernicke, believed auditory word memories were stored. In this scheme, the reason why Mr. L appeared to be deaf was because acoustic speech information was blocked from reaching the zone where memories of words were stored. And the reason why Mr. L could speak, read, and write without difficulty was because both language centers (*A*, the auditory word center, and *M*, the motor word center) and all their other connections were intact. Indeed, Lichtheim did not even consider isolated speech-deafness to be a form of aphasia. Today, we call this disorder *pure word deafness* or just *word deafness*, a nod to the fact that it is usually not so pure, often cooccurring with difficulty recognizing environmental sounds or aspects of music (Buchman et al., 1986).

In modern times, patients with word deafness have described their perceptual experience as follows:

> Like a great noise all the time . . . you think you can catch it and it fades away, like foreign folks speaking in the distance. . . . When people speak loudly or quickly the words just run together. (Klein & Harper, 1956)

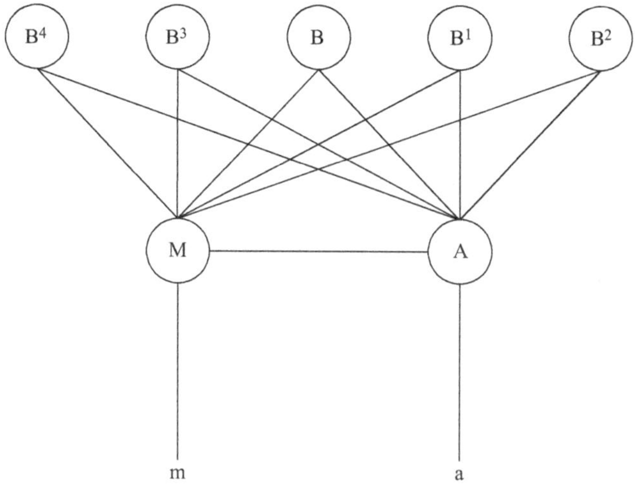

Figure 3.1
Lichtheim's diagram of the architecture of language. A, auditory speech center; a, acoustic periphery; M, motor speech center; m, motor periphery; B, distributed conceptual ("Begriffe") network. Redrawn from Lichtheim (1885).

> Meaningless noise, garbled sound, or a foreign language. (Mendez & Rosenberg, 1991)

> It's as if there were a bypass somewhere, and my ears were not connected to my voice. (Saffran, Marin, & Yeni-Komshian, 1976)

We can only infer that Mr. L must have experienced something similar.

Lichtheim did not report the location of Mr. L's brain damage; perhaps he never had the opportunity to examine his brain postmortem. But subsequent similar cases did undergo autopsy, which revealed a unique pattern: many of these people had *bilateral* lesions in and around auditory cortex. In retrospect, the fact that Mr. L had two episodes is at least consistent with bilateral strokes.

A dramatic and rather sad early example of the effect of confirmed bilateral lesions in auditory regions comes from the story of a young woman, E.H.M, who lived in a suburb of London in the 1890s. The description of her case (Mott, 1907, p. 310) suggests a complicated family history: "Father was a drunkard, and lived a fast life. Sister and brother were said to be hysterical." One brother was reported to have "died of fright" at age twenty-seven (likely a stress induced cardiovascular attack). In the early 1890s, at about age twenty, E.H.M. had her first attack after suffering her own fright due to a

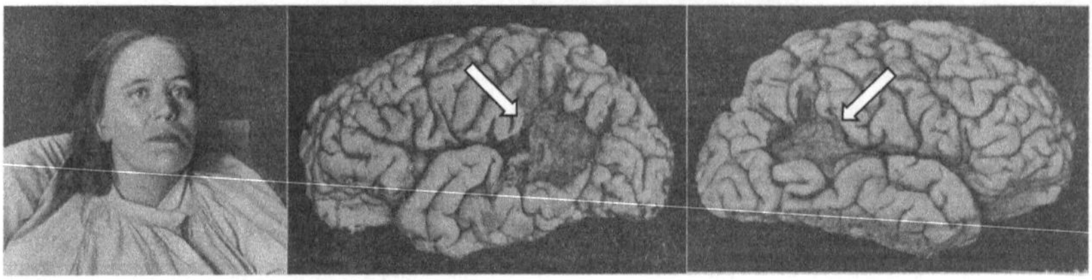

Figure 3.2
Patient E.H.M. Bilateral lesions shown in the posterior superior temporal region. Reprinted with permission from Mott (1907).

"supposed ghost." She was unconscious for three hours and lost her speech for two weeks, but she made a full recovery. The episode was written off by her physician as hysteria, botching the diagnosis.

E.H.M. was married in July 1894 and enjoyed good health for a few months. Then in November of the same year, she had another "fright followed by some family differences." A few weeks later, her husband found her helpless one morning, unable to understand what was said to her, although she could nod appropriately to written questions. Hospital examination revealed a right-sided paralysis. Her motor symptoms later improved, but her inability to comprehend speech, or indeed to hear much of anything at all, did not. She died in 1903 of tuberculosis. An examination of her brain revealed bilateral lesions in the superior temporal lobe corresponding to her two episodes, a stroke in the right hemisphere followed by a similar stroke in the left. Hers was not a case of pure word deafness, but rather *cortical deafness*, also called *central deafness*. Unlike Lichtheim's case, E.H.M. appeared not to be able to hear, period.

Cases such as Mr. L and E.H.M. accumulated in the literature sufficiently to enable Swedish neurologist Salomon Eberhard Henschen (1847–1930) to publish a detailed study of cortical deafness and word deafness in 1918. His analysis led him to a patently modern-sounding theory in which the cortex of the temporal lobe was organized hierarchically. I'll let him explain it: "There exist, in the temporal lobe at least 3 different centres above each other: the primary acoustic centre, the word-sound-centre, the word-comprehension-centre" (Henschen, 1918, pp. 442–443).

The primary acoustic center, which he correctly localized to Heschl's gyrus bilaterally, was sufficient for processing basic features of sounds but did not house the higher-order sound patterns that composed words. Making sense

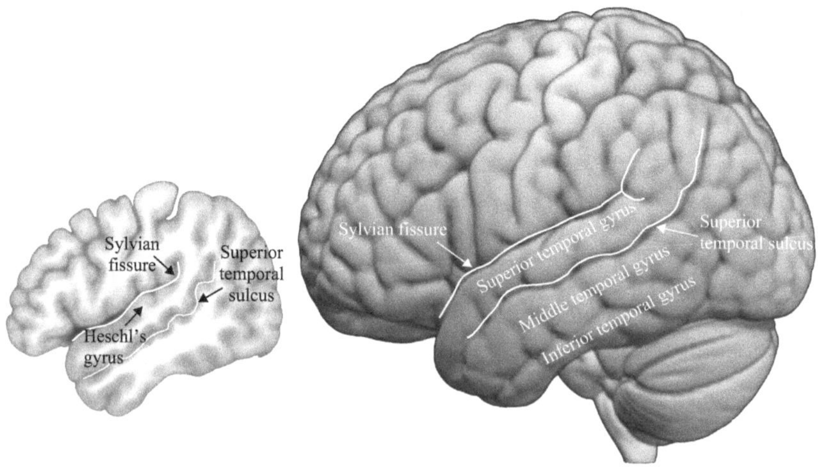

Figure 3.3
Temporal lobe anatomy shown on a rendered three-dimensional image (right) and a sagittal slice (left). Heschl's gyrus is not visible on the rendered image, as it runs medial to lateral on the dorsal aspect of the temporal lobe (the supratemporal plane) inside the Sylvian fissure. A portion of it is visible in the sagittal slice.

of those "word-clangs" (*wortklänge*), as he called them,[1] was the job of a separate downstream area in the posterior superior temporal gyrus (STG), which he called T^1. Figure 3.3 provides a visual guide for these regions. As Henschen (1918, p. 450) said, "The posterior part of T^1 is educated to receive the clangs of the language. Here the representations about the psychical explanation of these clangs are formed and here are the recollections [deposited]."

Although Henschen localized his word-clang center to Wernicke's area (roughly the posterior portion of the superior and middle temporale gyri), his theory differed from Wernicke's (and Lichtheim's) in its hierarchical arrangement. Wernicke did not distinguish primary auditory cortex from higher-order zones; for him, it was one undifferentiated auditory region. Henschen (1918) localized his third level of the hierarchy to T^2 and T^3, or the middle temporal and inferior temporal gyri, a zone that enabled a *meaning* to be linked to the word-clangs. From there, he speculated that word meanings could be connected to perceptions from other senses and ultimately "welded" into thoughts (p. 480). He suggested the frontal lobe as the welder.

Eighty-two years later, the cover of the May 2000 issue of the scientific journal *Cerebral Cortex* featured a map of the hierarchical organization of the temporal lobe for auditory word processing (Binder et al., 2000). It was generated using functional magnetic resonance imaging (fMRI) by Jeffrey Binder, a pioneer in its application to auditory speech research, and his colleagues at the Medical College of Wisconsin. Although rightfully considered groundbreaking in 2000, the map largely confirmed Henschen's hierarchical model: basic sound perception mapped to the dorsal STG in and around primary auditory cortex (roughly, Heschl's gyrus), phonological-level perception mapped to the superior temporal gyrus, and lexical/meaning-related word processing mapped to the middle and inferior temporal gyri (among other regions—stay tuned).

Bilateral or Unilateral Word-Clang Decoder?

Henschen made an exceptionally strong case for cortical deafness resulting from bilateral damage: "Only through bilateral lesions of the hearing centre, complete deafness may be provoked" (1918, p. 456), a statement that is widely accepted today. Yet he still seemed to adhere to the dogma of an entirely left-dominant language system, including his word-sound center, despite the fact that the majority of cases of word deafness that he evaluated in the literature and clinic setting had bilateral lesions. This included the case of Clara Nilsson, whom he himself presented in his 1918 report and diagnosed with "pure word deafness": she could recognize sounds and music, but not words. Although Henschen acknowledged the bilateral destruction of T^1 in Ms. Nilsson, he still referred to the "word hearing centre in the left T^1" and states without qualification, "Trough [sic] the destruction of W. [Wernicke's area, a left-hemisphere region] one looses the perception of words" (1918, p. 469). More directly, in a subsequent publication, he flatly stated, "Word-deafness is the result of disease of the first left temporal convolution" (Henschen, 1926, p. 113).

In hindsight, it appears that these early, influential authors missed an important clue. Lichtheim correctly noted that the receptive speech problem in Mr. L. was more severe than that found in sensory (Wernicke's) aphasia. Yet he failed to categorize the perceptual symptom as something qualitatively different, referring to the receptive problems in both sensory aphasia and in Mr. L. as "word deafness": "Patient gives the impression of an absolutely deaf man, and differs in this respect from *other* cases of word-deafness" (Lichtheim, 1885, p. 462, emphasis added).

Perhaps if Lichtheim had been aware of the bilateral lesions that were the likely source of Mr. L.'s deficit, he would have rethought his position. In his defense, however, only a few cases of sensory aphasia had been described, so it was probably premature to make a fuss about a difference in degree.

Henschen, for his part, *did* know about the underlying lesion pattern of pure word deafness, but he similarly did not make much of the difference in word recognition severity compared to that found in sensory aphasia, if he noticed it at all. From our vantage point, we can put two and two together even using only the information that was available in the 1910s: severe deficits in word understanding are, for the most part, associated with bilateral superior temporal lobe lesions, whereas only a milder receptive deficit, found in Wernicke's aphasia, results from left unilateral lesions to the temporal lobe.

Nonetheless, the language-equals-left-hemisphere dogma persisted into the 1960s, when neurologist Norman Geschwind (1926–1984) famously revived the Wernicke-Lichtheim model, which had faded from prominence in the wake of more holistic and behaviorist approaches to the mind/brain in the early twentieth century (Rutten, 2017).) Along with the Geschwind-led revival of the classical model came a renewed interest in word deafness and its neurological correlates. Geschwind articulated an elegant account that, following Henschen's model to a tee, placed the center for word-sound recognition in Wernicke's area in the left hemisphere and downstream from primary auditory cortex (A1), which was, of course, bilateral.

This left two pathways for word-clangs to get to Wernicke's area: one directly, from the left primary auditory cortex, and the other indirectly, from the right auditory cortex via the corpus callosum, the white-matter superhighway connecting the two hemispheres. Interrupt one pathway, such as by damaging left primary auditory cortex or anywhere in the right hemisphere, and word recognition is fine. Damage both pathways, though, and you have full-blown word deafness. Now notice that there are two lesion patterns that might damage both pathways. One involving bilateral damage to the superior temporal lobe, which snips the transmission near its source in each hemisphere, or a single left hemisphere lesion, which simultaneously snips both the within left-hemisphere pathway and the right-to-left crossing pathway. Geschwind (1965, p. 285) noted, "There are . . . many more cases recorded of this syndrome from bilateral lesions," but he emphasized the existence of the rarer unilateral examples to anchor this theory.[2] Perhaps this was the kind of model that Henschen had in mind to reconcile the bilateral lesion pattern with a unilateral word-clang decoder.

The Henschen/Geschwind model of pure word deafness makes a clear prediction: damage to left superior temporal lobe should cause the *symptom* of word deafness (a dense speech comprehension deficit), even if it coocurs with the other symptoms of Wernicke's aphasia (e.g., disrupted speech output). Geschwind (1965, p. 286) writes, "Pure word-deafness as the preceding discussion suggests probably results from the disconnexion of Wernicke's area from auditory stimulation. The normalcy of the patient's speech testifies to the intactness of Wernicke's area. With a lesion in Wernicke's area proper not merely is verbal comprehension impaired [as in word deafness], but speech is also impaired."

Wernicke's aphasia is word deafness plus impaired speech, according to Geschwind. But the receptive deficits in the two syndromes are not the same, as Lichtheim noted: "Patient gives the impression of an absolutely deaf man, and differs in this respect from other [i.e., Wernicke's aphasia] cases of word-deafness" (Lichtheim, 1885, p. 462). Geschwind, as Henschen had done before him, missed this critical clue. Years later, a junior colleague of Geschwind's, Harold Goodglass (1920–2002), noted another logical flaw in the disconnection analysis of word deafness. In his book *Understanding Aphasia,* Goodglass (1993, p. 125) writes, "One puzzle is why word-deafness is not seen more frequently in cases with extensive left subcortical lesions affecting both the left temporal isthmus and auditory pathways from the corpus collosum."[3]

To be fair to both Henschen and Geschwind, this was an easy clue to miss. Quantitative differences in severity are common in neurological disease and amounted to small potatoes in the context of the bigger problems that they were trying to solve in the neurology of speech and language. Further, quantification of linguistic abilities was not a strong feature of early research on aphasia (to say the least, as it relied mostly on clinical impressions). In fact, one might even make the case that the Henschen/Geschwind interpretation of the data was appropriately conservative. Nonetheless, their interpretation of the evidence helped cement the dogma of the strong left-dominance of speech perception.

Evidence against a Strongly Left-Dominant Word-Clang Decoder

The empirical landscape changed substantially in the 1970s and 1980s. During this time, several experiments yielded convincing evidence that

Henschen and Geschwind were wrong about word deafness and the idea that the word-clang decoder was strongly left dominant. Yet the standard theory—that Wernicke's area is the word-clang decoder—emerged unscathed through those decades and persists today. What is the evidence against the standard theory, and why didn't it change the theoretical landscape?

After the cognitive revolution of the 1950s and 1960s, when psychology shifted from behaviorism to cognitivism, psychologists and linguists flocked to the study of language and aphasia. They brought with them much more sophisticated experimental methods that were up to the task of providing detailed tests of neural models of language. One model that they tackled was the Wernicke/Lichtheim/Henschen/Geschwind claim that the source of the comprehension problem in Wernicke's aphasia is an inability to convert acoustic input to phonological word forms. Russian neurologist Alexander Luria (1902–1977) deserves mention here as well for his clearly articulated hypothesis that people with damage to Wernicke's area cannot carry out "phonematic analysis" (Luria, 1970; Luria & Hutton, 1977, p. 113), which today we might call a phonological-level impairment in the ability to perceive speech.

The clearest test was carried out by Errol Baker, Sheila Blumstein, and Harold Goodglass at the Aphasia Research Center in Boston (Baker, Blumsteim, & Goodglass, 1981). These authors recruited a group of patients with Wernicke's aphasia and gave them a simple test. It involved listening to a spoken word, such as "bear," and then pointing to the matching image from an array of four pictures. The pictures included the target, a phonological distractor (a pear), a semantic distractor (a moose), and an unrelated distractor (grapes), which was semantically related to the phonological distractor to remove asymmetry cues in the array (figure 3.4). If the comprehension problem in Wernicke's aphasia were attributable to a phonological deficit, patients should make errors where they choose the phonological distractor; if they were profoundly word deaf, they should select items randomly.

The results were *not* consistent with a primary deficit in speech-sound decoding as the basis for comprehension problems in Wernicke's aphasia. Overall, performance was impaired as expected (68.4 percent correct) but still well above chance (25 percent). That's the first clue: the patients were not behaving as if they were word deaf. The second, more telling clue came from the error-type breakdown. In contrast to the speech sound deficit theory, nearly half (47.6 percent) of the patients' errors were semantic (choosing the moose, in our example). Phonemic errors were less frequent (38.7 percent of all errors)

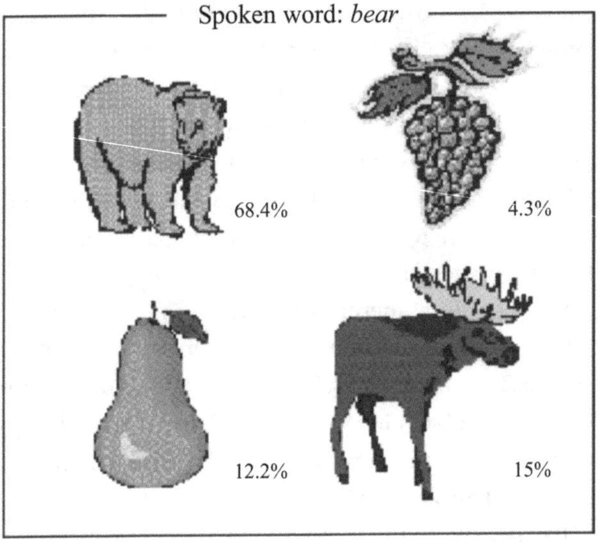

Figure 3.4
Stimulus example for the word comprehension test employed by Baker et al. (1981). Percentages indicate the average rate of selection of each picture type (target versus unrelated, phonemic, and semantic distractors) in the study. The stimulus array here is an actual item from Rogalsky et al. (2022).

and unrelated errors were relatively rare (13.7 percent of errors). In terms of overall performance, the breakdown is as follows: correct = 68.4 percent, semantic = 15 percent, phonemic = 12.2 percent, and unrelated = 4.3 percent. The implication is clear. The source of the comprehension problem in Wernicke's aphasia is at least as much semantic as it is phonemic, if not more so.[4] The century-old hypothesis of Wernicke, Lichtheim, Henschen, Geschwind, and Luria appeared to be incorrect, or more accurately, the speech-sound decoding theory isn't the whole story.[5]

Evidence from an entirely different source was already available in the 1960s, shortly after Geschwind resuscitated the classical models and proposed his disconnection analysis of word deafness. In the early 1960s, neurosurgeons Joseph Bogen and Philip Vogel were pioneering the modern human split-brain surgery for the treatment of epilepsy. The procedure, which involves cutting the corpus callosum to prevent the spread of seizures from one hemisphere to the other, had been tried in the 1940s by Dr. William P. Van Wagenen (Van Wagenen & Herren, 1940), but the medical community did not consider it successful. However, upon closer examination and in light of recent animal work on split brain surgery by Roger Sperry (1913-1994) (Sperry, 1961), Bogen thought otherwise and operated on three cases (Oransky, 2005). These individuals were studied intensively by Sperry and neuropsychologist Michael Gazzaniga.

Medically, the surgeries were a success: the seizures were under control and the patients appeared to be doing quite well, with "a remarkable absence of functional impairment in nearly all ordinary behavior" (Bogen, Fisher, & Vogel, 1965, p. 1329). Sperry and Gazzaniga, however, had been able to document several, now familiar neuropsychological peculiarities due to the lack of communication between the hemispheres (Gazzaniga, Bogen, & Sperry, 1962). Of concern to us here, though, are the results of language testing (Gazzaniga & Sperry, 1967).

It was clear that expressive language was possible only by the left hemisphere of all three patients, further confirming Broca's claim that we speak from our left hemisphere (Berker, Berker, & Smith, 1986). But receptive language testing revealed something different. Upon initial examination, it appeared that the right hemisphere could not comprehend language, but this was only because the tasks that were used initially required verbal output. When Gazzaniga and Sperry modified the tasks so responses could be made by nonverbal means, a different picture emerged. In two of the three initial cases, Gazzaniga and Sperry (1967, p. 143) clearly demonstrated the ability of the isolated right hemisphere to comprehend speech, at least at the word level: "Samples of some of the more complicated test words for which some recognition and understanding seemed to be present in the right hemisphere are listed as follows: rough, smooth, round, cone, rectangle, cylinder, square, pyramid, flashlight bulb, tweezers, safety pin, measuring instrument, door latch, clown, Indian, etc." The third case was reported to have "negligible" receptive language ability in the right hemisphere, but it was also stated that this individual had suffered considerable brain damage prior to surgery, "especially in the minor hemisphere." (p. 132).

Neuropsychologist Eran Zaidel also studied the receptive language abilities of split-brain patients, including tests of word comprehension using a bear-pear-moose-grapes-type task (Zaidel, 1985). He writes, "The disconnected RHs [right hemispheres] have large auditory vocabularies. . . . Also, standardized auditory-comprehension tests with multiple-choice pictures that include both semantic and phonological decoys usually lead to semantic errors." (p. 210).

In a collaboration with Kathleen Baynes, then a colleague of Gazzaniga's, I had the opportunity to assess a split-brain patient on a modified version of the bear-pear-moose-grapes task. We presented a spoken word to the patient ("bear," for example), which would have been heard by both hemispheres due to the bilateral projection of the ascending auditory pathways. Then we briefly presented a picture—either the matching target, a phonological foil, a semantic foil, or an unrelated foil—in either the left or right visual field, which guaranteed that only the right or left hemisphere, respectively, would see it. The task for the participant was simply to indicate whether the picture matched the word. We found that both hemispheres performed remarkably well in discriminating the target from the various foils. Here is the data showing the percent correct for each stimulus type in the right and left hemispheres:

- *Left hemisphere:* Target (100 percent), phonological foil (98 percent), semantic foil (80 percent), unrelated foil (100 percent)
- *Right hemisphere:* Target (90 percent), phonological foil (96 percent), semantic foil (84 percent), unrelated foil (100 percent)

Consistent with evidence from aphasia, this split-brain patient had the most trouble with semantic foils (but still well above the 50 percent chance level) and did quite well with phonological foils in both hemispheres.

Taken together, these studies show that the right hemisphere of split-brain patients is reasonably well equipped for word-clang decoding, disproving the standard model of a strongly left-dominant speech perception system. Indeed, from their own early work, it did not escape the notice of Gazzaniga and Sperry that their findings contradicted Geschwind's claims. In the summary of their 1967 report (p. 147), they state the following: "Our evidence that the disconnected minor hemisphere perceives and comprehends both the written and spoken word . . . contrasts with previous reports of 'word blindness' and 'word deafness' following callosal lesions (Geschwind, 1962)."

But the standard theory persisted despite some authors, like David Poeppel, arguing for a more bilateral organization of speech perception (Poeppel, 2001), a view that I eventually came to favor (Hickok & Poeppel, 2000). Clearly, further evidence would be needed to convince the field to give up on the deeply ingrained dogma.

In 2008, I teamed up with three of my graduate students at the University of California, Irvine, Kayoko Okada, Judy Pa, and Corianne Rogalsky, along with neurologist Arthur Grant and his team at New York University, to study the comprehension ability of the isolated left or right hemisphere using a different method, the Wada test. This procedure was developed by the Japanese neurologist Juhn Wada to assess the language lateralization of neurosurgical patients prior to surgery (Wada & Rasmussen, 1960). It involves injecting sodium amytal into the left or right carotid artery, which supplies the left or right cerebral hemisphere, respectively, effectively putting one or the other side of the brain to sleep for several minutes. While one hemisphere is out of commission, clinicians can assess the language function of the awake hemisphere. When the clinical work is done, and with the patient's permission, researchers can step in and run some experimental protocols.

Our experiment involved the bear-pear-moose-grapes word-to-picture matching test used by Baker, Blumstein, and Goodglass. Two things made our study different from theirs, however. One was that the *entire* left (or right) hemisphere was deactivated, rather than just a portion damaged by stroke. The second was that the deactivation was acute (it happened in seconds once the drug was injected), which allowed no time for the brain to reorganize or compensate for lost function. From a research standpoint, this means that our experiment was a particularly strong test of both the claim that speech perception is strongly left dominant and the hypothesis that the comprehension problems in Wernicke's aphasia results from a speech sound decoding deficit. Based on the previous research described here, we expected that the right hemisphere alone would be capable of comprehending words reasonably well and when errors occurred, we would see at least as many semantic mistakes as phonemic. We studied twenty people, most of whom were under evaluation for surgical treatment of temporal lobe epilepsy (Hickok et al., 2008).

Our predictions were confirmed by the results, shown in table 3.1, which provides the percentage of each response type during left-hemisphere anesthesia. This is compared with the data from the Baker et al. (1981) study discussed previously, as well as another study that used a similar task on a group of unselected people with chronic aphasia (Gainotti, Caltagirone, & Ibba, 1975). As in aphasia, our Wada study revealed that word comprehension was abnormal following left-hemisphere deactivation, and to a similar degree. But again, the errors were more often semantic than phonemic. In fact, in our Wada study, there were more than twice the number of semantic errors as phonemic errors, which is similar to the pattern in left-hemisphere aphasia in general and the split-brain patient discussed previously.

Table 3.1
Similar rates of performance on word-to-picture matching tests with phonological, semantic, and unrelated distractors from three samples of participants in three studies

	Left-Hemisphere Anesthesia	Wernicke's Aphasia	Unselected Aphasia
Correct responses	73.8%	68.3%	72%
Semantic errors	17%	15%	19.7%
Phonemic errors	7.50%	12.20%	5%
Unrelated errors	<1%	4.30%	3%

Another study by Argye Hillis and our group tested 237 acute left hemisphere stroke patients on a modified version of the bear-pear-moose-grapes test and found that only about 12 percent of patients performed worse than 90 percent correct in discriminating a target picture from a phonological foil and only 4.6 percent of patients performed worse than 80 percent (Rogalsky et al., 2008). At all levels of severity, patients had more trouble with semantic distractors than with phonemic distractors and the discrepancy grew as the severity increased. Thus, we see a skewed distribution of word comprehension abilities in acute stroke, with most people doing quite well and only a few having substantial difficulties. Moreover, most of these patients will recover, some substantially, leaving very few people with chronic moderate or severe word comprehension problems following unilateral damage.

To provide yet another examination of this question for this book, and with the help of my collaborator and former student Corianne Rogalsky, we probed our stroke dataset, identifying twenty-four cases of left unilateral temporal lobe damage, most with damage involving classical Wernicke's area. The standard model predicts that we should see significant deficits in this group of patients with damage to the presumed word-clang decoder. Yet the average score on the bear-pear-moose-grapes test in this sample was 97.9 percent correct; sixteen people had a perfect score, six got 95 percent, and the two lowest scorers were at 90 percent accuracy.

The data from cases of Wernicke's aphasia, unselected chronic and acute aphasia, split-brain, Wada procedures, and people with chronic temporal lobe lesions tell a coherent story. Wernicke, Lichtheim, Henschen, Geschwind, and Luria were wrong. The word-clang decoder is not to be found exclusively in the left posterior temporal lobe. Damage to the posterior temporal lobe does not regularly cause the symptom of word deafness as it should if these early theorists were correct. In contrast, the majority of cases of word deafness have *bilateral* superior temporal damage. There is one simple explanation for this pattern: the word-clang decoder is, to a first approximation, bilateral.

Yet unilateral cases of word deafness still exist, even if they are rare. What do we make of these? Previous theorists, such as Geschwind, accounted for their rarity by assuming that the lesion had to be small and exquisitely placed to doubly disconnect Wernicke's area from the two acoustic pathways leading to it. Given that strokes are not smart bombs, these cases are rare. But this logic works only if word deafness also results from the total destruction of Wernicke's area due to stroke, the total isolation of the right hemisphere via callosal surgery, or the total deactivation of the left hemisphere via the Wada test. We've seen that this is not the case. The logic doesn't hold. We need a different explanation. The only explanation that I can think of is one proposed by Goodglass (1993):[6] unilateral cases occur in people with atypically strong left dominance. They are the exception that proves the rule.

Summary

What we've seen in this chapter is that from the start of modern neurology in the nineteenth century, the empirical record never supported the widespread view that speech perception is strongly left dominant. It was a reasonable assumption in the early days, given that speech production does seem to be strongly left dominant and comprehension deficits were observed following left posterior lesions. But the source of the comprehension problems was only *assumed* to hold at the phonemic perception level. When researchers finally tested that assumption in the 1970s and 1980s, it was found to be faulty. Add to this the bilateral lesion pattern associated with word deafness, the receptive speech abilities of the isolated right hemisphere in split-brain and Wada studies, and the relatively good receptive speech perception abilities of large samples of acute and chronic stroke patients, and one wonders why the view persists that "speech perception and production are left lateralized in the human brain" (Rauschecker & Scott, 2009, p. 720).

Why is there so much resistance to the idea that this part of the receptive language system is more bilaterally organized? I suspect that the reason can be found in the long obsession with the duality of mind and brain, which was heightened first with the discovery of the left-hemisphere dominance for expressive speech, and later with split-brain research (Bogen, 1969). We turn to this issue next, in chapter 4.

4 Left Brain, Right Brain: Wrong-Minded

> In summary, the two halves of the encephalon, being perfectly identical from an anatomical point of view, cannot have different functions . . .
> —Paul Broca, 1865

> The human brain is actually two brains, each capable of advanced mental functions.
> —Michael Gazzaniga, 1967

These quotes appear to reflect two very different viewpoints on the nature of the cerebral hemispheres. Broca seems to promote a more symmetric, wholistic view, while Gazzaniga champions a more asymmetric, dualistic picture. Most of us would probably side with Gazzaniga. After all, isn't that the major lesson we learned from split-brain research? A close read, however, reveals that there is room for both perspectives. The two sides can each have advanced mental functions, but this does not necessarily imply a strong asymmetry or complementarity. Gazzaniga's viewpoint resonates more with us these days because we have inherited the split brain–inspired duality of mind philosophy as a replacement for, rather than a refinement of, the older view. My goal in this chapter is to bring back some balance between the two, and even nudge the needle a bit more toward Broca. It will help to start by tracing the early development of ideas on asymmetry of function.

Wrestling with Asymmetry of Function

The functionally lateralized view of the brain and mind is younger than you might think, particularly if you are a bit puzzled by the introductory quote from Broca (Bogen, 1969; Harrington, 1985). For most of the 1800s and well into the 1900s, the idea that the two hemispheres might support different mental functions was considered "patently absurd, an insult to the wisdom of Nature" (Harrington, 1985, p. 620). This is because such an idea violated the "laws of symmetry," which held that the limbs, sensory organs, and cerebral hemispheres "necessarily function identically and in unison" (Harrington, 1985, p. 619). When the two sides of the brain failed in their state of functional unity as a result of pathological breakdown, insanity was the result, or so it was proposed (Wigan, 1844).

In this context, Broca's observation in the 1860s that the control of articulate speech is a function of the *left* (not right) inferior frontal gyrus (IFG) challenged this deep-seated understanding of the relation between mind and brain and, understandably, led to heated and sometimes indignant discussion. For example, one commentator, Paul Briquet, quipped, "Would it be possible for the right eye to see only blue, black, and red, while the left eye would be made for seeing only green, yellow, and blue? Could the right ear hear in music nothing but do, re, mi, fa, and the left ear so, la, ti?" (quoted in Harrington, 1985, p. 620).

Broca himself was keenly aware of the conundrum that resulted from his findings. Writing in 1865, after even more evidence had been revealed from the unpublished work of Marc Dax (1770–1837) decades earlier, Broca acknowledged the implication of the proposition as follows: "Is there then a functional difference between the two halves of the encephalon? If this proposition should be accepted, it would disrupt our understanding of physiology. . . . Now there is one physiologic law, which everywhere else [in the body] is without exception, namely, that two organs that are equal and symmetrical have the same attributes, and it would be quite strange that this law should present here a marked exception" (from the translation in Berker et al., 1986, p. 1067).

So he proposed a reconciliation between the physiological law of symmetry and the observed left-dominance of the capacity for articulate speech, allowing both principles to coexist. Drawing on the analogy of left-hemisphere control of the right-dominant hand (in most people), Broca concluded (quoted from Berker et al., 1986, p. 1070):

> In summary, the two halves of the encephalon, being perfectly identical from an anatomical point of view, cannot have different functions, but the more precocious development of the left hemisphere makes us prone, in our first groping ways, to execute with that half of the brain the manual and intellectual actions that are

> most complex. Among such actions, one must certainly include the expression of ideas by means of language and, more particularly, articulate speech.

In short, the two hemispheres are organized in functionally identical ways and are indeed symmetrical at birth, preserving the law of symmetry. It is just that the left hemisphere is more precocious in its development, leading to its dominance in complex functions like speech and language.

Over time, Broca's developmental argument for the left-dominance of speech morphed into the view that the left hemisphere is inherently endowed with a general superiority to the right for human intellectual ability (Harrington, 1985). We see this clearly in Henschen's review of left- versus right-hemisphere function (1926, p. 120, p. 122, respectively):

> We have seen that in speech, music and calculation, the left hemisphere plays a predominant or exclusive role. . . .

> All these facts seem to be in favour of the theory of a primary organization [a basic organizational principle] of the brain as the original cause of the predominance of the left side of the brain. In nature there seems to be a law that in order to attain full development of higher faculties originally distributed over two organs, a concentration of the capacity in only one of them is necessary.

It wasn't long before textbooks crystalized the idea for the next generations of scientists, as here from Strong and Elwyn's text, *Human Neuroanatomy*: "It is an interesting but unexplained fact that in man the higher cortical functions are vested principally in one cerebral hemisphere, the left one in right-handed individuals, a phenomenon known as 'cerebral dominance'. The various agnosias and apraxias discussed above occur only when the dominant hemisphere is damaged, lesions of the other hemisphere producing as a rule no recognizable disturbances." (1948, p. 393).

To be sure, there were a few theorists, such as John Hughlings Jackson (1835–1911), who promoted complementary equality of function in the left and right hemispheres: "If, then, it should be proved by wider evidence that the faculty of expression resides in one hemisphere, there is no absurdity in raising the question as to whether perception—it's corresponding opposite—may not be seated in the other" (Jackson, 1864, p. 604).

But "the tide was running against Jackson," as Joseph Bogen put it in his review of the history of thought on cerebral lateralization (1969, p. 136). When the right hemisphere was mentioned at all during these early days of neurology, it was often to point out its inferior ability in the same higher mental functions carried out more perfectly in the left, or its possible involvement in less intellectual functions, such as emotion or "animality" (Harrington, 1985, p. 622).

Slowly, however, researchers opened their minds to the possibility that the right hemisphere might have some superior talents of its own. Notably, the 1930s saw the emergence of research on the effects of right-hemisphere damage on nonverbal tasks, such as those developed as part of IQ scales (Bogen, 1969; Weisenburg & McBride, 1935). Even so, while these new tests confirmed a left-hemisphere dominance for verbal task performance, differences on nonverbal tasks produced equivocal results, with some studies showing that right-hemisphere patients were worse than left-hemisphere patients, while other studies showed no difference (Heilbrun, 1959).

Even as more evidence accumulated regarding the type of deficits that one could or couldn't document following right-hemisphere damage, nagging questions remained regarding whether the results indicated a true right-brain dominance for some abilities. Théophile Alajouanine and François Lhermitte's evaluation of the various types of deficits that one finds after left versus right parietal lobe lesions exemplifies the complexities of interpretation: "A lesion of the right parietal lobe, far from revealing any particular function in this hemisphere, demonstrates on the contrary its minor role. It gives rise to a behavioral pattern whose main feature derives from the activities of the major hemisphere" (quoted in Bogen, 1969, p. 141).

A (literally) shocking experiment published in 1968 in the journal *Science* illustrates one attempt to clarify the situation (Cohen et al., 1968). The study used strong electroconvulsive shocks, then a common treatment for psychiatric disorders, to induce seizures in anesthetized patients, and then measured the disruptive effects on previously learned tasks. The investigators trained a group of eight patients on two paired-associate learning tasks, one involving memorizing arbitrary pairs of words and one involving memorizing shape patterns. They then shocked the participants' brains with 150 volts delivered either to the left, right, or bilateral cerebral hemispheres in separate treatments. The authors reported that a shock to the left hemisphere impaired memory for the verbal task more than a shock to the right hemisphere, and vice versa for the nonverbal task. Bilateral shocks, however, were reported to be the most disruptive to both tasks. While this result showed a clear interaction between task and hemisphere, lending support to the view that the right hemisphere may in fact be better than the left on nonverbal tasks, an examination of the data still reveals a justification for caution. I've plotted

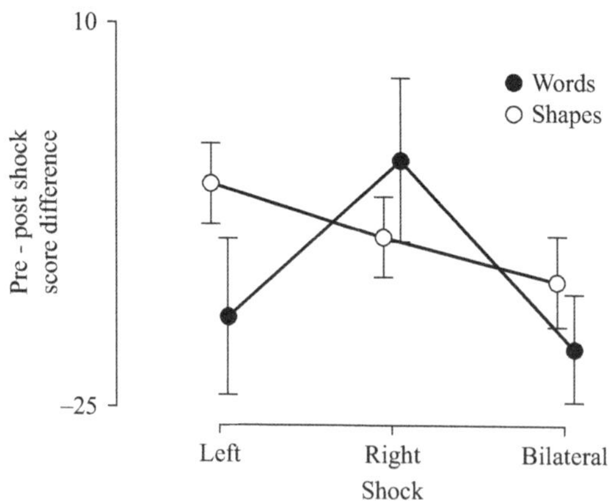

Figure 4.1
The effect of electroconvulsive shock to the left, right, or bilateral cerebral hemispheres on a word versus shape task. Plotted from data in Cohen et al. (1968). The vertical axis indicates performance level, with lower being worse. The error bars are confidence intervals.

the averaged data in figure 4.1 to show the result more clearly—namely, most of the effect is driven by hemispheric differences on the word task, not the nonverbal shape task.

What changed the game by erasing all (or at least most) doubt was split-brain research. When that work started, it took some effort just to demonstrate that the isolated right hemisphere in split-brain patients wasn't completely impotent, as its lack of expressive verbal abilities seemed to imply. But once Michael Gazzaniga and his team switched from verbal to nonverbal response tasks, it became astonishingly clear that the right hemisphere was quite capable of understanding and problem solving and had some level of consciousness quite independent of the left hemisphere (Gazzaniga, 2015). Soon it was shown that it even outperformed the left hemisphere on some tasks, such as arranging blocks to match a pattern. Gazzaniga's astonishing conclusion framed the new revolutionary understanding of hemispheric asymmetries: "The human brain is actually two brains, each capable of advanced mental functions" (Gazzaniga, 1967, p. 24).

With such a proclamation, and with a clear confirmation that there were indeed different abilities in the two hemispheres, the obvious research program was to uncover precisely what the various abilities are. So that was the focus of much research, many books, conferences, and even new journals for the next three decades. The trajectory of interest in and research on hemispheric asymmetries was steep. In the 1960s, only a handful of studies were published with "hemispheric asymmetries" and related terms in their titles or abstracts. The 1970s saw that number rocket up more than forty times and then increase another 400 percent in the 1980s over the 1970s.

Geschwind, for his part, made another major contribution in the late 1960s that further stoked enthusiasm for a strong theory of lateralization. He reported an anatomical asymmetry in the vicinity of Wernicke's area: the planum temporale is larger on the left than the right in most people, which was interpreted as a reflection of that area's left dominance for linguistic function (Geschwind & Levitsky, 1968). The classical assumption that the hemispheres were essentially identical from an anatomical point of view was shattered.

There followed a parade of abilities for which the right hemisphere seemed to have an advantage: face recognition (Leehey et al., 1978), music (Bever & Chiarello, 1974), emotion (Ley & Bryden, 1979), visual-spatial attention (Heilman & Van Den Abell, 1980; Vallar & Perani, 1986), and discourse-level language processes (Beeman, 1993; Hickok et al., 1999; Winner et al., 1998), to name just a few. These findings provided strong support for the idea that there is no "minor hemisphere," that each had its own set of proclivities. I also participated in the lateralization excitement as an eager young scientist studying sign language aphasia, adding to the literature of left-right hemisphere dominance patterns (Hickok et al., 1995; Hickok, Bellugi, & Klima, 2001; Hickok, Klima, & Bellugi, 1996).

Thus, the historical trend in thinking about brain (a)symmetry went from a dominant view of complete symmetry and unitary function in the pre-1860s era, to functional symmetry with augmentation of one or the other hemisphere through skill development (Broca), to functional symmetry with augmentation built into a dominant left hemisphere (Henschen), to full-blown functional *asymmetry*, at least for some higher functions (Gazzaniga, Bogen, indeed almost everyone), which morphed into the overblown popular view of two entirely different minds, abilities, and ways of thinking.

Indeed, by the 1990s, the scientific obsession with cerebral asymmetries had fully inebriated the public. This was due, in part, to theoretical musings sparked by split-brain observations. Bogen himself wrote about the "duality of the mind" and "two different ways of thinking" supported by the two hemispheres. The left hemisphere, he argued, was propositional, logical, and analytic, while the right hemisphere was appositional, intuitive, and gestalt oriented (Bogen, 1969). He viewed these as a kind of yin and yang–style duality of thought, which resonated with the Eastern-looking philosophical mindset of the 1960s and 1970s. Psychologist Michael Corballis suggested that the popular book *The Psychology of Consciousness*, by Robert E. Ornstein (1972), which pursued Bogen's speculations with vigor, is largely responsible for the popularization of dualistic interpretations of the left and right brains (Corballis, 1980). We could also point to other pop science books such as Springer and Deutsch's *Left Brain, Right Brain* (1981), which delivered the basic science to the masses, where it achieved mythical status.

Asymmetries in Receptive Speech

We ended chapter 3 with a question: Why the resistance to the idea that speech perception is bilaterally organized? The answer that I'm developing is that we in the field have focused so much on what's different about the two hemispheres that we have failed to fully appreciate the similarities. I've sketched the backstory of hemispheric asymmetries generally, showing how today, we tend to think of them as much more dramatic than previous generations of neurologists and cognitive neuroscientists did, largely because of the remarkable work on the split brain. Now let's situate research on receptive speech in this context.

Given the obsession with asymmetries in the 1960s through 1990s, it is no surprise that left-right asymmetries in the perception of speech and other kinds of acoustic information were investigated intensely and discovered. Most of the research buzz in the early days of this period centered on a technique called *dichotic listening*. The method, developed in the 1950s by British psychologist Donald Broadbent (1926–1993), involved presenting two auditory signals simultaneously, one in each ear. In a typical experiment, three or four pairs of words were presented in sequence, and the listeners' task was to report back as many words as they could remember. The object of study in Broadbent's hands was to assess the effects of multiple spatially distributed streams of information on attention and working memory (Broadbent, 1954). A few years later, Brenda Milner, a pioneer and titan in modern neuropsychology, whose career remained active past her 100th year, applied dichotic listening to study the effects of left versus right temporal lobe resections on auditory perception (Milner, 1958). A student of Milner's at the Montreal

Neurological Institute, Doreen Kimura (1933–2013), developed the method into a workhorse for the field.

The initial observation was that the number of words that could be recalled using the dichotic task was lower in patients with left temporal lobe disease than right temporal lobe disease, and lower still after the diseased portion of the left temporal lobe had been removed. Right-hemisphere resection had no significant effect (Kimura, 1961, 1967). But Kimura noticed something else in the data. In the presurgery testing, more items were recalled from the sets of words presented to the right ear than to the left ear. This *right ear superiority* was true for all patient groups and held for healthy controls as well. What was going on?

To solve this puzzle, Kimura drew on animal work suggesting that even though there were connections from each ear to both hemispheres, the contralateral (opposite-side) connections were stronger than the ipsilateral (same-side) connections. She reasoned, "The explanation for the right-ear superiority on the digits [number word] test, then, was that the right ear had better connections with the left hemisphere than did the left ear, and since the left hemisphere was the one in which speech sounds were presumably analyzed, the right ear sounds had the advantage of having better access to these speech centres" (Kimura, 1967, p. 164).

Now researchers interested in cerebral asymmetries had a ready tool for studying them in healthy people. (Recall that this was before the advent of functional imaging.) In short order, dichotic listening methods were used to assess hemispheric asymmetries for melodies (Kimura, 1964), rhythm (Gordon, 1978), consonants versus vowels (Shankweiler & Studdert-Kennedy, 1967), prosody (Blumstein & Cooper, 1974), syntax (Zurif & Sait, 1970), semantics (Heeschen & Jürgens, 1977), and more. Special issues of journals were published on the topic, as were bibliographies of the mushrooming body of work (Richardson & Knights, 1970) to help pre-Internet researchers track the rapid progress.

For our purposes, the two important nuggets that emerged from these studies were a right ear/left hemisphere advantage found for words (more specifically consonants) and a left ear/right hemisphere advantage for melodies and prosody. This was taken to indicate a "left hemisphere dominance in speech perception . . . at the level of speech sound structure" (Shankweiler & Studdert-Kennedy, 1967, p. 62), that "the processing of spoken nonsense sounds is carried out by the left hemisphere" (Kimura, 1967, p. 176), and that "the right hemisphere may thus be especially important for the perception of melodic patterns" (Kimura, 1964, p. 357). This is a good example of a theory positing very different functions for the two hemispheres. But these dichotomous interpretations (pun intended) overstepped the data by zeroing in on just the differences and ignoring the overlapping abilities of the two hemispheres. Let me elaborate.

The validity of dichotic listening was anchored from the beginning in traditional neuropsychology. In the foundational studies, left- versus right-ear advantage was correlated with language lateralization as determined by the Wada test (right hemisphere–dominant patients on Wada showed a left-ear advantage) and the speech versus music ear advantage pattern found in healthy people mirrored the pattern found after left versus right temporal lobectomies (Kimura, 1967). This means that to understand the nature of the dichotic ear advantage results, we should look closely at the data from the neuropsychological anchors. When we do, a more nuanced picture emerges.

Consider the correlation with the Wada test. Language dominance as revealed by the Wada test primarily measures expressive speech. Receptive speech at the phonological and word levels is more bilaterally organized, as we saw in chapter 3. So either dichotic listening is picking up on much more subtle perceptual asymmetries, or it reflects the expressive reporting task that was customary of dichotic testing, or working memory, or perhaps asymmetries in attentional mechanisms—Broadbent's original motivation for inventing dichotic listening.

The left versus right temporal lobe resection data paint a similarly nuanced picture. I noted at the beginning of this section that left resection patients recalled fewer words than right resection patients on the dichotic task. But fewer by how much? Kimura reports a 7-percentage-point pretest difference between right- and left-hemisphere diseased patients (91 percent versus 84 percent recall, respectively) and a 4-percentage-point difference between pre- and post-left-hemisphere resection (84 percent versus 80 percent recall, respectively). Pre- and post-right-hemisphere cases showed a statistically nonsignificant, 2-percentage-point decrement (Kimura, 1967). To summarize, the asymmetries amount to differences of less than 10 percentage points, which indicate asymmetries of diminutive degree, not black-and-white dichotomies. In other words, there are vastly more similarities of ability in the two hemispheres than differences.

Dichotic listening wasn't the only source of data on hemispheric asymmetries in the acoustic domain. As a complement to Kimura's findings that word recall was affected more by left than right temporal lobectomy, others reported that the perception of tonal patterns or melodies was more affected by right than left temporal lobectomy (Shankweiler, 1966). But again, the effects were subtle. On a recognition memory test for melodies, left-hemisphere diseased patients correctly recognized 23.7 melodies (out of a total of 40) preoperatively and 23.9 melodies postoperatively after resection of the left anterior temporal lobe. Right-hemisphere diseased patients got 24.1 items correct preoperatively and 21.9 postoperatively (about 9 percent difference) (Shankweiler, 1966)—worse to be sure, but not hugely so.

In the mid-1970s, a study of the ability of left- or right-hemisphere-damaged stroke patients to comprehend emotional prosody in speech appeared to provide dramatic evidence for a strong right-lateralized ability. Patients were asked either to select among pictures that matched the literal content of a sentence or select among line drawings of facial expressions that matched the emotion of the sentences (angry, happy, sad, indifferent). All the participants performed perfectly on the content-matching task, but on the emotion-matching task, the right hemisphere patients did no better than chance, scoring only 26 percent correct; the left hemisphere group performed at 64 percent correct (Heilman, Scholes, & Watson, 1975), a pretty large difference.

Not long thereafter, Elliot Ross and Marcel Mesulum reported a disturbance of *expressive* speech prosody following right-hemisphere damage (Ross & Mesulam, 1979). Ross followed this up in 1981 with a paper that presented ten cases in which he assessed the ability to produce, comprehend, and repeat emotional prosody information. He reported that right-hemisphere damage could yield *aprosodias* that looked a lot like the aphasias caused by left-hemisphere damage, and with similar lesion distributions: expressive aprosodias were caused by right frontal damage, receptive aprosodias were caused by right temporal lobe damage, and so on. He proposed a mirror language system in the right hemisphere, but controlling emotional speech information rather than linguistic content. He concluded his article by saying, "It has been my recent experience, as attending neurologist, that the aprosodias are as commonly encountered on the clinical services as are the aphasias" (Ross, 1981, p. 569).

Subsequent research has supported Ross's observation that emotional prosody dysfunction can occur after right-hemisphere damage (Durfee et al., 2021). But as more data accumulated, questions arose regarding whether emotional prosody was strongly right dominant (Cancelliere & Kertesz, 1990; Pell & Baum, 1997). To quote from a 2008 paper, "aprosodic deficits may occur after either RBD [right brain damage] or LBD [left brain damage], thus bringing into question the concept that affective prosody is a dominant and lateralized function of the right hemisphere" (Ross & Monnot, 2008, p. 64). To be sure, differences in the patterns of deficits can be found following left- and right-hemisphere damage (Ross & Monnot, 2008), but, again, the asymmetries that exist are much more nuanced than was initially suggested.

Nuances aside, differences *were* found, even if small, and several theories popped up to explain their source. Cognitive neuroscientist Paula Tallal put forward an influential proposal that the left and right hemispheres differ in their ability to handle rapid changes of information in time, the left side being better at the fast changes (Tallal, Miller, & Fitch, 1993). Speech is full of rapid changes, such as the acoustic transition between a consonant and a vowel, which is measured in a few tens of milliseconds. Music, on the other hand, is substantially slower. This provided a plausible explanation of the left advantage for speech and also jibed with claims about a left-right difference in visual processing, wherein the left hemisphere was thought to be better at high-spatial frequency information (fine-grained or local-level information) and the right hemisphere better at low spatial frequency information (course-grained or global-level information) (Sergent, 1982). Robert Zatorre and colleagues weighed in with a similar proposal, which emphasized the left hemisphere's facility with fine-grained temporal information (discrimination over time) versus fine-grained spectral information (discrimination over frequency or pitch) (Zatorre, Belin, & Penhune, 2002; Zatorre et al., 1992). A bit later, David Poeppel (2003) added his own remix of these proposals with his *asymmetric sampling in time* hypothesis: the left hemisphere samples and integrates the acoustic stream over relatively shorter time windows (on the order of a few tens of milliseconds), whereas the right hemisphere samples and integrates over relatively longer time windows (a couple hundred milliseconds). Poeppel added a further nuance, explicitly proposing that the asymmetry was one of degree, with the left and right hemispheres exhibiting overlapping distributions of computational biases instead of dichotomies.

Poeppel and I reevaluated the available data a few years later and arrived at a hybrid conclusion: "Neural mechanisms for integrating information over longer timescales are predominantly located in the right hemisphere, whereas mechanisms for integrating over shorter timescales might be represented more bilaterally" (Hickok & Poeppel, 2007, p. 396). I'm not convinced that we got it right.

What's interesting about these proposals is that they all moved away from framing asymmetries in terms of distinct representational domains (such as speech sound categories versus melodies) and toward more subtle differences in perceptual analysis. So we have subtle asymmetries with effect sizes in the 10 percent difference range (90 percent similarity) being explained by subtle processing differences. The field was regressing to the law of symmetry.

Speech and the Brain-Mapping Revolution

Functional imaging crashed the party in the late 1980s through early 1990s. The functional imaging revolution had two effects on the study of hemispheric asymmetries. One was to dethrone it as *the* hot topic in cognitive neuroscience. Prior to positron emission tomography (PET) imaging in the 1980s and functional magnetic emission imaging (fMRI) in the early 1990s, most cognitive neuroscientists had to settle for mapping the human brain at the level of the hemisphere. However, once we had the tools to map activation patterns with centimeter resolution, hemisphere-scale differences seemed instantly old-fashioned (although studying asymmetries at the scale of regions remained prominent). The other effect was to fossilize the doctrine of asymmetry in the mind of many researchers. *There were differences,* and that was (and remains) a major focus when researchers look at left- and right-brain regions, regardless of the magnitude of the effect. The data streaming off the shiny new scanners would be viewed, for the most part, through the lens of dichotomy.

What did these early functional imaging studies of speech perception show? Were there dramatic asymmetries in the activation pattern in left and right auditory areas when people listened to speech? No. The response to speech was robustly bilateral and largely symmetric in the superior temporal lobe compared to a resting baseline (Demonet et al., 1992; Petersen et al., 1988; Petersen et al., 1989; Wise et al., 1991; Zatorre et al., 1992). Despite this, and despite all the evidence that we've reviewed so far showing that receptive speech function is mostly bilateral, the three most influential reports led by Petersen et al., Zatorre et al., and Demonet et al. outright claimed a unilateral phonological system despite the observed bilateral response to speech,[1] citing other evidence for their interpretation.

Petersen and colleagues noted that activations are bilateral and symmetric around the superior temporal gyrus, but they noticed a more posterior, left-lateralized focus in the vicinity of the angular and supramarginal gyri. They cited two neuropsychological studies of phonological *agraphia* (output disorders of writing) that implicate this area and concluded that it is a good candidate for "phonological coding." Zatorre and colleagues also cited neuropsychological evidence implicating the left supramarginal gyrus region in phonological short-term memory, not straight-up speech perception ability. Demonet and colleagues' interpretation was influenced in part by the asymmetry in their own data when they compared their phonological task (monitoring for the phoneme */b/* in nonwords following the phoneme */d/* in a previous syllable) to their baseline tone task (monitoring for a rising pitch in a sequence of tones). Activation was bilateral, but more widely distributed in the left hemisphere. They read these neural tea leaves as an indication of left dominance in the task[2] and attributed the right-hemisphere activation to "a role for this area in the analysis of complex sounds, as suggested by Zatorre" (Demonet, et al. 1992, p. 1762). The Zatorre study that Demonet et al. refer to here (Zatorre, 1988) specifically implicated the right Heschl's gyrus on a pitch perception task. But this misaligns with Demonet's findings because speech-related activation did not implicate Heschl's gyrus, but rather the higher-order auditory association cortex. The findings simply didn't line up, a fact noticed by David Poeppel early in his career (Poeppel, 1996). Further, none of these authors included a discussion of pure word deafness, split-brain findings, or the work on the source of the comprehension deficit in Wernicke's aphasia to inform their conclusions regarding hemispheric asymmetries for speech perception.

In contrast to the other three groups, Wise and colleagues (1991) begin their discussion by stating that they are not surprised by the bilaterally symmetric activation to speech: "The involvement of both superior temporal gyri when listening to words, real or otherwise, was expected" (p. 1809). And they even state that "the auditory analysis system and input lexicon . . . considered together appear to be distributed throughout the auditory association cortices of both hemispheres" (p. 1809), referring to work on the bilateral basis of pure word deafness to back up their statement. But still, the overall tone of their commentary slants toward dichotomies. After stating that bilateral activation was expected, they give the reasons why:

> Psychophysical methods (dichotic listening tasks) have demonstrated a right ear advantage for consonant-vowel combinations, which has been interpreted as evidence that the left hemisphere acoustic analysers are better able than those on the right to identify the rapid changes in frequencies (formant transitions) that permit the discrimination of individual words . . . There is no ear advantage for vowel sounds, which are the most constant sounds in the English language, and there is evidence that the right hemisphere is involved in nonlinguistic aspects of language processing, such as comprehension of the affective intonation of speech. (p. 1809)

Such a statement, read through the dichotomy spectacles of most readers, translates to the conclusion that the right hemisphere is responding to vowels, which one can conceptually reframe as nonspeech complex tones, and affective intonation. Even their discussion of the bilateral basis of pure word deafness opens itself to a left-dominant speech hemisphere interpretation: "The problem in many of these [pure word deaf] patients seems to be a defect in an early stage of speech comprehension, at the level of the acoustic analysis system . . . and it is often observed only when there have been bilateral temporal lobe lesions involving the superior temporal gyri" (p. 1809). Dichotomy glasses translation: acoustic analysis is bilateral, but phonological analysis is in the left hemisphere.

Let me be clear—I'm not suggesting ineptness on the part of any of these scientists. They interpreted their data perfectly in line with the current thinking about the state of the evidence. I would have done the same. In fact, I did exactly that when I ran into claims of bilateral speech perception by David Poeppel, then a graduate student at the Massachusetts Institute of Technology (MIT). I'll tell that tale in chapter 5. For now, though, let me summarize and elaborate what I have been trying to convey in this chapter.

Summary, Conclusions, and a Path Forward

The concept of brain asymmetry between the two hemispheres of the cerebrum has evolved dramatically in the last 200 years. Prior to the 1860s, hardly a soul could fathom the idea that the two hemispheres were *not* symmetric in function. When evidence came to light in the 1860s that there is, in fact, a substantial asymmetry in the probability of speech problems following left versus right inferior frontal lesions, Broca proposed what I'll call *symmetry with modification*, which was the dominant view until the 1960s. The 1960s game-changer was split-brain discoveries with results so astonishing that the shackles of the law of symmetry were cast off, allowing an unconstrained view of the kind of asymmetries that could exist, including dichotomies such as verbal versus visuospatial. It was in this later, unconstrained milieu that new evidence for linguistic asymmetries emerged from neuropsychological and dichotic listening research. Although the differences between the hemispheres were subtle for the most part, the interpretations tended to be more categorical, more dichotomous. And even as researchers who spent portions of their careers trying to understand hemispheric asymmetries regressed a bit toward the *symmetry with modification* view, the broader field tended to see the tidal wave of new functional imaging results through the lens of dichotomies and whopping left-right differences that were never supported by the evidence for most abilities. Theories of the hemispheric organization of speech recognition, one of the least lateralized linguistic functions, simply got swept into the left hemisphere by the tidal surge, where it largely remains.

Where do we go from here? I think we need to reset our theory of the left-right organization of the brain and return to something closer to Broca's *symmetry with modification*. Specifically, I would argue that to a first approximation, the two hemispheres are symmetric. What I mean by this is simply that the two sides of the brain are vastly more similar to each other than they are different. In defense of this claim, consider the following four facts:

- The gyrification (bump-and-groove) pattern of the brain is strongly related to the underlying cellular and functional architecture of the cortex (Ronan & Fletcher, 2015), and the major morphological landmarks of these patterns are highly consistent across individuals. Put differently, even though there is individual variation in the details, the brains of any two humans are more similar in their basic anatomy and functional organization than they are different.[3] No big revelation here. How else would we be able to draw functional anatomic maps of the human brain showing the location of various functional areas in relation to gyral and sulcal landmarks? But now consider the fact that the gyrification patterns of the left and right brains are also highly similar—so similar, in fact, that it inspired the law of symmetry. Even Geschwind's famous leftward asymmetry of the planum temporale is subtle, holding for only 65 percent of the brains in his sample. If neural similarity among *individuals* leads us to the conclusion that their different brains are more similar than different, then similarity between *two hemispheres of one brain* leads to the same conclusion: they are more similar than different.

- Confirming the logic sketched so far, damage to different regions within a hemisphere (e.g., visual cortex versus auditory cortex versus motor cortex versus the medial temporal lobe versus prefrontal cortex) will reliably produce radically different deficits, whereas damage to homologous regions in the left versus right hemispheres will tend to produce similar families of deficits (visual, auditory, motor, memory, cognitive), even if some of the details differ.
- Indeed, the most profound discovery from split-brain research, the poster child for hemispheric asymmetry, was the existence of two well-functioning, self-sufficient minds in one head, both with its own will, desires, consciousness, attentional spotlight, reasoning ability, memory, perceptual, and motor skills. The differences are a matter of degree. Even speech production, which is one of the most strongly lateralized abilities, has been found to emerge in the isolated right hemispheres of some split-brain cases (Gazzaniga et al., 1984), and it does so rather often in cases of childhood left-hemispherectomy (Nahum & Liegeois, 2020), although not to fully normal levels in many instances.
- Bilateral lesions to homologous brain regions tend to produce the most severe deficits in a given domain. We've seen in chapter 3 that this is clearly the case for cortical deafness and word deafness. Visual agnosia is another example, typically caused by bilateral damage to the ventral visual stream (chapter 2). Medial temporal lobe memory loss is yet another example, as was shown clearly by the bilateral resections in the famous case of H. M. (Corkin, 2013; see also chapter 8 of this book). Even speech production deficits are much more severe following bilateral lesions, as in *opercular syndrome* (Silveri et al., 2017; Walker et al., 2021). If unilateral lesions cause less severe deficits than bilateral lesions, this indicates that both hemispheres play a role in the ability in question.

These four points provide a sketch of the case for the symmetry part of the *symmetry with modification* hypothesis. I don't think that anyone would deeply disagree with these arguments, but it is worth making them explicit because many of us are trained to focus on the differences (the modification part of the theory) and not think much about the similarities. This leads to potential problems, as we'll see momentarily.

Against this background of broad symmetry, evidence strongly supports the existence of asymmetries between the hemispheres, even if they are subtle in some cases. We must, therefore, admit to left-right modifications to symmetry that arise through the effects and interactions of experience, genetic endowment, and development (Jansen et al., 2015; Mitchell, 2020). These modifications are matters of degree, not kind, such that homologous regions in the two hemispheres can vary in the *way that information is processed* or in their *patterns of connectivity* at the network level. Further, asymmetric modifications occur to different extents for different systems, with some more asymmetric than others. There is also individual variation, with different people having different patterns and degrees of asymmetry.

Consider now how different views of cerebral lateralization can lead to different interpretations of experimental results. Suppose that we adopt the standard hypothesis that speech perception and word recognition are predominantly left-lateralized to the superior temporal lobe. To test this hypothesis, we might recruit a sample of 100 patients who have poor auditory language comprehension on a clinical aphasia screening test. We then administer an experimental test of word comprehension, such as the pear-bear-moose-grapes word-to-picture matching test (chapter 3) and collect MRI data to map the associated lesions. We find (hypothetically) that 95 percent of the patients in our sample have left-hemisphere lesions, and when we correlate the severity of their performance on the word comprehension test with lesion location, we find an association between damage to the left superior temporal lobe and the number of errors made on the comprehension test. This correlation holds (hypothetically) whether we look at the overall score, phonemic foil error rate, or semantic error rate. Such a result would confirm our hypothesis straightforwardly, and we would conclude that speech perception and word recognition are left-dominant activities. This approach is not uncommon today, as many groups (including some of my own collaborators) recruit study participants on the basis of the presence of aphasia.

Now suppose that we take the symmetry with modification perspective and ask a different question: To what extent are speech perception and word recognition asymmetrically organized in the two superior temporal lobes at the individual and population levels? To measure this, we might recruit 200 people, 100 with damage restricted to the left superior temporal lobe and 100 to the right superior temporal lobe (and cases of bilateral lesions, if we can find any). This recruitment must include all cases with the lesion pattern that we are interested in, whether or not they have language problems; otherwise, we'd miss cases of people who have damage to the critical zones but don't show symptoms. Now suppose that we find that the vast majority of people in our sample, even the left hemisphere-damaged cases, perform at or near ceiling on the task. Perhaps we find that of those cases with notable deficits, most of them are left-lesioned but some have damage to the right. In light of such results, would it make sense to claim that speech perception and word recognition is left dominant? No. It is left dominant for some people and right for a smaller fraction of others, and shows no asymmetry in most cases. For a more detailed, quantitative answer to the question, we could simply plot the distribution of performance (or asymmetry) on our test. The result would be a graphic estimation of the (a)symmetry of speech perception and word recognition.

For this book, I generated such a graph from the bear-pear-moose-grapes word comprehension task that we administered to people undergoing a Wada procedure (Hickok et al., 2008). Although this is a relatively small sample, it provides a nifty way to estimate the degree of (a)symmetry for speech perception because one can assess complete suppression of each hemisphere in turn. The Wada method also induces an acute functional disruption, precluding the argument that the remaining ability is due to compensatory plasticity of the nondominant hemisphere.

I made two graphs (figure 4.2), one tabulated from phonological errors (pointing to the picture of the pear rather than the bear) and the other from semantic errors (pointing to the moose), and did so separately for left- and right-hemisphere anesthesia. Unrelated errors were rare, occurring in only 1 percent of the trials, so I did not include those in the graphs. I then calculated the difference in proportion of errors separately for each type in order to derive a simple measure of the degree of symmetry, where a score of zero indicates complete symmetry and scores between zero and 1 indicate degrees of asymmetry, with 1 being complete asymmetry in one direction or the other.[4] I then calculated the percentage of people at each degree of asymmetry.

The overall phonological error rate was low, at 7.5 percent following left-hemisphere deactivation (right = 1 percent). In the phonological error asymmetry graph, you can see that perfect symmetry is the most common

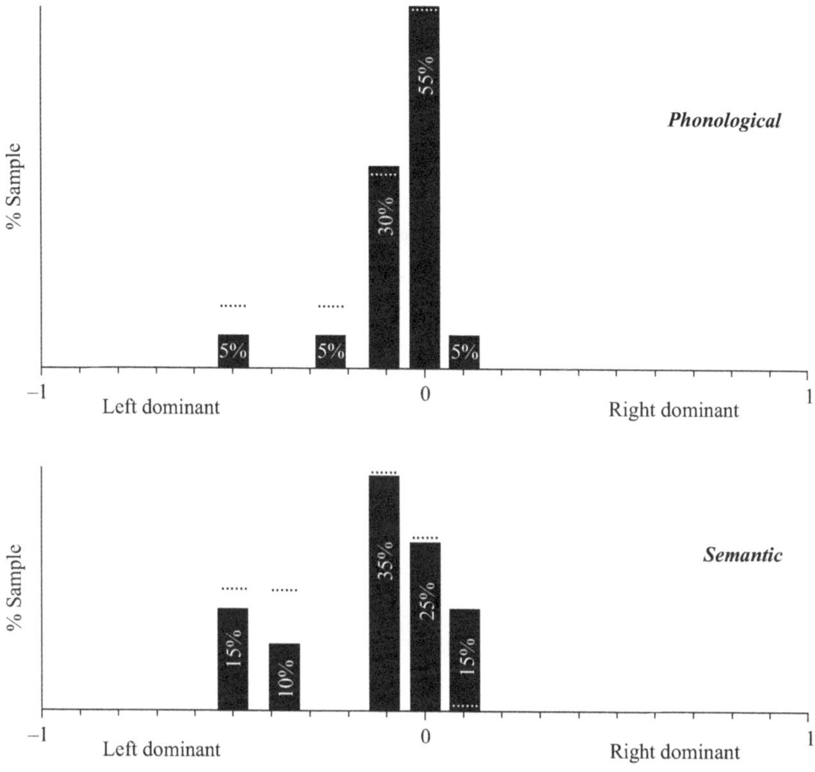

Figure 4.2
Distribution of (a)symmetry in word comprehension during left- versus right-hemisphere deactivation in the Wada procedure, using data from Hickok et al. (2008). The horizontal axis plots the degree of left-right asymmetry expressed as a difference score ranging from –1 to 1, with perfect symmetry corresponding to zero. For the most part, the ticks on the horizontal axis correspond to 10 percent differences in both absolute error rate (0 = no errors, –1/1 = 100 percent errors), as well as 10 percent left-right difference in error rate. For example, the bar on the middle tick between –1 and 0 indicates a 50 percent error rate during left deactivation and 50 percent worse during left versus right deactivation. The only two exceptions to this are one case in the zero bin in the semantic error analysis, who made one error with left deactivation and one error with right deactivation; and another case in the 0.1 bin, who made one semantic error with left deactivation and two semantic errors during right deactivation. For these two cases, their performance was worse (i.e., farther from zero) than their asymmetry score. The vertical axis represents the percentage of the people in the sample (specified by the values in each bar) who exhibit (a)symmetry at a given degree.

pattern, representing just over half the sample. None of these zero asymmetry participants made any phonological errors during left or right deactivation. Another 30 percent had a slight left asymmetry, making phonological errors at a rate of about 10 percent during left-hemisphere deactivation; and another 10 percent of participants made errors at a more substantial rate, between 25 and 50 percent. Finally, 5 percent had a slight right bias, making phonological errors at a rate of about 10 percent during right-hemisphere deactivation and scoring perfectly during left-hemisphere deactivation. The dashed lines in figure 4.2 represent the results of the same analysis but include only those cases with a healthy left hemisphere (no evidence of epilepsy in that hemisphere); this analysis shows a highly similar (a)symmetry pattern.

The overall semantic error rate was higher, at 17 percent following left-hemisphere deactivation (right = 3 percent), consistent with the pattern found in poststroke aphasia. You can see this increase in the leftward asymmetry in the graph. Perfect symmetry was found in 25 percent of cases, while the largest fraction of people, 35 percent, had a slight left-hemisphere bias, making semantic errors at a rate of about 10 percent following left deactivation and no errors following right deactivation. Another 25 percent had stronger left-hemisphere asymmetry, with an error rate ranging between about 35–50 percent following left but not right deactivation. Finally, 15 percent showed evidence of a weak right-hemisphere bias. Again, the dashed lines represent only cases with a healthy left hemisphere and show a similar pattern.

To summarize, even though most of the errors in this experiment were caused by left-hemisphere deactivation, the majority of people are either symmetric or very mildly biased to one hemisphere or the other (typically the left) on the word comprehension task. The tendency toward symmetry is stronger at the phonological than the semantic level. Again, this is a small sample (only twenty people), but given all the other evidence for more or less bilateral organization of speech perception, I think it is a decent approximation.

What this analysis indicates, along with the other evidence reviewed here, is that clinical observations of severe word comprehension problems following unilateral left-hemisphere lesions, even acute ones, are not representative of the lateralization of speech perception in the population as a whole. Instead, these cases reflect an asymmetry pattern present in only a minority of individuals—somewhere, I estimate, in the 10–20 percent range. This result also suggests a new perspective on leftward asymmetries that show up in functional imaging studies of speech perception and word comprehension.

Rather than indicating a population-level leftward asymmetry, they reflect the strong asymmetries present in a minority of people, the weak asymmetries present in a larger fraction of individuals (but still a minority), and the slightly more left-leaning effects of lexical-semantic processing.

It is worth reemphasizing that this discussion of (a)symmetry applies to speech *perception*, not speech production, which is much more strongly left dominant. Indeed, every one of the participants in the Wada study that we have just discussed were rendered completely mute during left-hemisphere deactivation, even while they were comprehending words well. I don't think that this violates the symmetry with modification theory, for the reasons mentioned earlier. It's just that there is apparently a lot more modification in the case of expressive speech.

The difference in the degree of asymmetric modification for speech perception versus production is important to keep in mind because the asymmetries in production can alias into measurements of speech perception for some tasks, which can yield paradoxical results. In chapter 5, we start with a story from my own scientific journey that highlights how this can happen and how understanding why it happens has led to a deeper understanding of the neural architecture of speech processing—namely, the dual stream model. After that, we will be in a position to understand other apparently contradictory results in the realm of speech perception, and finally to home in a bit more precisely on the underlying functional neuroanatomy.

5 Toward a Neural Architecture of Speech Perception

> The functional neuroanatomy of speech perception has been difficult to characterize.
>
> —Gregory Hickok and David Poeppel, 2000

When I first started thinking about the neural foundations of speech perception in the late 1990s, I thought it would be an easy nut to crack. After all, speech perception is a form of auditory perception, and the location of auditory cortex was well established. Yes, there were questions about the degree of lateralization, but there was abundant evidence in the literature that should allow one to sort it out. Yet around the turn of the millennium, there were several wildly different claims in the literature and little consensus. The regions that were implicated in speech perception included the following (figure 5.1):

- Left posterior superior temporal gyrus (STG; Geschwind, 1970)
- Left and right hemispheres (Gazzaniga & Sperry, 1967; Zaidel, 1985)
- Left inferior frontal gyrus (IFG/Broca's area; Fiez et al., 1995; Gallese et al., 1996)
- Left supramarginal/angular gyri (Caplan, Gow, & Makris, 1995; Petersen et al., 1989)
- Left dorsal precentral gyrus (Wilson et al., 2004)
- Various combinations of all these (Demonet et al., 1992; Zatorre et al., 1992)

This mishmash of proposals is why David Poeppel and I started our 2000 paper saying that speech perception has been difficult to characterize. "Part of the difficulty," we continued in the second line, "stems from the fact that

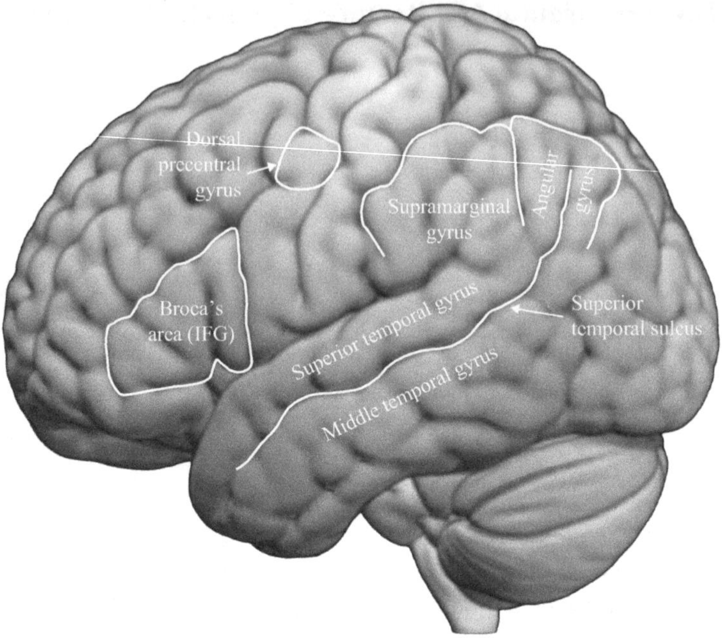

Figure 5.1
Regions implicated in "speech perception" using syllable discrimination/identification (dorsal stream regions) and word comprehension (ventral stream regions) tasks.

the neural systems supporting 'speech perception' vary as a function of the task" (Hickok & Poeppel, 2000, p. 131). Allow me to tell the story of how we arrived at that conclusion.

The Seeds of the Dual Stream Model of Speech Processing

I met David Poeppel during my stint as a postdoctoral fellow in Steven Pinker's lab at the Massachusetts Institute of Technology (MIT) between 1991 and 1993. David and I were both funded by the new McDonnell-Pew Center for Cognitive Neuroscience, me as a postdoc and David as a graduate student. The prospects for the field at that time were dynamic and exciting. The first positron emission tomography (PET) studies of language and other cognitive abilities were emerging, and functional magnetic resonance imaging (fMRI), which offered the hope of a completely noninvasive and widely available method for measuring neural activity with high spatial resolution, had just been invented. At the same time, magnetoencephalography (MEG) was also being developed, which held promise for high-temporal-resolution measurements of auditory and speech processes. It was a good time to be in one of the world's first cognitive neuroscience programs.

David was working on the neural basis of phonological processing and had taken a long, hard look at the PET studies discussed in chapter 4, along with a couple more that involved written stimuli. He basically called bullshit on their localizations of phonological processes, which, he noted, didn't converge with each other due to task design issues (Poeppel, 1996). This gained him substantial attention—which sometimes included the glare of disdain—from the senior guard of the field.

He had his own theory about the neural basis of speech perception, which I learned about when he gave a talk at the "McPew Center," as we called it. Arguing primarily from the bilateral distribution of pure word deafness and the bilateral distribution of speech activation in functional imaging, Poeppel claimed that speech perception was bilaterally mediated. He was just starting out on his career, but I was a seasoned scientist—or so I thought, having commenced my graduate training way back in 1988 (it was about 1991)—and was well schooled in hemispheric asymmetries, aphasia, and the Geschwind model. In fact, I had just finished my PhD under Edgar Zurif, a big shot at the Boston V.A. Aphasia Research Center, which was founded by Norman Geschwind. I had attended grand rounds regularly and hobnobbed with the likes of Harold Goodglass, Margaret Naeser, Sheila Blumstein, and other prominent figures in aphasia research. (Well, I saw them in the halls now and then.) After his talk, I informed David that, while the idea was interesting, the evidence that he presented for it was not strong. Geschwind, I explained, had a theory about why bilateral lesions can cause word deafness; and, by the way, you can get word deafness from unilateral lesions too. There was also evidence, I continued, from dichotic listening and lesion data that the right hemisphere is specialized for tonal, prosodic, or emotional aspects of speech. Finally, I concluded my mini-lecture by explaining that you get aphasia from left- but not right-hemisphere damage in the typical case. Yes, I was one of those people who was fully hypnotized by hemispheric dichotomies.

I can't remember his counterarguments. Perhaps I was unimpressed or just too dichotomy drunk to take them seriously. We did our own separate thing for the rest of our overlap at MIT, interacting now and then I suppose (I honestly don't remember) until I left to work on hemispheric lateralization of sign language with Ursula Bellugi and Ed Klima at the Salk Institute in La Jolla, California.

The critical issue circulating around sign language aphasia, my new research topic, was its lateralization pattern (Poizner, Klima, & Bellugi, 1987). Was it left hemisphere organized because it is a language? Or right hemisphere organized because it has a strong visuospatial component (then thought to be a right-hemisphere function)? I realized that we could home in on more specific and trendy ideas, such as Paula Tallal's fast temporal processing theory of left-hemisphere specialization for language. Spoken language contains such fast temporal signals, but signed language is much slower and more spatial. Therefore, Tallal's model predicted that signed language should not be strongly left dominant. It turns out, however, that the hemispheric asymmetry of signed language is pretty much identical to that of spoken language, as revealed by our sign aphasia work, thus disproving fast temporal processing as the basis of left dominance for language (Hickok et al., 2001, 1996).

Our conclusion was questioned, though, when the first fMRI studies were carried out on signed language, comparing the perception/comprehension of American Sign Language (ASL) with the perception/comprehension of written English. Reading English yielded mostly left-dominant activation, whereas viewing ASL produced a more bilateral pattern (Neville et al., 1998). This pushed me to look more closely at the activation patterns of spoken language (a more natural comparison to ASL), which was largely bilateral, as we've seen in the previous chapters. And this led me to consider the possibility that language lateralization was not all or nothing; rather, it varied across tasks. We concluded that the right hemisphere is indeed involved in receptive sign language processing, but so is it for spoken language, and to a similar degree (Hickok, Bellugi, & Klima, 1998). Exactly what it was doing, though, was unclear.

Thus, my interest was piqued in the possible involvement of the right hemisphere in aspects of receptive language, both signed and spoken, and David and I started talking at conferences and by phone and email about ways to further test the dominant theories. We toyed with a few ideas but never really hit on a critical test. In the meantime, while preparing a lecture on the neuroanatomy of speech perception for my undergraduate "Language and the Brain" course, I came across a surprising piece of evidence.

This was around 1997, so I was well aware of the essentially bilateral, symmetric activation in the superior temporal lobe during speech listening.

The mainstream interpretation of these results at the time, as detailed in the previous chapter, was that left-hemisphere activation reflected low-level auditory plus linguistic-phonological processes, whereas right-hemisphere activation reflected low-level auditory plus some other higher-level nonlinguistic process, such as emotional prosody. This seemed to be broadly (if indirectly) supported by lesion data—aphasia is typically caused by left-, not right-hemisphere damage—but I wanted to show my students more specific evidence for this claim.

The most straightforward test would be to examine the effects of lesions to the left- versus right-hemisphere regions that activated during speech listening in functional imaging studies. In the era of functional neuroimaging, I had an advantage over previous theorists interested in this question because I knew exactly which areas of the brain activated to speech, the superior temporal lobe, and could focus my attention on the effects of lesions in that zone. In the past, researchers like Geschwind couldn't be sure whether the critical zone was just the superior or middle temporal gyrus, whether it was the anterior, middle, or posterior parts, or the angular gyrus or supramarginal gyrus, or some combination of these. Since I knew where to look, I could take a different tack. I consulted the literature and tried to find cases of patients with damage to the left STG. If that was the critical zone for speech perception, as the dominant interpretation of the functional imaging data suggested, then I should find examples where damage in that area caused at least moderate receptive phonological deficits, if not full-blown word deafness.

I succeeded in finding cases with damage in the critical left STG zone, but they had no trace of receptive speech problems. Instead, affected individuals had speech *production* problems, conduction aphasia in particular (Damasio & Damasio, 1980; Hickok, 2000). I dug deeper, probing the literature for evidence that more extensive damage in the left temporal lobe might cause receptive phonological problems. This led me to the work described in chapter 3 showing that even patients with substantial receptive language problems (such as Wernicke's aphasia) don't have predominantly phonological receptive speech perception deficits. Finally freed from my hemispheric dichotomy goggles, I called David and told him that I had found some evidence that really clinched his case for bilateral speech perception. We decided that we should write a simple, short paper in which we lay out the evidence and then move on to more interesting things. But it turned out not to be so simple.

Tsk-Task

So far in this book, we've been evaluating the neural basis of receptive language from the perspective of auditory comprehension tasks, the pear-bear-moose-grapes task (discussed in chapter 3). There is good reason for this. Receptive language as it is used in the wild is all about comprehension. That is what makes language useful after all: the communication of meaningful information. It stands to reason, then, if we want to understand the neural circuitry for wild-type language, we should study comprehension. In the technical terminology, comprehension tasks have *ecological validity*.

But there are problems with studying comprehension. If I ask a participant to "point to the red square" and the person picks up a black circle, how can I tell what's gone wrong? Is it difficulty perceiving the speech sounds? Accessing the right word meanings? Deriving the compositional meaning of the phrase? Planning the response? It's hard to tell.

Speech scientists have devised methods for isolating a process of interest. The bear-pear-moose-grapes task uses the approach of allowing participants to make different types of errors, which can be used to distinguish failures at the phonological versus lexical-semantic levels. A favorite alternative approach to isolate phonological processes is to strip speech of its meaning completely by using nonsense syllables like "ta," "da," "ba," or "ga." Such syllables are typically used in two types of tasks. In *syllable identification*, a syllable is presented to a listener, who reports what was heard by repeating the syllable or selecting a match from an array of written options. In *syllable discrimination*, pairs of syllables are presented, and the task is to decide whether the two are the same (as in the pair "ba"-"ba") or different ("ba"-"pa"). These sorts of tasks have served as staples in speech perception labs the world over since the 1950s and have found their way into the ears of many people with aphasia. What came out, though, was a paradox.

We learned earlier that people with unilateral left-hemisphere disruption do not typically have serious impairment in receptive phonological ability as measured by comprehension tasks, even when they have severe aphasia or complete acute deactivation of their left hemisphere. Under these conditions, word-level comprehension is reasonably good, and when participants do make errors, they are more likely semantic confusions. What does this fact predict regarding the ability of people with aphasia to decide whether "ba" and "pa" are the same or different? Well, if they can correctly match the words "bear" or "pear" to their corresponding pictures more often than not, they will certainly be able to hear the difference between "ba" and "pa," right? Wrong. A number of studies over the years have shown that unilateral left-hemisphere damage can severely impair performance on syllable identification and discrimination tasks (Baker et al., 1981; Basso, Casati, & Vignolo, 1977; Blumstein, Baker, & Goodglass, 1977; Caplan et al., 1995; Miceli et al., 1980). Consider the results of a 2012 study of eleven people with Wernicke's aphasia, which reported a group average of 54 percent correct on a syllable discrimination test (range = 44 percent–63 percent; healthy control average = 98 percent) and with only two people performing statistically better than random guessing (Robson et al., 2012). Quite severe deficits, indeed.

Even more puzzling, when researchers compared the same patient's performance on syllable discrimination tasks and word-level comprehension tasks, the abilities doubly dissociated. Some people were good at syllable discrimination but bad at word comprehension, while others were bad at discrimination but good at comprehension (Miceli et al., 1980).

I'll illustrate this paradoxical result with a modern, large-scale replication and extension that our team recently published (Rogalsky et al., 2022). The study was led by my former graduate student and longtime collaborator, Corianne Rogalsky. We utilized the familiar bear-pear-moose-grapes comprehension test and also constructed two discrimination tasks, one involving word pairs (e.g., "page"-"cage") and a matched set of nonword syllables (e.g., "pag"-"kag"). With the help of several collaborators at aphasia research centers in Iowa, California, South Carolina, Wisconsin, and Arizona, we assessed a group of 158 individuals with focal brain damage, including 113 with left-hemisphere stroke, 19 with right-hemisphere stroke, 18 with left anterior temporal lobectomies, 8 with right anterior temporal lobectomies, and 26 neurologically intact controls. The results for the left-hemisphere stroke group are shown in figure 5.2. There are several notable findings.

The first is that overall word comprehension as measured by the bear-pear-moose-grapes task was remarkably good, as past work led us to expect. The average performance level was 97 percent correct in the left-hemisphere stroke group;[1] the other groups did even better. The distribution was highly skewed, with only a few patients (five to be exact) performing worse than 90 percent, as figure 5.2A shows. This replicated previous studies that showed minimal word comprehension deficits following unilateral damage.

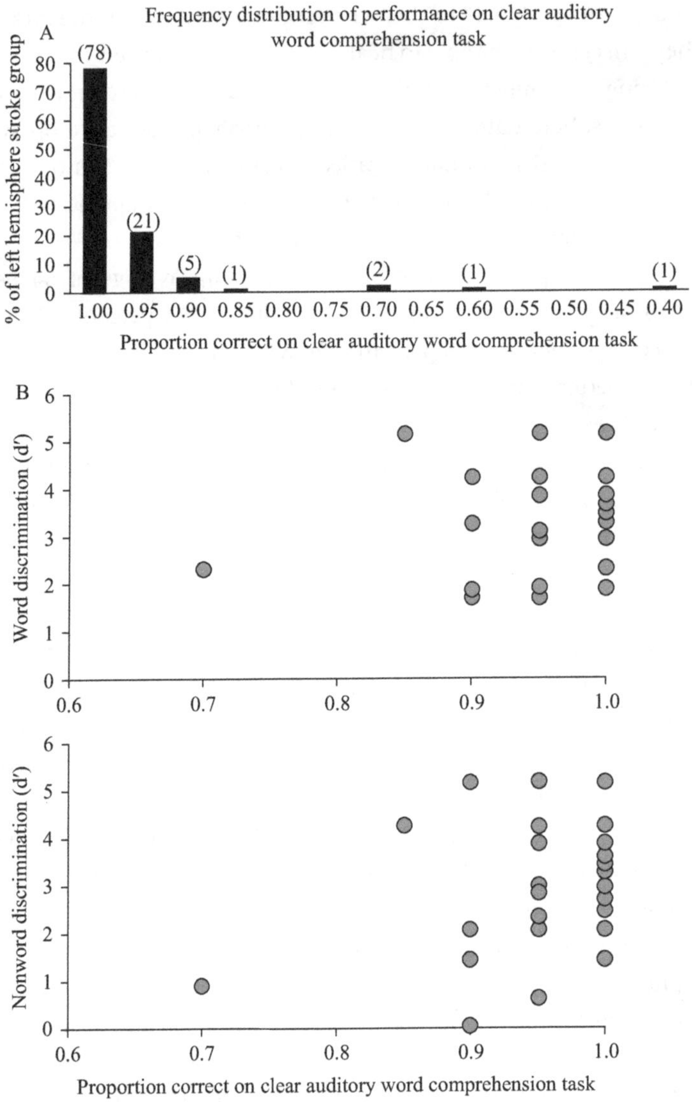

Figure 5.2
Performance on three tasks requiring speech perception in people with left-hemisphere stroke. A. Distribution of the percentage of correct scores on the pear-bear-moose-grapes word-to-picture matching task. B. Scatterplots of performance on word (top) and nonword syllable (bottom) discrimination (vertical axis) compared to the pear-bear-moose-grapes task in the same patients. Reprinted with permission from Rogalsky et al. (2022).

The second notable result was that there was substantial variation in performance on both discrimination tasks in the left-hemisphere stroke patients (vertical axis), as the graphs in figure 5.2B show,[2] even though nonword discrimination was statistically significantly harder than word discrimination. We can explain the word-advantage effect in discrimination in terms of the fact that real words allow two levels of comparison, at the phonological and semantic levels, whereas nonwords allow comparison only at the phonological level. It's interesting, though, that word *discrimination* proved to be much more difficult than word *comprehension* for many patients. This indicates that the task itself is to blame rather than the ability to comprehend words. There's something particularly difficult about comparing two sequentially presented words, which apparently isn't involved in comprehending one word and comparing it to a set of pictures.

The third notable result from this study is that it strongly confirmed the dissociability of syllable discrimination and word comprehension. You don't need a statistical analysis to recognize this fact—just look at the group of patients who perform perfectly on the word comprehension test and note the spread of their performance on the discrimination tasks: it spans a wide range. This is the paradox: how can a person struggle to tell the difference between /*b*/ and /*p*/ on a discrimination test and yet readily tell the difference in the context of a comprehension test? We'll return to this study later in the chapter. For now, the data illustrate the puzzle that stood in the way of the straightforward bilateral story that Poeppel and I wanted to tell.

Some clues for the solution came from functional imaging research that had been published in the 1990s. The study by Zatorre and colleagues (1992) that I introduced earlier had found bilateral activation when participants passively listening to words compared with passively listening to noise. However, when the research team asked their subjects to listen to pairs of words and decide whether the last phoneme was the same or different—a kind of discrimination task—they found that left lateralized activation was localized not to auditory areas, but to Broca's area, the posterior parietal lobe, and some other regions when compared to passive listening to speech. Other studies reported similar findings using what Poeppel and I called "sublexical tasks" (various sorts of tasks requiring explicit attention to sublexical phonological features) (Burton, Small, & Blumstein, 2000; Demonet et al., 1992), and which appeared to be linked to a left lateralized, frontoparietal network, unlike the bilateral superior temporal lobe network, which supports speech comprehension. Adding more support to this idea was a lesion-based study of syllable discrimination that included lesion localization information, unlike previous studies that reported the aphasia subtype as the only clue to the location of the damage. This study localized the source of syllable discrimination deficits to the left posterior inferior parietal lobe, not to the STG (Caplan et al., 1995). Figure 5.1 shows the location of these regions.

At the same time, there was a flurry of high-profile scientific reports on the cortical networks supporting verbal short-term memory as assayed using functional imaging. Verbal short-term memory is the system that allows us to retain a short grocery list or phone number for a brief period. The dominant theory of verbal short-term memory at the time was Alan Baddeley's *phonological loop model*. This model is comprised of a passive, temporary storage component called the *phonological store* and an active rehearsal component called *articulatory rehearsal,* which can refresh the store to keep its content active for as long as one rehearses it (Baddeley, 1992). When researchers used functional imaging to localize these components, they reported a left lateralized frontoparietal network: articulatory rehearsal was linked to left frontal motor speech areas, including Broca's area and more dorsal premotor areas, while the phonological store was linked to the posterior parietal lobe (Awh et al., 1996; Jonides et al., 1998; Smith & Jonides, 1997). These were the same regions implicated in sublexical speech tasks.

These facts suggested a solution to the paradox: sublexical speech tasks rely on the frontoparietal phonological working memory circuit, whereas comprehension does not. A stroke victim with frontoparietal damage, then, could end up with a syllable discrimination deficit while leaving word comprehension intact. This was the solution that Poeppel and I proposed in our paper:

> The functional neuroanatomy of speech perception has been difficult to characterize. Part of the difficulty, we suggest, stems from the fact that the neural systems supporting "speech perception" vary as a function of the task. Specifically, the set of cognitive and neural systems involved in performing traditional laboratory speech perception tasks, such as syllable discrimination or identification, only partially overlap those involved in speech perception as it occurs during natural language comprehension. (Hickok & Poeppel, 2000, p. 131)

It was useful to point out the task-dependent pattern in the lesion and functional imaging data because, in one sense, this was enough to allow us to move on to the claim that speech perception, as assessed using ecologically valid tasks, is approximately bilaterally organized. Sublexical tasks, in contrast, because of their reliance on working memory systems that are not typically used during normal perception for comprehension, could give misleading results. In other words, such tasks fall short of the ecological validity standard and should not be used to assay the neural organization of speech perception.[3]

But this was unsatisfying. We wondered what it is about discrimination versus comprehension that led to the recruitment of different processes. Relatedly, we also wondered why processing phonological information in working memory should be more lateralized to the left than processing phonological information during comprehension. The answer that we came up with was the dual stream model of speech processing. Here's how we got there.

I had recently read Milner and Goodale's book *The Visual Brain in Action* (1995), in which they described their dual stream model of the visual system, with one stream involved in perception for comprehension and the other involved in perception for guiding action. It occurred to me that speech processing could be conceptualized in exactly the same way. If you hear a word, you can comprehend it (ventral stream) or you can reproduce it with your motor speech system (dorsal stream). With this as the starting point, we could then reason using homology about the organization of speech processing. The train of thought went something like this:

- Assessing speech perception using comprehension tasks requires a mapping from sound onto meaning, a ventral stream function.
- The neuroanatomy of syllable discrimination implicates frontoparietal areas, which is the part of the brain that generally supports sensorimotor function.
- The two speech tasks are dissociable and map to different regions because, for some reason, one (comprehension) involves the ventral stream and the other (discrimination) involves the dorsal stream. Why would this be?
- Syllable discrimination and phonological working memory are closely aligned, and one can conceptualize phonological working memory as a sensorimotor (dorsal stream) system: a motor-related rehearsal mechanism refreshing the contexts of a sensory-related store (Wilson, 2001).
- Maybe the phonological loop is an emergent property or a special use of the normal dorsal stream auditory-motor speech production system. Supporting this idea is the observation that working memory functions

have also been proposed for dorsal visual stream areas (Rizzolatti, Fogassi, & Gallese, 1997).

- Maybe deciding whether two syllables are the same requires articulatory recoding to enable both conscious awareness of the sounds (as in learning to read and sounding out words) and facilitates keeping them active long enough to compare and make a decision (as in working memory tasks).
- Speech production, a dorsal stream ability, is more strongly left dominant than receptive speech, explaining why syllable discrimination and phonological working memory are left dominant: these tasks rely on the dorsal stream, auditory-motor speech production system.

So we concluded that

> there is a fronto-parietal network predominantly in the left hemisphere, which functions to interface auditory and articulatory representations of speech. This system plays a key role in the development of speech (i.e. learning to articulate the phonetic structure of the target language) and comprises the network underlying phonological working memory in the adult. Because this system provides a means for explicit access to sub-lexical speech segments, it is also recruited during the performance of tasks requiring explicit operations on sub-lexical information. (Hickok & Poeppel, 2000, p. 137)

We took the logic a bit farther, arguing from homology that the dorsal stream should involve at least three components: sound-based representations of speech in the superior temporal sulcus (STS) bilaterally that is shared by the dorsal and ventral streams, motor-based representations of speech in the left IFG (Broca's area) for articulation planning, and an auditory-motor interface in the inferior posterior parietal lobe (which is near homologous visuomotor interface areas). We then redefined the functional neuroanatomy of the phonological loop, arguing that what others had called the "store" in the inferior posterior parietal lobe was actually the auditory-motor interface system:

> We hypothesize that activations attributed to the functions of the phonological store reflect the operations of the proposed auditory–motor interface system in inferior parietal cortex. That is, inferior parietal cortex is not the site of storage of phonemic representations per se (or more precisely, it does not store the type of sound-based representations that are used during lexical access), but rather serves to interface sound-based representations of speech in auditory cortex with articulatory-based representations of speech in frontal cortex. (Hickok & Poeppel, 2000, p. 136)

We proposed that the store was localized to the auditory system, the part shared by the dorsal and ventral streams.

Within a year, fMRI research from my lab, led by one of my graduate students, Brad Buchsbaum, had confirmed activation of the superior temporal lobe in a phonological short-term memory task and refined the localization of the auditory-motor integration area. We dubbed the auditory-motor integration area *Spt,* which stands for "Sylvian parietal-temporal," reflecting its location at the posterior end of the Sylvian fissure at the parietal-temporal boundary (Buchsbaum, Hickok, & Humphries, 2001; Hickok et al., 2003). Spt's location did not line up perfectly with our proposed "inferior parietal lobe" location, but it was very close, and more interesting in some ways. Spt turned out to be in the posteriormost part of the *left* Sylvian fissure involving parts of the planum temporale, a part of the classical Wernicke's area, and was in the middle of the zone where lesions produce speech *production* problems—namely, conduction aphasia, as we later determined (Buchsbaum et al., 2011). This was a good sign and suggested that we were on the right track. We return to area Spt and the auditory dorsal stream in chapter 10.

The struggle to explain the paradox of the dissociability of syllable discrimination and word comprehension tasks ultimately led to the dual stream model of speech processing, schematized in figure 5.3, from our 2000 paper. The model enabled us to account for the different lateralization patterns and dissociability of the two types of tasks, and it brought the speech network in line with (i.e., homologous to) the dual stream organization of the visual system. As a bonus, the phonological loop fell out of the model for free (Buchsbaum et al., 2011; Buchsbaum & D'Esposito, 2008). I thought that we had clarified the state of confusion (as I saw it) that the field of speech perception research had found itself in during the 1990s. But other challenges emerged even while we were working on solving the task paradox.

Challenge 1—The Pathway for Intelligible Speech

Just a few months after Poeppel and I published our 2000 paper, Sophie Scott, Richard Wise, and colleagues published a PET study on receptive speech that added yet another hypothesis to the long list of regions claimed to be the basis of speech perception. The report, titled "Identification of a Pathway for Intelligible Speech in the Left Temporal Lobe" (Scott et al., 2000), starts by acknowledging the bilateral and generally symmetric response to speech that was reported consistently in functional imaging studies. It also notes that the pattern is inconsistent with aphasia: "A

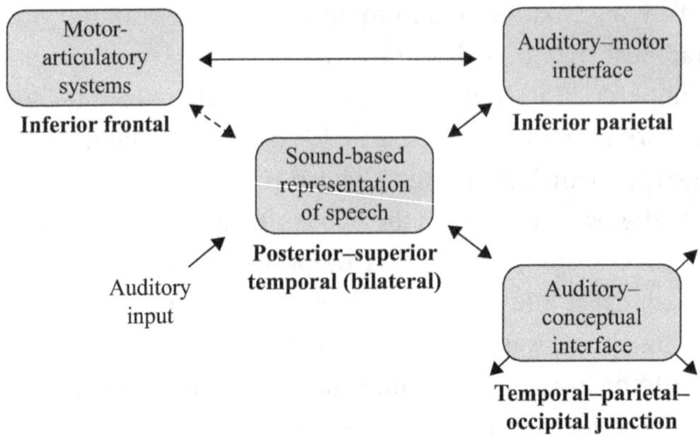

Figure 5.3
The original Hickok & Poeppel dual stream model. Reprinted with permission from Hickok and Poeppel (2000).

distinctly lateralized temporal lobe system for speech perception has not been demonstrated previously with functional imaging techniques, despite a wealth of clinical evidence that aphasia arises from damage to the left, not the right, hemisphere. Instead, hearing speech, contrasted with silent rest, simple tones or noise, results in activations in the left and right superior temporal cortices, sometimes associated with relative interhemispheric asymmetry" (Scott et al., 2000, p. 2400). We now understand that the association between aphasia and left-hemisphere damage is not a good indicator of the lateralization of speech perception. But that fact was not appreciated by most researchers at that time, so it did seem like a significant puzzle.

The group designed a clever experiment that attempted to disentangle the brain's response to language from the response to the acoustic signals that delivered it. They did this by creating four types of stimuli. One was clear, intelligibly spoken sentences. They then created unintelligible versions of these sentences by digitally flip-flopping the high and low frequencies of the signal, a process called *spectral rotation*. They noted that spectrally rotated speech is similar acoustically to regular speech in terms of the frequency range covered and the temporal characteristics. The result of these manipulations is something that sounds like "alien speech." Therefore, the comparison between these two stimuli roughly—very roughly—holds acoustics constant while manipulating intelligibility. The third type of speech stimulus was intelligible, but it was substantially degraded acoustically using a process called *noise vocoding*. Here, speech is divided into a small number of frequency bands (they used six), and then the spectral information within each band is replaced by noise that has the same *amplitude envelope*, which is a technical name for the loudness fluctuations of the signal. Noise-vocoded speech sounds very gravelly, but with a little practice, it can be readily understood. A comparison between clear speech and noise-vocoded speech allowed the team to manipulate the acoustics while holding intelligibility constant. The final condition was unintelligible noise-vocoded, spectrally rotated speech.

A total of eight participants listened passively to several examples of each type of stimulus while their brain activity was monitored using PET. The aim was to determine whether there was a region or network that responded equally to the two intelligible sentences, despite the acoustic differences, and not as well to the unintelligible stimuli, despite their rough acoustic similarity to the intelligible sentences. There was a single area that met that criteria. It was in the anterior superior temporal lobe of the left hemisphere. The authors concluded, "This novel observation demonstrates a left anterior temporal pathway for speech comprehension" (Scott et al., 2000, p. 2400).

This was news. Not only had a left-dominant receptive language area been identified with functional imaging, but it was in the anterior temporal lobe, outside the traditional language areas. The authors asserted that "this lateralization is consistent with the neuropsychology literature" (Scott et al., 2000, p. 2404), which is true if one looks only to the left-dominance of aphasia as the literature representative. But the finding was quite inconsistent with the neuropsychological literature on receptive speech function, which indicated a more bilateral distribution, one that implicated the posterior, not the anterior, temporal lobe. Even when unilateral left-hemisphere lesions cause impairments in speech understanding, as in Wernicke's aphasia, the lesions tend to be posterior in the temporal lobe (Bates et al., 2003; Damasio, 1992). Moreover, surgical resection of the anterior temporal lobe was regularly performed for the treatment of medically intractable epilepsy without destroying the ability to understand speech (Hermann, Wyler, & Somes, 1991). The facts didn't add up.

In the study led by Corianne Rogalsky that I introduced earlier, we had an opportunity to test the left anterior temporal lobe hypothesis (Rogalsky et al., 2022). Eighteen people with left anterior temporal lobectomies, which included Scott et al.'s region, and eight people with right anterior temporal

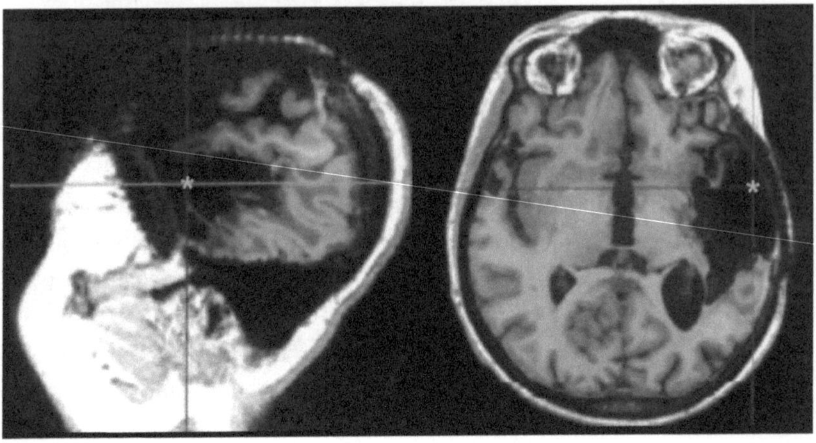

Figure 5.4
Magnetic resonance imaging (MRI) scans of two participants with left anterior temporal strokes. The left image shows a sagittal slice of one participant, and the right image shows an axial slice of the other. The asterisks indicate the coordinate for the peak of the left anterior temporal intelligibility effect from the Scott et al. (2000) study, which is well within the lesioned zone.

lobectomies took the pear-bear-moose-grapes test. Not a single error was made by anyone in either group. One might argue that anterior temporal lobectomies are not a strong test of the hypothesis because they are removing tissue that is diseased; perhaps speech perception flopped to the other hemisphere, leaving no residual ability in the left. A stronger test would be strokes restricted to the relevant zone. Anterior temporal lobe strokes are rare, but we did have two cases in our sample who had substantial damage to the zone activated in the Scott et al. study (figure 5.4). Both individuals scored perfectly on the bear-pear-moose-grapes task. Recall, too, that the isolated right hemisphere of split-brain patients and people undergoing Wada procedures can process heard speech quite well at the phonological level. The neuropsychological evidence does not support the left anterior temporal lobe theory of the pathway for intelligible speech.

Why, though, was the proposed pathway for intelligible speech found only in the left hemisphere? Kayoko Okada, along with several members of my lab, decided to have a look by attempting a replication of the original PET study (Okada et al., 2010). We used fMRI and a larger group of participants in hopes of revealing a clearer picture of the extent of activation. We indeed replicated the anterior temporal lobe intelligibility effect but found that it wasn't restricted to that region at all. Instead, we found that the effect included both anterior and posterior temporal regions—a finding that Scott's group had also reported in 2003 (Narain et al., 2003)—and did so in both hemispheres, which was new information. Further, using a more sensitive machine learning analysis technique, we found that even primary auditory cortex responded differentially to intelligible, clear speech and unintelligible, spectrally rotated speech: the *pattern* of activation was different even though the averaged magnitude was not. The "pathway for intelligible speech," then, involves much of the superior temporal lobe of both hemispheres.

Challenge 2—Mirror Neurons and the Motor Theory of Speech Perception

In 1996, four years before Poeppel and I published our claim that "cortical fields in the posterior–superior temporal lobe, bilaterally, constitute the primary substrate for constructing sound-based representations of speech" (Hickok & Poeppel, 2000, p. 131), Giacomo Rizzolatti and colleagues speculated that "neurons with properties similar to that of monkey 'mirror neurons', but coding phonetic gestures, should exist in human Broca's area and should represent the neurophysiological substrate for speech perception" (Gallese et al., 1996, p. 607). Mirror neurons are cells in the macaque monkey area F5—claimed to be a homologue of the human Broca's area—that respond both during the execution of motor actions, such as grasping with the hand, and during the observation of the same or similar actions (Gallese et al., 1996; Rizzolatti & Craighero, 2004). Because of this property, they were believed to support action understanding, which, as the quote shows, was quickly generalized to include speech perception. This theory is in opposition to the view that I've argued here, so it deserves some attention. I won't dwell on it, however, because it was the topic of my previous book, *The Myth of Mirror Neurons* (Hickok, 2014b).

Based on what we've covered in this and preceding chapters, we can already question the idea that the motor speech system plays an enabling role in speech perception. We know that speech production is rather strongly left hemisphere dominant, while speech perception is more symmetrically organized. Given just this information, we can conclude that the motor system is, at the very least, not the whole story. But let's work through some of the details.

The *motor theory of speech perception* was proposed by Alvin Liberman (1917–2000) and colleagues in the 1950s and 1960s (Liberman, 1957; Liberman et al., 1967). These researchers made important discoveries about the nature of the speech signal and how it is perceived. A major finding was the *lack of invariance problem*, the fact that the acoustics of a phoneme, say /*d*/, is different depending on its context, as in the words "deep," "dot," and "doodle." The relation between our percept of an individual speech sound and the physical signal that gives rise to it varies (i.e., there is a lack of invariance). Liberman and his team proposed a solution: we don't perceive the unruly acoustics of speech directly, rather we perceive the motoric *gestures* that produce them, which are well behaved and invariant.[4] This is the motor theory of speech perception: you need the motor system to perceive speech because the objects of speech perception are motoric gestures.

Decades of research eventually disproved this strong version of the theory (Galantucci, Fowler, & Turvey, 2006; Hickok, 2014b; Liberman & Mattingly, 1985; Massaro & Chen, 2008). Some of this evidence included experiments showing that you don't need a speech production system to perceive speech. This conclusion is supported by evidence from people with Broca's aphasia (who by definition have impaired production with spared word comprehension), people with bilateral lesions to motor speech regions resulting in mutism (opercular syndrome) (Walker et al., 2021), people with congenital inability to produce speech (Lenneberg, 1962), prelingual infants (Eimas et al., 1971), and nonhuman animals (Kuhl & Miller, 1975), all of which can perceive speech sounds quite well. Findings like these forced a retreat to a weaker version of the motor theory by some researchers, including Liberman (Liberman & Mattingly, 1985), and also to its complete abandonment by others (Diehl & Kluender, 1989). But the discovery of mirror neurons in the 1990s resuscitated both strong and weak versions of the motor theory. Soon a flurry of new studies, published in high-profile journals, showed that listening to speech activated the speech motor system and speech perception could be disrupted by stimulating it using transcranial magnetic stimulation (TMS)[5] (D'Ausilio et al., 2009; Meister et al., 2007; Watkins, Strafella, & Paus, 2003; Wilson et al., 2004).

But none of these new observations erased the old facts: people with damaged, undeveloped, or absent speech production systems can still be shown to perceive speech quite well (Bishop, Brown, & Robson, 1990; Hickok et al., 2011; Lotto, Kluender, & Holt, 1997; Rogalsky et al., 2011). Take, for example, the Wada experiment that I described in chapters 3 and 4. I used it to argue for bilateral speech perception capacity: deactivate the entire left hemisphere, and the right hemisphere still comprehends words very well. Recall too that while the right hemisphere was merrily comprehending words, the Wada study participants were completely mute as a result of their left-hemisphere deactivation. Ditto for the mute but comprehending right hemisphere of split-brain patients. So we can shut down speech production control completely and still demonstrate unimpaired word-level speech perception and comprehension in most people. Facts like these have led most speech scientists to conclude that the strong version of the motor theory is just plain wrong.

What remains debatable is whether the motor speech system plays a weaker role. The most viable hypothesis is that it provides an advantage in noisy listening conditions (Wilson, 2009). In fact, every one of the TMS studies that showed an effect of motor stimulation on speech perception did so using speech that was masked by noise. If TMS is applied to the motor system while listening to *clear* speech, performance is unaffected. This seems to support the noisy listening claim. There are two caveats, however. The first is effect size, which is small in these studies. A study led by one of my former graduate students, Ryan Stokes, estimated that the motor speech system turned up the perceptibility knob by no more than 1–2 decibels (Stokes, Venezia, & Hickok, 2019). That's not a lot. Then there's the second caveat: the tasks in these experiments are typically syllable discrimination or identification, not comprehension, which we know rely more extensively on the dorsal auditory-motor stream and do not correlate with normal comprehension. There was one study that used a comprehension task, and it found no effect of TMS on word comprehension, only a reaction time difference for some (but not all) of the stimuli (Schomers et al., 2014).

To summarize, claims that the motor system is critical, or even helpful, under some conditions are either wrong or overstated. The motor system adds about 1–2 decibels of assistance under noisy listening conditions and only using nonecologically valid tasks that are known to implicate the motor system. This is a modest contribution of the motor system to speech perception indeed. My estimate is that it is essentially zero in real-world communication.

The Auditory Phonological Area

Having dispensed with challenges to the bilateral speech perception hypothesis, I will focus on making the case for a more precisely defined superior temporal region, which we will call the *auditory phonological area (APA)*.

A good starting point is to look to one of our best methods for spatial localization of function in the healthy human brain: fMRI. And a good way to get a quick snapshot of the anatomy of speech perception is to look to meta-analyses, statistical reductions of the central trends that exist across many studies. There have been a few in the last decade. I carried out an informal analysis with Poeppel in one paper (Hickok & Poeppel, 2007), the aim of which was to identify temporal lobe regions involved in perceiving syllable-length, sublexical chunks of speech. (Note that while I've argued that the sublexical speech perception task can implicate dorsal stream networks, perceiving chunks of speech will also necessarily involve the bilateral speech perception network, as has been observed repeatedly. Thus, by focusing on the superior temporal regions, we can identify critical subareas therein.) I identified several studies that contrasted listening to speech (usually consonant-vowel syllables) against various acoustic controls and then displayed the locations of each blob across all the studies on a standardized brain image. Figure 5.5A shows the distribution. The blob locations were roughly symmetrically distributed in the left and right hemispheres and centered around the middle part of the superior temporal sulcus (STS).

Two subsequent and more formal meta-analyses of neural activations associated with listening to sublexical units found nearly identical distributions (DeWitt & Rauschecker, 2012; Turkeltaub & Coslett, 2010). Figure 5.5B shows DeWitt and Rauschecker's meta-analysis at the sublexical speech level and figure 5.5C shows their meta-analysis at the word level (DeWitt & Rauschecker, 2012). It is worth noticing that the distribution of individual activation foci shows that the right hemisphere distribution is more spread out than in the left hemisphere. Turkeltaub and Coslett (2010) quantified and confirmed that this left-right difference in activation distribution is statistically significant. This affects the across-study averaging in meta-analyses (and probably individual studies too), making the more distributed right-hemisphere activations appear weaker.

What I'd like to focus on here, though, is the consistency of location within the hemispheres, which is centered on the middle STG and STS just lateral to Heschl's gyrus. From there, activations tend to extend more posteriorly than anteriorly. At the word level (figure 5.5D), there are a few more anterior foci, but so are there more posterior and ventral foci.[6] The picture from functional imaging is clear: perceiving speech implicates the middle-to-posterior STG and STS bilaterally.

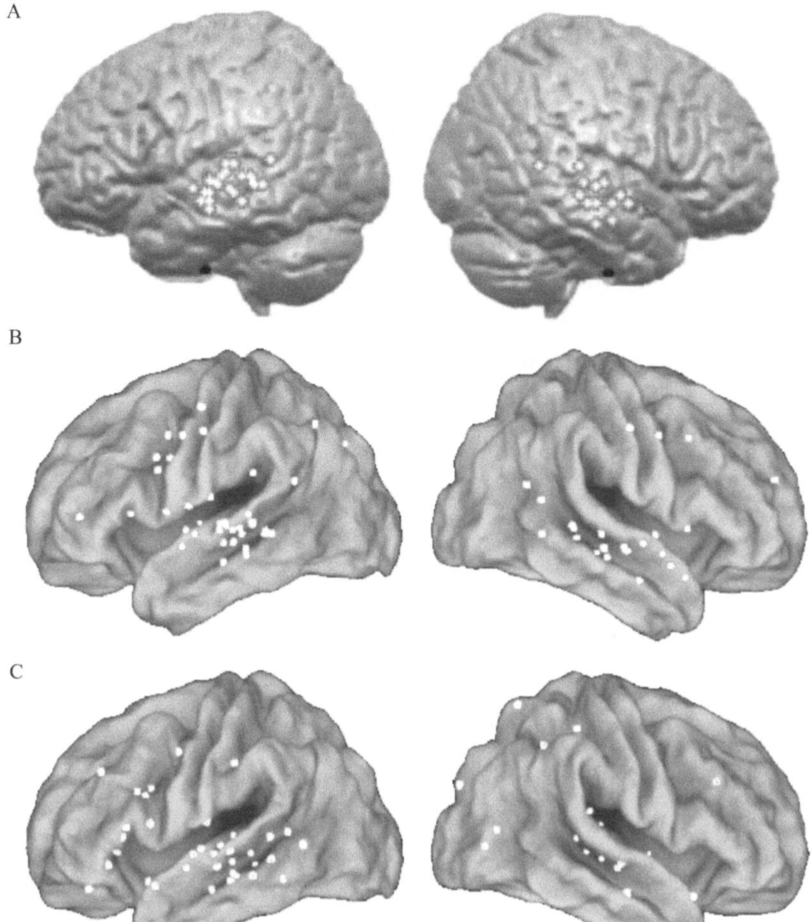

Figure 5.5
Meta-analyses of functional activation studies of speech listening at the sublexical (A, B) and word (C) levels. A (from Hickok & Poeppel, 2007) shows only temporal lobe activation foci; B and C (from DeWitt & Rauschecker, 2012) show all activation foci. Reprinted with permission.

If you glance back to figure 2.2 and notice where we drew the "phonological network" on the schematic brain map, it is a little posterior to the distribution suggested by these meta-analyses. The reason for this is that we were taking account of data from two other sources. One was from work by Pascal Belin, Robert Zatorre, and colleagues, who identified a region in the STG/STS just lateral to Heschl's gyrus that is relatively selective for vocal sounds generally, not just speech—a *voice area* (Belin et al., 2000). It's likely that many studies of speech perception are showing activation associated with these more general voice-related activations. Thus, the more anterior part of the middle-to-posterior distribution of responses to speech may not be related to phonological-level processes, but something more general.

The second source of evidence was an fMRI study that my then-graduate student, Kayoko Okada, and I carried out to identify, as precisely as we could, the receptive phonological network. We were not convinced that the typical approach of comparing speech with nonspeech acoustic controls is an optimal localization method. Here's why:

- This so-called *subtraction method* assumes the existence of a phonological faculty in the brain that is spatially distinct from other networks involved in processing acoustic information. There are two reasons why this assumption may be wrong. One is that the brain may not need any specialized neural machinery to process speech sounds, as opposed to other kinds of nonspeech acoustic signals. Indeed, some have argued that the human auditory system did not evolve anything special to handle speech (e.g., Fitch, 2018). The other is that even if something special is needed for decoding word-clangs, it may not be housed in its own island of gray matter. To give a simple example, our vocal apparatus—including the lips, tongue, jaw, pharynx, nasal cavity, larynx, and lungs—is clearly necessary (and some would say adapted) for producing speech, and yet it is hardly specific to it, given the role of this anatomical constellation in eating, drinking, and breathing. The subtraction approach to localizing phonological networks, then, may subtract out the very things that we seek to find.
- Even if the assumptions of the subtraction method hold true, things can still go awry if the nonspeech control is suboptimal. The hitch is that there is no foolproof way to know what *is* optimal. We have to rely on our educated guesses.

- Even if we hit on just the right acoustic control, we are still not guaranteed to nail down the phonological network. The reason stems from another assumption of the subtraction method—namely, that one can cleanly subtract out a level of analysis with no collateral change in processing (and therefore activation). This is referred to as the *pure insertion assumption*, which may not hold (Friston et al., 1996). Another related problem is that even when a sensory signal is just some random pattern, we can't help but try to analyze it as something meaningful. The number of Jesus sightings in everything from socks to toast is sufficient to make this point. What, then, is to stop the brains of fMRI participants from attempting a phonological analysis of an unintelligible signal, especially when they are in a speech study? The title of this report by Azadpour and Balaban (2008) indicates that indeed, there is nothing to stop them: "Phonological Representations Are Unconsciously Used When Processing Complex, Non-speech Signals." This could mean that phonological areas are stripped right out of a study that was carefully designed to map phonological areas!

There are various ways to get around these problems. Tobias Overath, Josh MacDermott, Jean Zarate, and David Poeppel published a good example of the parametric feature-based approach, where a feature of interest is varied in its degree of change (Overath et al., 2015). A functional activation map is then generated by identifying regions that correlate with the degree of change. In Overath and colleagues' study, they started with natural speech stimuli, chopped them into various-sized time chunks, and then shuffled the order of the chunks. This modification to the speech signal resulted in the preservation of various-sized units of analysis, from the speech sound or syllable to prosodic contours. The STG bilaterally was the region that correlated with this manipulation.

Kayoko Okada and I adopted a similar approach in an attempt to specifically identify phonological-level stages of word recognition. Our approach was to vary phonological-level features of words during fMRI and look for areas with brain activity that varied in kind. We chose to manipulate *phonological neighborhood density* (Vitevitch & Luce, 1999; Vitevitch et al., 1999). It turns out that words vary in terms of how many similar-sounding words exist in a language. As any poet can tell you, "cat" has lots of phonological neighbors: "mat," "rat," "sat," "fat," "hat," "pat," "cap," "can," "cut," "kit,"

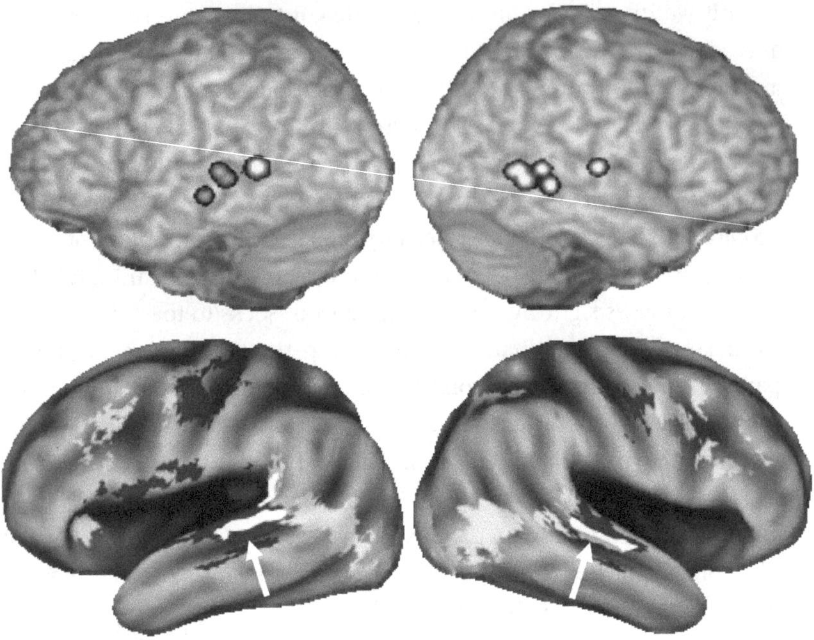

Figure 5.6

Functional activations associated with phonological-level processing during receptive speech tasks. Top row: activations associated with listening to words with high-density phonological neighborhoods compared to low-density phonological neighborhoods. From Okada and Hickok (2006) with permission. Bottom row: activations during audiovisual speech processing. White zone (arrows) are regions that are active both during auditory and visual speech perception *and* more active for visual speech perception than perceiving nonspeech facial gestures. Dark gray zones are more active for visual speech and nonspeech facial gestures. Reproduced from Venezia et al. (2017) under Creative Commons copyright.

"cot," and so on; "orange," not so many. It also turns out that it is harder to recognize words with more dense phonological neighborhoods. Presumably, the brain has to work harder to discriminate, say, "cat" from its many partial lexical doppelgängers.

Now, if the brain is working harder to distinguish one phonological pattern from its similar-sounding neighbors, it should be more active. We tasked our participants simply with listening to lists of words. Unbeknownst to them, some came from high-density neighborhoods and others from low-density neighborhoods. Naturally, we controlled for length and other nuisance features. Our listeners didn't notice the neighborhood density difference, but their brains did: we found higher activation levels to high- compared to low-density words in and around the posterior STS bilaterally, a bit more posterior than the meta-analyses indicate. Perhaps this was because we weren't picking up on the voice area (figure 5.6, top row) (Okada & Hickok, 2006).

Another source of evidence for the localization of the APA comes from functional imaging studies of audiovisual speech perception. It is well known that the visual speech information that comes from watching someone speak can lead to substantially better perception of what they are saying (Erber, 1969; Ross et al., 2007; Sumby & Pollack, 1954). Moreover, when audio and visual speech cues conflict, the visual cues can completely change the phonemic information that is acoustically perceived in some cases, a phenomenon called the *McGurk-MacDonald effect* (McGurk & MacDonald, 1976). Assuming that audio and visual speech interact at a phonological level of analysis (a common view), mapping the location of that interaction would provide a clue to the location of receptive phonological networks. This is an interesting approach because it gets around the problem of trying to separate acoustic from the phonological features in the signal. Instead of trying to subtract out the low-level acoustic information from a speech signal, we *add* a high-level phonological cue from vision and look to see where there is overlap between auditory and visual activation.

Several studies have been carried out on audiovisual speech with consistent results implicating the mid-to-posterior STS, particularly its upper (dorsal) bank, bilaterally (figure 5.6, bottom row, white regions indicated by arrows) (Beauchamp et al., 2004; Nath & Beauchamp, 2012; Venezia et al., 2017). This is not to say that this region is the only site of audiovisual interaction. There is good evidence that signals from different modalities can and do interact at multiple levels of the sensory hierarchies, including in or near the primary auditory cortex in the case of speech (Calvert et al., 1997; Hickok et al., 2018; Luo, Liu, & Poeppel, 2010; Okada et al., 2013). But the mid-to-posterior STS seems to be where the bulk of the interactions take place.

Yet another fascinating source of evidence comes from studies that attempt to decipher information coded in brain activity—a literal form of mind reading. The basic idea is this: in the typical functional imaging experiment, participants are exposed to a stimulus or given some task, and we look to see what pattern of brain activity correlates with that stimulus or task. A decoding experiment does the reverse, in a sense: the experimenter looks at the brain activity and tries to read what the participant was experiencing or doing.

In an elegant study published in the journal *Science*, Elia Formisano and colleagues (2008) asked volunteers to listen to three different vowel sounds spoken by three different talkers. In standard subtraction-type analyses, they found that their stimuli activated portions of auditory cortex bilaterally, as expected, but no difference between the various sounds. They then used machine learning to determine whether the brain response *pattern* could discriminate among the different vowels (with talkers varying) and among different talkers (with vowels varying). It did. Next, they mapped which regions in the brain enabled the discrimination among vowels and which enabled the discrimination among talkers. They were not the same areas. Regions enabling vowel discrimination, or decoding, included a swath of cortex covering the mid-to-posterior STG and STS, bilaterally. Regions enabling talker decoding were organized more sparsely, more asymmetrically in the hemispheres (right more than left) and extended a bit more anteriorly, especially in the right temporal lobe. There was some patchy overlap between the two decoding networks as well. The main result, then, is consistent with what we've seen from other methods: speech sound recognition involves the mid-to-posterior STG/STS region, nicely overlapping the zones highlighted in figure 5.6. Similar decoding studies have been carried out using surface electrodes implanted in human cortex as part of clinical surgery planning. Again, in these studies, electrodes recording from the mid-to-posterior STG/STS yielded the best decoding of speech information (Mesgarani et al., 2014; Moses et al., 2016; Pasley et al., 2012).

If the mid-to-posterior superior temporal region is critical for speech perception, disrupting it should cause speech perception deficits. This is complicated to test because of the bilateral organization of this system, and word-deaf cases are too rare and lesions too variable to get a clear picture of the precise location implicated. However, we do have some relevant neuropsychological evidence that confirms the involvement of this region.

Earlier in this chapter, I discussed the large-scale lesion study of speech perception and word comprehension led by Corianne Rogalsky. Recall that word comprehension performance in the pear-bear-moose-grapes task was at or near ceiling for nearly all the participants. This precludes the possibility of mapping the lesions associated with phonological-level word comprehension deficits because there were effectively no such deficits. However, we also

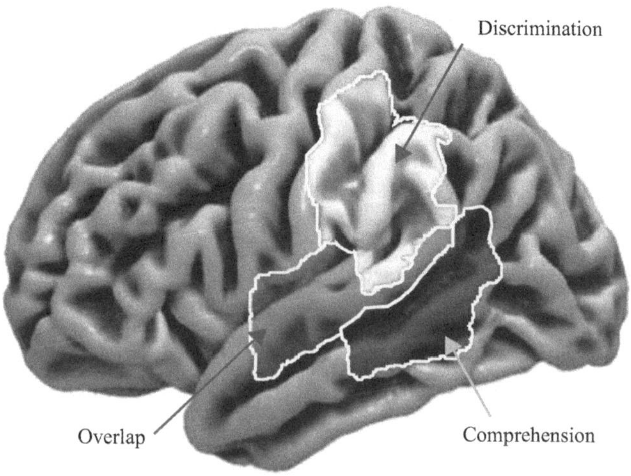

Figure 5.7
Lesions associated with word comprehension in noise, word and nonword discrimination, and their overlap. Modified with permission from Rogalsky et al. (2022).

presented the same task with the words masked by noise so we could get performance off the ceiling and then map the lesions associated with deficits.

Figure 5.7 shows the lesion pattern associated with impaired word comprehension performance in noise, along with the lesion pattern associated with impaired syllable discrimination. The results show that the STG is implicated in both tasks, including posterior and anterior zones, while word comprehension uniquely implicates only more posterior STG/STS regions. Discrimination ability, on the other hand, uniquely involves more dorsal areas in the planum temporale and inferior parietal lobe. The more anterior involvement of the STG in both tasks does not quite reach the zone identified by Scott et al. (2000), but it is more anterior than I would have predicted. This could be an artifact of lesion distributions, mechanisms associated with separating voice signals in noisy environments, or some other effect. Motor-related areas were not implicated in contrast to the hypothesis mentioned earlier that the motor system is important for speech perception in noise. Overall, these results support the localization of the APA to the mid-to-posterior STG/STS, as well as revealing the neural basis of the double dissociation between word comprehension and syllable discrimination.

Corroborative and spatially more precise neuropsychological evidence regarding the location of the APA comes from direct cortical simulation mapping studies of speech perception. In 1998, Dana Boatman and her colleagues reported a study of a twenty-three-year-old man who was undergoing evaluation for epilepsy surgery. The team performed a Wada test and assessed syllable discrimination ability, among other functions. No impairment whatsoever was noted with left- or right-hemisphere deactivation. However, when syllable discrimination was tested during direct cortical stimulation of the left hemisphere, the authors report, paradoxically, that performance was severely disrupted by electrical interference between "a single electrode pair on the left posterior superior temporal gyrus" (Boatman et al., 1998, p. 461). A more recent and much larger study (90 patients) of direct cortical stimulation found that word deafness was induced by stimulation to the left mid-to-posterior STG in the same distribution of the regions shown in figure 5.6 (Roux et al., 2015). The researchers inferred a word-deaf state from the experience of the patients during stimulation: "I really don't know; I hear something but I can't say what it is" (Roux et al., 2015, p. 404), reported one patient. Evidence from direct cortical stimulation, then, lines up very nicely with traditional functional imaging studies, as well as decoding studies: the relevant zone is the mid-to-posterior STG/STS.

It is worth taking a short digression to resolve the paradox that focal stimulation in just the left hemisphere can severely disrupt speech perception, while complete deactivation of the *entire* left hemisphere can leave speech perception intact, even in the same patient. It's likely that the explanation is to be found in the different manners of disruption: stimulation injects an electrical storm of energy into the brain that not only paralyzes the directly stimulated cortex but also can desynchronize and interfere with the entire network to which it is wired (including regions in the opposite hemisphere). In contrast, stroke or amobarbitol injection causes an electrical blackout, allowing the rest of the network to function as well as it possibly can without the added neural noise. Thus, focal stimulation to left-hemisphere speech perception networks can add noise to what is perceived in the right hemisphere, causing more severe disruption than the complete silencing of left-hemisphere activity.

Resolving Disagreements

There have been some surveys of this same body of literature that come to a different conclusion—namely, that the critical region for speech perception is mid-to-*anterior* STG/STS (DeWitt & Rauschecker, 2016). Some of this disagreement can be chalked up to terminological semantics: what I call "midposterior," others call "midanterior." For example, DeWitt and Rauschecker (2016) examined the same large-scale electrical stimulation study and concluded that the zone yielding transient word deafness is "mid-to-anterior STG." I'll let you examine the figure in that paper (or DeWitt & Rauschecker's meta-analysis maps) and come to your own conclusion about the best terminology. Either way, we are talking about the same zone.

Some of the disagreement, however, stems from the relative weight given to studies that use the rotated speech subtraction method, which, as we saw earlier, has implicated more anterior regions in the STG. I have to admit, this is a robust effect that I don't fully understand. From the functional imaging data alone, it does seem clear that the anterior temporal lobe is part of the story. But evidence from other sources strongly constrains our interpretation of the anterior activation as discussed earlier. My suspicion is that a good chunk of the magnitude of the effect is related to phrasal-level processing (but not morphosyntactic ability per se, as we'll see) with some contribution from voice- and possibly prosody-related processing (Humphries et al., 2005), not phonological processing of the sort that we've been discussing.

In chapter 8, we will discuss the role of the anterior temporal lobes in semantic knowledge. Perhaps the anterior superior temporal lobe intelligibility effect is tied more to semantic processing than speech processing. More work is needed to clarify what is going on.

Specificity and Differentiability of the APA

Is the APA specific to phonological processing? This is a rather loaded question that is best answered if we reframe it as an issue of differentiability. I'll try to clarify the distinction here.

It would seem that cognitive neuroscience has a vestigial phrenological tendency to carve the cortex into distinct functional zones. The fusiform face area, the parahippocampal place area, and the visual word form area are prominent examples, not to mention my own labeling of the APA. This tendency has led some authors to brand the whole field as *neophrenology* (Uttal, 2001) and to argue that it is misguided (Landreth & Richardson, 2014). Others defend the stance that the cortex is parceled into discrete areas (Petersen et al., 2024).

Both sides have at least some valid points, in my opinion. The idea that there is an area for faces or places or words, isolatable from the rest of the brain and handling nothing but faces or places or words, is surely incorrect.

Even defenders of the specificity of these areas would agree that such an extreme claim is not tenable. These areas are receiving and require already highly processed information from earlier stages of sensory analysis, and no one is surprised to find face-, place-, or word-*like* stimuli resulting in partial activation of their respective brain areas. At the same time, few question the functional and anatomical distinctions among the many structural and functional cortical areas such as primary visual, auditory, somatosensory, or motor cortex, among many others. Surely, then, the cortex is functionally specialized to some degree. Why, then, shouldn't similar specializations exist for higher-level functions? But, to flip-flop the argument yet again, we know that even primary cortex of one ilk can receive inputs from sensory signals of another. If even primary cortex isn't strictly selective, higher-level zones must be even less so.

While writing the first draft of this section in October 2018, I raised the question of speech specificity on Twitter (now X), mentioning specifically the regions that respond more to speech than nonspeech control stimuli. Language neuroscientist Marc Joanisse, responded, "I'm unsure how to interpret speech > nonspeech findings—in plenty of cases STG/STS subregions show activity for both, it's just that the effect is significantly bigger for speech. Is that 'specificity' in a strict enough sense?" He makes an important point. Whether a region is specific to some function or stimulus type boils down to how you define specificity. If we adopt a hyperstrict sense, not even primary cortex can be defined as specific, and that seems like the wrong result. But then where do we draw the line on the selectivity continuum before we call an area specific to *x*?

For this reason, I like to ask the question in terms of *differentiability*. Does V1 differentiate between visual and auditory input? The answer is yes, even if A1 receives some visual input and V1 receives some auditory input. Likewise, the face, place, and word areas may not be strictly specific to faces, places, or words (according to some), but do they differentiate between stimulus types in some way? If they do, then we can use that as a clue to working out how the zone in question contributes to the network's computations. And this is really what we are after: What is an area doing, computationally, such that it is really good at (or activates strongly to) speech? Whether we call it "specific" is just a reflection of a decision regarding an arbitrary boundary.

So does the word-clang zone differentiate between phonological-level speech and nonspeech? The results of the three meta-analyses that I described previously strongly suggest that it does. Listening to meaningless syllables activates the APA more than listening to various nonspeech control stimuli. Perhaps this reflects their tuning to different phoneme-related classes of sounds such as fricatives, plosives, and nasals (Mesgarani et al., 2014). Nonspeech controls might activate the network less strongly because those sounds or sequences of sounds don't hit the sweet spot of the network.

I would not want to rely on functional imaging evidence alone as an argument for the existence and localization of an APA, but bundled with evidence from studies that manipulate phonological variables directly, audiovisual speech studies, speech decoding, and direct cortical stimulation, we have a pretty strong case.

Having identified the word-clang zone, the natural next questions are: What is it coding, and how does the brain extract that code? We turn to these questions in chapter 6. But before we do, I want to button up our discussion about localization and lateralization.

A Remaining Puzzle

Here's a question that sometimes gets thrown my way: If speech perception is bilaterally organized as you say, why doesn't right hemisphere damage cause any problems? This is a reasonable question. I've presented data showing that left-hemisphere damage or deactivation produces only relatively mild speech perception deficits, even in people with significant auditory comprehension problems (Wernicke's aphasia) and complete destruction or complete deactivation of the left superior temporal lobe results in little or no receptive speech deficits in most people. Why don't we see similar mild speech perception problems following right-hemisphere damage, particularly to the superior temporal lobe?

One answer is that no one has really looked carefully. Mild comprehension deficits are often subclinical, meaning that they don't interfere much with everyday life and require sensitive tasks to measure. In fact, we probably go through much of our day missing some proportion of the phonological information that tickles our eardrums. Our brains readily fill in any missing or ambiguous bits using phonological (Warren, 1970), semantic (Ganong, 1980), and any other knowledge context we can bring to bear on the conversation (Sohoglu et al., 2012). So a right-hemisphere damaged patient with a subclinical speech perception problem and no other signs of

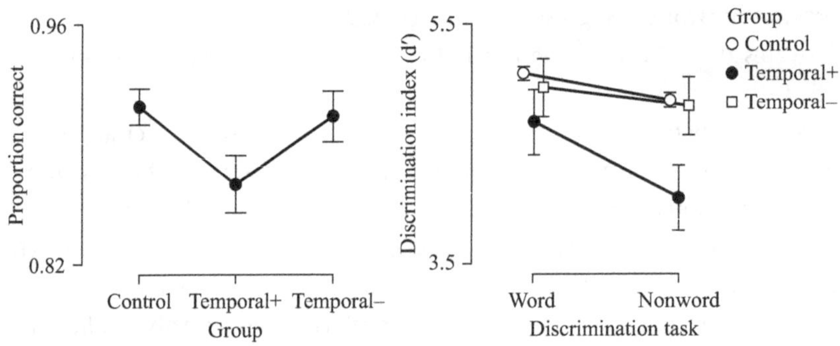

Figure 5.8
Performance on speech perception tasks following right-hemisphere stroke, grouped by the presence (+) or absence (–) of temporal lobe involvement. Left: word comprehension in noise. Right: word and nonword discrimination. Reprinted with permission from Rogalsky et al. (2022).

aphasia would typically not be assessed in depth for such a problem, nor would they typically be enrolled in a study investigating the neural basis of word-clang decoding.

Again, I teamed up with Corianne Rogalsky and our colleagues to test this possibility (Rogalsky et al., 2022). We assessed nine patients with damage to the right superior temporal lobe, eight patients with right-hemisphere damage that spared the temporal lobe, and a group of 26 nonbrain-damaged control participants on four tasks: two versions of the pear-bear-moose-grapes task, one with clear speech and one with the target word presented in noise, and two discrimination tasks, one using words and the other using nonwords. Unsurprisingly, we found that all participants performed well on the clear speech pear-bear-moose-grapes task. Recall that the vast majority of the left-hemisphere-damaged patients have no trouble with this task. Only two right-hemisphere-damaged patients made an error on the task, and both were in the temporal lobe group. However, performance on the other three tasks revealed differences between the groups, with the right temporal lobe–damaged patients performing significantly worse that the others, as the graph in figure 5.8 shows (Temporal+ is the group with temporal lobe damage). When we look carefully, then, we indeed find evidence of mild speech perception deficits following right-hemisphere damage.

6 Cracking the Speech Code

> What's all this fuss I keep hearing about violins on television. Now why don't parents want their children to see violins on television? I thought the Leonard Bernstein concerts were just lovely.
>
> —Gila Radner as Emily Litella on *Saturday Night Live*, 1976

The physical speech signal, the stuff that travels through the air from talker to listener, is nothing more than waves of air pressure. Think about that for a second. We can take a potentially infinite number of complex thoughts, code them into sound wave patterns as a talker, and then decode them back into complex thoughts as a listener. Maybe this doesn't sound so impressive to us inhabitants of the digital age where we can code anything in patterns of zeros and ones, communicate with someone anywhere on the globe nearly instantaneously, and even send pictures of Pluto back to Earth through billions of miles of space. But consider the fact that scientists and engineers have been working for decades to build automated speech recognition systems and, despite reasonable success in recent years, still haven't matched the ability of the system built by biology. Nor do we yet understand fully how humans do it. Apparently, the speech code is a bit more complicated than sending photos from Pluto. Why?

The Speech Signal

Figure 6.1 shows what the phrase "From waves to thoughts" looks like in graph form, where we represent time on the horizontal axis and air pressure level on the vertical axis.

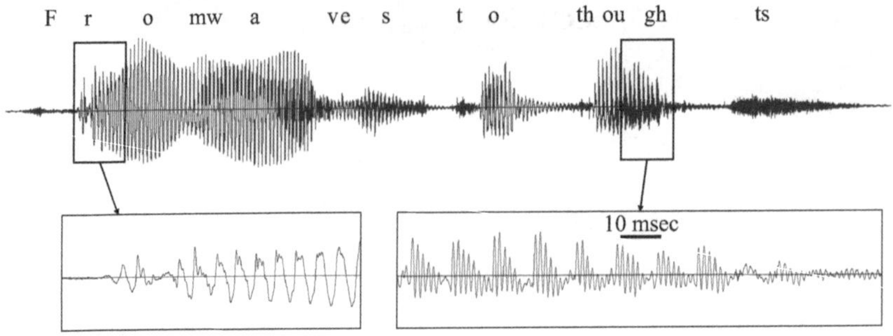

Figure 6.1
A sound pressure wave plotting time on the horizontal axis and amplitude on the vertical axis. The lower boxes zoom in on portions of the wave pattern.

The written sentence above the graph roughly shows the portion of the wave that corresponds to the different sounds of the words in the sentence. From this perspective, the most noticeable feature of the waveform is that there are mountains and valleys of wave action; that is, changes in the *amplitude envelope*. Notice that the valleys do not necessarily correspond to word boundaries. The two graphs on the bottom show a zoomed-in view of the boxed sections on the upper graph. Here, we can clearly see individual waves as the pressure goes up and down around the horizontal line. Moreover, we can see waves embedded in waves: in the lower right box, the highest-frequency (i.e., they are the closest together) waves cycle at about 800 Hz (Hz = cycles per second), going from peak to peak in about 1.25 msec. These high-frequency waves form larger wave clusters, which cycle at about 100 Hz, a cluster peak-to-peak time of approximately 10 msec. And now we can appreciate that these waves themselves cluster into the lower-frequency waves (typically in the 3–5 Hz range), which we already saw as the "mountains" in the upper graph.

We can see all the frequencies more directly by plotting the speech sample as a *spectrogram* (figure 6.2). Here, time is again on the horizontal axis, but now frequency is plotted on the vertical axis instead of sound pressure amplitude. The darker or lighter portions represent the strength or energy of the signal (height of the wave) at the various frequencies, with darker denoting more energy. Notice that the energy tends to concentrate in horizontal bands that are kind of stacked on top of one another. These are called *formants,* and the ratios between the bands in a stack provide cues to

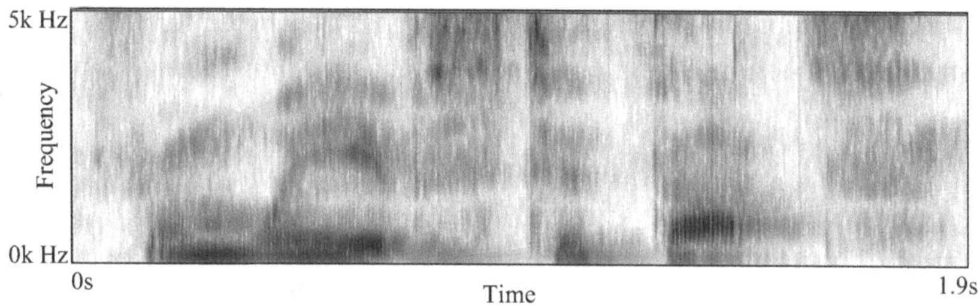

Figure 6.2
A speech spectrogram.

the speech sounds that we hear, particularly vowels. Notice too that the formants are not always straight horizontal bands; they often drift up or down over time, sometimes rather abruptly. These are called *formant transitions,* and they result from transitioning from one vocal tract position to the next during speech articulation.

Hopefully, now you can appreciate the *spectrotemporal* nature of the speech signal. It is composed of a spectrum of frequencies (waves upon waves upon waves) that change over time. A spectrogram, then, describes the particular spectrotemporal pattern associated with a chunk of speech. A core goal in the neuroscience of speech perception is to figure out how these spectrotemporal patterns are represented in neural signals and how those neural signals are collapsed onto categories of speech sounds and words. We've made respectable progress in understanding how the brain codes spectrotemporal patterns, but it is not so clear how such codes are transformed into speech recognition. Let's consider the general problem with speech before we dig into what we know about how the brain tackles it.

The Problem with Speech

The reason why it has been so difficult to crack the speech code is *not* that those complex waveforms are hard to reproduce by a machine or to decompose precisely into their component frequencies and amplitudes. You've seen in the figures in this chapter so far that we can do that quite readily. The problem is acoustic variability coupled with perceptual stability. More than one spectrotemporal pattern—many, in fact—can lead to the same percept of speech. We encountered one example previously: the spectrotemporal pattern corresponding to the /*d*/ in, say, "dog" is not the same as the /*d*/ in "dig," even though we hear /*d*/ sounds in each case. But the problem goes well beyond such phonological context effects, as we'll see.

The human larynx and vocal tract differ anatomically from person to person, which changes the fundamental frequency of the speech waveform. We hear these differences in the pitch of a talker's voice, from the squeaky highs of a young child (average 250–300-Hz range) to the middle-range of an adult female (average 200–250 Hz), to the lower sound of an adult male (average 120–150 Hz) (Hillenbrand et al., 1995; Lee, Potamianos, & Narayanan, 1999). This alone means that the absolute values of the formants associated with a speech sound will differ from talker to talker. Yet, despite all of this between-talker variability in the acoustics of speech, we still manage to understand everyone we talk to. This ability is called *talker normalization*.

Beyond this normal voice-to-voice variation, we also encounter speech under a variety of conditions including accents, rates of speech, slurred speech, false starts, sung speech, and the speech of parrots or even Hoover the talking seal (Google it). We still manage to understand most of it.

The fact that we can understand speech despite substantial variation in speech acoustics should remind you of the lack of invariance problem discussed in chapter 5. It's part of a pervasive family of perceptual problems that plagues not only speech, but other aspects of perception as well, such as visual object recognition. We recognize a chair as a chair, for example, even though chairs can vary indefinitely in size, shape, color, orientation, position on the retina, viewing angle, and degree of occlusion. Vision scientists typically describe the situation using the term *perceptual constancy*, emphasizing what the brain achieves rather than the problem that it faces. Some speech scientists also use "perceptual constancy" to describe the speech situation (Shankweiler, Strange, & Verbrugge, 1977). I like *perceptual constancy* better than *lack of invariance*, so I'll mostly do the same.

Perceptual constancy in the face of acoustic variation is the reason why automated speech recognition has been such a hard nut to crack. If human speech was robotlike, with no acoustic variation, the tech giants wouldn't have to employ entire teams to work on automated speech recognition. One way around some of this variability is to train your personal digital assistant on your own voice, but that doesn't help much with variations in phonological context, speech rate, and the like. Another kind of crutch is to restrict the system's guesses to known words. Ask your digital assistant to spell or recognize a nonsense word like "blicket." You'll likely find that it guesses a similar-sounding real word or name. Now try it with a five-year-old child, as psychologist Jean Berko did in her famous "wug test" (Berko, 1958), and you'll likely get a correct response. We humans readily hear pseudowords because we *have* to in order to learn new words, which children do prodigiously (Brysbaert et al., 2016). Another trick that engineers have discovered is that instead of trying to detect a couple dozen individual phonemes from their many-faced, context-dependent forms, they build in a few thousand *triphone* detectors (Dahl et al., 2012). So instead of trying to figure out that the first phonemes in "dog," "dig," "dope," and "dad" are all /*d*/, they just built detectors that can recognize "dog," "dig," "dope," and "dad" more directly, sidestepping the problem. Some scientists have proposed that humans do this too, claiming that the syllable, not the phoneme, is the principal unit of speech recognition (Greenberg, 2005; Massaro, 1972). I agree with them.

What's the Code?

Now that we are aware of some of the problems that speech presents, let's return to the issue that we started with: What kind of information is coded in those air pressure waves? It's useful to consider analogies, like Morse code, where we know the code precisely. In Morse, different sequences of dits and dahs correspond to different letters of the alphabet, A • —, B — •••, C — • — •, and so on. A message such as Mark Twain's 1897 telegram, "The reports of my death are greatly exaggerated" can then be coded as the following sequence: "-••••• / •-••• —•—•-•— •••/—••-• /—•—/ -•••• —••••/ •- •-•• /—••- ••• —•-•• -•—/ • -••- •—•— ••• -• •—• -••" And, like speech, there can be physical variability in the dits and dahs, which can be electrical pulses, light flashes, horn blasts, or other stimuli. We can go a step further and imagine dit/dah patterns that look like - — - versus — —— —, where the dah in one transmission is the same length as the dit in the other. If there was this much variation in Morse "talkers," it would be impossible to know, without a larger context, whether an individual signal element was a dit or a dah because the categorization of one element is relative to the other. In more technical terms, this is a lack of invariance problem that requires "talker" normalization to achieve perceptual constancy. Hopefully, you can see why finding the invariant acoustic signature of an individual speech sound is probably a quixotic quest, and why it makes good sense instead to look for acoustic patterns in a broader context (like triphones) that correspond to speech categories.

At this point, we might wonder: What are the speech equivalents of Morse dit/dah sequences, and what do they stand for? But before we think about that, let's consider another instructive lesson from Morse code. Notice that the message itself isn't directly coded in the sequence, only the letters. For example, if you know Morse code, you can accurately transcribe this message "•-••— ••• / ••— ••• -•—•-•—•••• /— ••• /—•• /—••- • •-•—• / ••• —• /—••—•—/ • -••- •— • • • -• •—••—•••• -•-•-" as "Los informes de mi muerte son muy exagerados," even if you don't understand the message. To get to the message, a further decoding is needed from letters to words and words to meaning. Samuel Morse didn't have to worry about this additional decoding because it was already built into the interpreter: a human mind equipped with language. But we have to worry about it when thinking about the speech code because we have to traverse the entire span between waves and thoughts.

The Language Code

The broader language code is actually a collection of codes in a complex transformation from meaning to sound and from sound to meaning. Linguistic theory represents our best understanding of a subset of these codes, although, as with any active domain of science, there is still much disagreement about the details. In traditional models, this includes some form of a phonology, a morphology, a syntax (or an integrated morphosyntax), a semantic system, and a lexicon of words. But there are more transformations that occur outside these traditional domains of linguistic theory. On the input side, for example, sound pressure waves are transformed into their component frequencies and tracked over time well before any contact with a phonological system, as we'll discuss shortly. And on the output side, phonological representations of speech must be transformed into sequences of articulatory gestures.

This way of framing the problem naturally leads one to think in terms of a funnel-like process. On the output side, thoughts are transformed into some sort of formalized semantic representation and a syntactic frame, which is populated by morphemes that contain syllables built of phonemes that are spelled out as articulatory gestures, which finally result in air pressure waves that create little Morse-like packets of coded information that represent the end product at the bottom of the funnel. On the input side, the Morse-like packets impinge on our ears and get transformed into neural signals from which the phonemes are recovered and then assembled into syllables, syllables into morphemes, and morphemes into syntactic hierarchies and meanings. Let's call this the *funnel model,* in which the goal of the computation is to transform a thought into tiny packets of information (e.g., phonemes) that can be coded sequentially in wave patterns.

I will argue instead for a *leaky funnel model.* Few speech scientists (if any) would disagree that the thought-to-sound-to-thought transformations involve funnel-like processes, or in linguistic terminology, *hierarchical levels of representation.* I myself concur. But the various levels of the funnel are not completely sealed off from the acoustic airwaves. Rather, information relevant to multiple levels of the funnel leak out from their own levels of the thought-to-sound transformation process and add to the wave pattern directly. Of course, I'm not suggesting some telepathic direct transmission of information from higher levels straight into the acoustic soup. No, everything has to pass through the motor-articulatory system. What I'm pointing out is that what gets communicated to the articulatory system is a *multiplex* signal that comes from more than one source in parallel.

To see why this is not a far-fetched idea, consider the full range of information contained in the speech signal. Not only do we hear words and sentences, but we also gain information about the emotional state of the talker, the person's gender, approximate age, approximate size, identity (if the talker is an acquaintance), where in the world the person is from (accent), the person's health status, personality traits, social receptiveness or threat level, interests and concerns, and more. This information has its origin in varied and far-flung sources ranging from emotional networks in the brain to the length of the vocal tract, and they all end up leaving their own stamp on the acoustic wave that listeners can readily decode, quite separate from the content of the linguistic message.

So multiplexing is a thing in the acoustic waveform of speech, but does it include information from multiple *linguistic* levels? Does the linguistic funnel leak, in fact? Let's start low and work our way up. Information about phonemes is certainly present in the speech signal, even if we don't understand precisely how it is recovered. Vowels are a good example. The spacing between the formants, particularly the ratios between them, as well as the length and transition pattern, are cues to vowel identity (Hillenbrand et al., 1995; Mi et al., 2016). Some acoustic correlates of consonant identity have also been identified (Stevens, 1980). It is, therefore, uncontroversial that phonemic information is somehow coded in the speech signal. What about higher-level information?

Syllables are built of sequences of phonemes with a vowel at the core. Vowels tend to have the highest-amplitude waves because they are voiced and are formed by relatively open vocal tract configurations. Consonants, in contrast, can be unvoiced (like */t/*, */p/*, or */k/*) and generally involve more constricted vocal tract positions. Look back at the speech waveform plotted in figure 6.1, and notice that the vowels in the sentence correspond to the high-amplitude lobes of the signal. What this means is that the low-frequency rhythm of speech—the 3–5-Hz "lobe frequency"—is an acoustic reflection of the rate of syllables. We'll return to this basic rhythm and its implication for speech perception soon. For now, though, we are just interested in the fact that syllable-level information is cued in the speech waveform, at least broadly so.

Overlaid on the backbone rhythm of the syllables is an orchestra of further modulations of the speech signal. These include pitch accents within and across words and phrases, pauses and sound lengthenings, emphases, tempos, and higher-order rhythmic patterns. Collectively, this is the *prosody* of speech, and it communicates a wealth of information through sound that transcends phonemes (Cole, 2015; Cutler, Dahan, & van Donselaar, 1997; Shattuck-Hufnagel & Turk, 1996; Wagner & Watson, 2010). The following definition from a review of the topic (Wagner & Watson, 2010, p. 906) captures the scope of this prosodic code:

> "Prosody" is often used to refer to those phonetic and phonological properties of speech that are crucially not due to the choice of lexical items, but rather depend on other factors such as how these items relate to each other semantically and/or syntactically, how they are grouped rhythmically, where the speaker places emphasis, what kind of speech act the utterance encodes, whether turn taking in conversation is being negotiated, and they can reflect the attitude and emotional state of the speaker. While these factors can also determine the choice of lexical material, they can affect the signal directly without any mediation by a lexical morpheme with segmental content, and it is this kind of information that is often referred to as the prosody of an utterance.

At the word level, prosody, or tone, can differentiate one word from another, such as in the related English words, *PRO*duce versus pro*DUCE*, and it does so more systematically in tone languages like Mandarin, where different pitch contours overlain on the same syllable can cue completely different words. At the sentence level, prosody can cue syntactic boundaries, as punctuation does, and, in some cases, it can disambiguate alternative parses of a sentence. Prosody can also substantially change the meaning of an utterance, even while the phonemes, words, and syntax are unchanged. Consider the sentence *I don't think you can eat the fish.* Depending on which word is emphasized (which will be **bold** in the examples to come), the meaning flops like, well, a fish out of water:

I *don't think you can eat the fish.*

Meaning: I don't think you can, but someone else might think you can.

I ***don't*** *think you can eat the fish.*

Meaning: It's incorrect that I think you can eat the fish.

I don't ***think*** *you can eat the fish.*

Meaning: I don't believe that you can eat the fish, but I'm not sure.

I don't think ***you*** *can eat the fish.*

Meaning: You can't, but someone else can.

I don't think you ***can*** *eat the fish.*

Meaning: You're not even capable of eating it.

I don't think you can ***eat*** *the fish.*

Meaning: You can't eat it, but you can do something else with it.

I don't think you can eat the ***fish****.*

Meaning: The fish is out, but you have other options.

The different emphasis patterns shift attention to different discourse topics. Is the utterance fundamentally about me, you, thinking, eating, or the fish? These shifts in prosodically cued meaning (and emotion) can lead to misunderstandings in emails or text messages: without the prosody, the meaning and tone of the message are left for the reader to impose. A humorous sketch showing how different prosodic orchestrations can drive radically different interpretations was performed by the comedy duo Keegan-Michael Key and Jordan Peele in "Text Message Confusion." (Check it out on YouTube.)

There are, of course, many other types of prosodic cues (Cole, 2015), like the sentence-final rising or falling tone patterns that signal questions, statements, and demands. There are prosodic cues that are related to turn taking in conversations. There are cues to indicate topic changes, and cues to sarcasm, which flips the literal meaning of a sentence on its head. I could go on, but I think I've made my point.

These facts argue against a strict funnel model of the mapping between thoughts and sounds and instead support a leaky funnel model. This is not a new or even controversial idea. In fact, the very existence of the entire subfield of linguistics that focuses on prosody is an acknowledgment that the funnel is perforated. Prosody *is* these higher linguistic, pragmatic, semantic, and emotional levels *affecting the acoustic speech signal directly,* without mediation by lower levels of the funnel, to paraphrase the previous quote. Nonetheless, it is worth motivating the leaky funnel model because it broadens our perspective on the speech code. We can now see that it's not as simple as Morse code, where only tiny bits of information (the bottom of the funnel) are contained in the signal. Instead, speech is a massively multiplexed signal coding information across multiple hierarchical (e.g., phonological, syllabic, lexical, syntactic, semantic) and parallel (e.g., emotional, prosodic) levels.

When we listen to speech, we are taking in all this information and somehow combining it to arrive at an interpretation of the intended message. This process adds further complexity beyond what we see in the multiplexed speech code because, while the speech code contains all the necessary information to get to the intended message in most cases, it doesn't give it to the listener for free. Despite all these cues, many of the details within those levels of representation must be *computed* from what's given, as opposed to simply being decoded. For example, while prosodic structure can cue syntactic phrase boundaries, it's only a loose relationship (Shattuck-Hufnagel & Turk, 1996). And while syllable timing information is available directly in the amplitude pattern of waveform, this does not tell you which syllables are part of the same versus different words. To compute the nitty gritty of a syllable, word, or syntactic parse, multiple sources of information must be combined. It's only when multiple sources of information are used in a mutually constraining fashion that we can resolve potential ambiguities in the possible parses and interpretations. In psycholinguistic jargon, this multisource combination process is referred to as *constraint satisfaction* and has become the dominant model of language comprehension (MacDonald & Seidenberg, 2006).

We've seen that the brain doesn't boil down thoughts simply into Morse code–like, phoneme-sized bits for airborne transmission. Rather, it leaks from multiple levels of the transformation from thoughts to sounds. So the problem with speech is not limited to the ever-vexing perceptual constancy problem. That's just one complication. Another is how to decode the parallel streams of information in the multiplexed signal and combine them in the right way to arrive at the intended meaning of the utterance.

The process of combining all these sources of information is itself rather complicated and worth the space of at least a few lines here. One important point is that the influence of one source on another is bidirectional. In one direction, lower levels can be used to build higher levels: phonological features must be combined into syllables, syllables into words, words into phrases, and so on. This is bottom-up combination. But top-down information from higher levels influences how the lower levels are combined. For example, word-level information can influence the perception of speech sounds (the *Ganong effect*) (Ganong, 1980); intonation and stress patterns can influence perceived word boundaries in a speech stream (Cutler, 2005; Cutler & Norris, 1988; Spinelli et al., 2010); sentence structure, compared to lists of words, can facilitate recognition of a target word (Simpson et al., 1989); and the particular low-frequency rhythmic structure of an utterance can be the difference between complete gibberish and a fully intelligible sentence (Zion Golumbic, Poeppel, & Schroeder, 2012).

To summarize, the speech signal is difficult to decode fully, and it's difficult in many ways. It's highly variable at the acoustic level, and yet we have perceptual constancy of phonological categories. It's a multiplexed signal with information from several levels of analysis coded directly in the waveform. Converting thoughts to waves, however, is not a lossless compression, as there are ambiguities to resolve across multiple levels. Thus, analyzing the signal requires combining information from the various multiplexed channels using both bottom-up and top-down mechanisms.

Now, hopefully, you can more fully appreciate why cracking the speech code is rather more difficult than beaming a picture through space.

How Does the Brain Decode Speech?

There has been a lot of progress in understanding how the brain decodes the speech signal, but most of the advances have been at the earliest stages, as the acoustic signal is transformed to a nervous signal and progresses up to auditory cortex. At the level of the cortex, we have a basic understanding of the organization of primary areas; beyond that, we have good clues but few details. I'll summarize some of the important facts in the following sections:

Acoustic signals are split into frequency channels.

Earlier in this chapter, we looked at a spectrogram of speech, which breaks the wave pattern into its component frequencies for us to visualize, and we learned that formants, energy in different frequency bands, are important cues in speech. It seems reasonable, then, that the auditory system performs some sort of frequency or spectral analysis of speech. Indeed, in the 1920s, Hungarian biophysicist Georg von Békésy (1899–1972) showed that the cochlea in the inner ear functions as a spectral analyzer, breaking complex sound waves into separate frequency bands or channels, as they are also called (von Békésy, 1960). He carried out detailed measurements on the mechanics of the cochlea in a variety of animals—from chicken to cow to elephant to human—and found that sound-induced vibrations of the eardrum and middle-ear bones were transformed by the cochlea into a wave that traveled along its basilar membrane. He further observed that the peak of the wave along the length of the cochlea was determined by the frequency of the sound hitting the ear. Higher-frequency sounds have maximum peaks closer to the base of the cochlea and lower-frequency sounds peak nearer the end.[1] Different bundles of nerve fibers connect to different locations along the cochlea, resulting in acoustic information being transmitted up to the brain through separate frequency channels (Narayan et al., 1998). This *tonotopic* organization is preserved all the way up to auditory cortex.

Acoustic signals are also coded in the time domain.

Spectrograms represent not only frequency information, but also how it changes over time. Accordingly, the auditory system also tracks temporal information. One way that this has been documented is to record the neural response to amplitude-modulated wideband noise, such as a "shhh" sound, that oscillates between quiet and loud (amplitude modulation). Research with such signals has found that some auditory areas prefer (i.e., respond more to) rapid modulations, up to 256 Hz, while others prefer slower modulations, down to 4 Hz or even less (Giraud et al., 2000). In general, there is a fast-to-slow trend in preferred temporal modulation rate as the signal progresses from the cochlear nerve up to the cortex (Creutzfeldt, Hellweg, & Schreiner, 1980; Giraud et al., 2000). We might think of this as a progressive broadening of the temporal receptive field of auditory neurons: earlier stages of processing care more about what's happening over a few milliseconds of time, whereas later stages are looking at the broader temporal patterns that occur over tens or hundreds of milliseconds. Some researchers call these *periodicity* or *temporal envelope* codes.

Spectral frequency and temporal modulation rate combine to yield spectrotemporal codes.

If frequency bands are maintained from cochlea to cortex in the form of tonotopic codes, and if different levels of the auditory system code temporal modulation at different rates, it is reasonable to hypothesize that these two codes intersect within each frequency channel, resulting in *spectrotemporal* codes. At a simple level, we can think of the intersection of a spectral frequency channel (say, one centered on 1,000 Hz) and a temporal modulation preference (say, one centered on 8 Hz) as defining a *spectrotemporal receptive field (STRF)* because a signal that contains that combination will maximally excite the cell or network that is tuned to it. Indeed, if you separately map preferred spectral frequency (tonotopy) and preferred temporal modulation rate (periodotopy) in auditory cortex, as my colleagues and I did, you can find evidence for such combinations of spectral and temporal fields (Barton et al., 2012).

Primary auditory cortex as a spectrotemporal modulation filter bank.

There is a growing consensus that primary (also called *A1* or *core*) auditory cortex functions as a *spectrotemporal modulation filter bank* (Chi, Ru, & Shamma, 2005; Sabin, Eddins, & Wright, 2012; Santoro et al., 2014; Santoro et al., 2017; Shamma, 2001). What does that mean? A *filter bank* is just a set of filters that separate an input signal into multiple components. We encountered this concept earlier when we talked about the cochlea, which breaks complex waves into multiple frequency channels. It functions as a *spectral* filter bank. A *modulation* filter bank just means that the filters apply to patterns of modulation (change) in some dimension. We've encountered a kind of modulation filter in the temporal domain, with different areas responding maximally to different pulse rates. A *spectrotemporal* modulation filter bank means that it is filtering modulation patterns in two dimensions, spectral and temporal.

While it's not too hard to conceptualize temporal modulation rate filters, each tuned to different rates of acoustic rhythms, spectral frequency modulation filters are less intuitive. A graphic will help. Figure 6.3 shows spectrogram

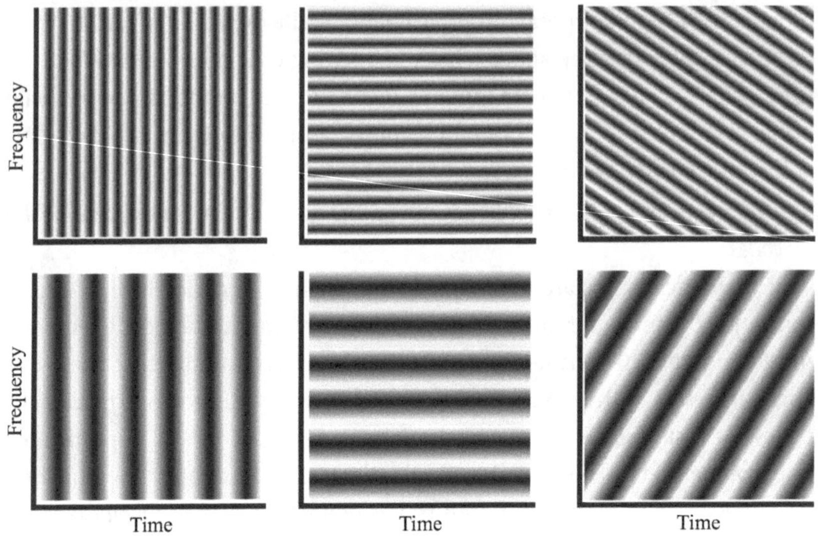

Figure 6.3
Examples of spectrotemporal modulation patterns. Left column: temporal modulation without spectral modulation. Middle column: spectral modulation without temporal modulation. Right column: spectral and temporal modulation. Top row: higher modulation frequencies. Bottom row: lower modulation frequencies.

plots with time on the horizontal axis, frequency on the vertical axis, and energy amplitude represented by the grayscale shading. The left columns show temporal modulation without spectral modulation: there is a lot of change in the amplitude as you move from left to right in the graph at every spectral frequency, but there is no change in amplitude as you move from bottom to top at a given point in time. Notice that the top and bottom rows differ in their rate of temporal modulation frequency: the top is more densely spaced and the bottom is more sparsely spaced. Now think of an array of detectors or filters tuned to a variety of temporal modulation frequencies. The middle column represents spectral modulation without temporal modulation: there is a lot of change in the amplitude as you move from bottom to top at every point in time, but no change in amplitude as you move from left to right. Again, the rows show different modulation frequencies, more densely/sparsely spaced, but now in the spectral dimension. Imagine, now, an array of filters tuned to a variety of spectral modulation frequencies.

Finally, the right column represents modulation in both dimensions, one falling (top) and one rising (bottom). To see that there is modulation in both dimensions, notice that if you pick any point on the horizontal axis and trace a line vertically from that point, there is spectral modulation, and if you pick any point on the vertical axis and trace a line horizontally from that point, there is temporal modulation. The amount of modulation that you see as you trace a vertical or horizontal line depends on how steep the rise or fall angle is. Now imagine an array of filters tuned to a variety of spectral and temporal frequencies from tightly to sparsely spaced at a variety of falling or rising angles. That is a spectrotemporal modulation filter bank.[2]

What a spectrotemporal modulation filter bank does computationally is that it breaks a complex signal into its component modulation frequencies, both temporally and spectrally. In mathematical terms, it performs a two-dimensional Fourier transform. Since the inverse of a Fourier transform is the time-domain signal itself (the waveform), what the filter banks derives is a representation of the acoustic signal, the brain's code for what hit the ear. Higher-level processing streams in the auditory system can then tune into different filters or combinations of filters—that is, portions of this spectrotemporal modulation code—to extract phonemic representations, talker voices, prosodic or melodic patterns, or other features.

Speech Codes beyond Early Auditory Cortex

For speech perception, the relevant portion of spectrotemporal modulation space—the *modulation power spectrum*—is *low pass* in both dimensions; that is, low-spectral and low-temporal modulation frequencies. Jonathan Venezia, a former graduate student in my lab and longtime collaborator, along with my University of California, Irvine colleague, Virgina Richards, developed a method called *auditory bubbles,* which allowed us to measure the speech intelligibility regions of modulation space (Venezia, Hickok, & Richards, 2016). The basic procedure is that a spoken sentence is digitally transformed into modulation space. Figure 6.4 shows what this looks like for speech. In this type of graph, we plot the temporal modulation frequency on the horizontal axis and spectral modulation frequency on the vertical axis, instead of time and frequency as in a spectrogram. Once in modulation space, we can then remove some spectrotemporal modulations in the original signal by applying a Swiss cheese–like filter mask, the "bubbles." Then the speech waveform is reconstructed, but now lacking modulation in some spectrotemporal frequencies. Volunteers listen to the sentence and indicate whether they understand what was said. After presenting lots of sentences with lots of different

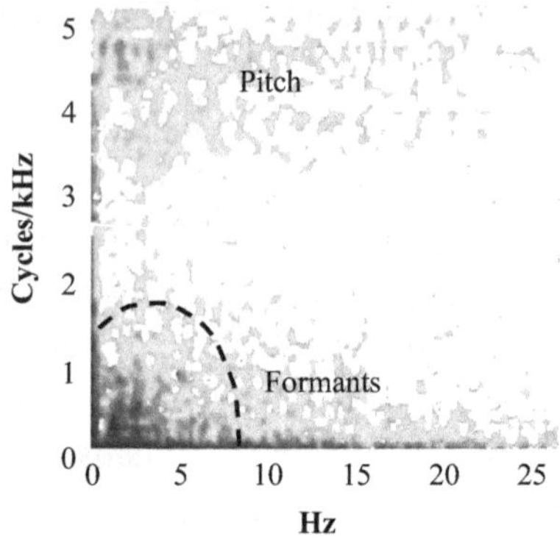

Figure 6.4
Average speech energy in the spectrotemporal modulation space (modulation power spectrum) from 452 sentences spoken by a female talker. The horizontal axis is temporal modulation and the vertical axis is spectral modulation. Energy (power) is indicated in grayscale, with darker shades indicating more energy (power). Most of the spectral energy is concentrated in two bands; the lower-frequency band (<2 cycles/kHz) is associated with speech formants, while the higher-frequency band (4–5 cycles/kHz) is associated with pitch. Most of the temporal energy is concentrated in lower frequencies (<8 Hz), which includes the syllable rate. The dotted line represents the intelligibility zone, the region of the modulation power spectrum that is critical for the perception of intelligible speech. For details, see Venezia et al. (2016, 2019).

regions of spectrotemporal space removed, it is possible to generate a map in spectrotemporal space that corresponds to the modulation frequencies that, when present in the signal, enable the listener to understand the sentence. This *intelligibility zone*, indicated by the dotted line in figure 6.4, is below 8 Hz in the temporal dimension and below 2 cycles/kHz in the spectral dimension. This makes sense because, in the spectral domain, low-modulation frequencies align with the spectral frequencies critical for vowel identification (below 2 cycles/octave) (Liu & Eddins, 2008) and, in the temporal domain, low-modulation frequencies align with the syllable rate in spoken language (between 2 and 8 Hz) (Poeppel & Assaneo, 2020).

An inspection of figure 6.4 shows that there is a notable pocket of energy in the high-spectral, low-temporal modulation range that falls outside the intelligibility zone. This region of modulation space is important for pitch perception, which is important for talker identification, prosodic aspects of speech, and music but isn't critical for intelligibility, at least for nontone languages.

The bubbles method can also be used for mapping SRTFs in human auditory-responsive cortex. In a study again led by Jonathan Venezia, we asked participants to listen to bubble-filtered sentences during functional magnetic resonance imaging (fMRI) scanning and examined the data to determine the regional tuning preferences to different patterns of spectrotemporal modulation (Venezia et al., 2019). The results showed a hierarchy of tuning preferences (figure 6.5). In and around Heschl's gyrus, tuning was found for a wide variety of spectrotemporal modulation patterns (black, pattern 1), consistent with this region functioning as a spectrotemporal modulation filter bank. Pattern 1 tuning was also evident in a more lateral posterior superior temporal gyrus (STG) location. There is some evidence that this posterior STG zone is involved in prosody (Turker et al., 2023), in which case sensitivity to pitch makes sense. The other tuning patterns that we observed were restricted to low-spectral modulations, the "formant" regions of figure 6.4, but they differed in terms of their temporal modulation tuning. Tuning to rapid temporal modulations (10–15 Hz, dark gray pattern 2) was mostly found in the planum temporale region and posterior STG, while more lateral and ventral regions of the STG were tuned to slower modulations, with one set of clusters responding best to temporal modulations in the 5–6-Hz range (light gray pattern 3) and the other in the 2–4-Hz range or lower (white pattern 4). You can see that the slowest spectrotemporal modulation receptive fields (pattern 4) look like the behaviorally determined intelligibility zone in figure 6.4; and, indeed, the brain locations with such tuning overlapped areas that are identified in fMRI studies of speech intelligibility. Presumably, the different patterns of temporal modulation tuning represent subregions of auditory-related cortex that are analyzing speech over different-sized temporal windows, which may correspond to different-sized linguistic units ranging from acoustic-phonetic features to syllables and words to phrasal chunks.

There are a couple of important caveats to keep in mind regarding these broad patterns of STRF maps. One is that these are general tendencies of averaged patterns rather than detailed individual maps. So, while we can see a general hierarchy of tuning in this kind of analysis, the detailed organization is obscured. A second, related caveat is that the tuning patterns (1–4) shown in figure 6.5 are representative examples of a class of tuning; there is variation within each class. For example, within pattern 1, some bits of cortex are tuned mostly to the pitch region of spectrotemporal space, some more so to portions of formant space, and others to both. Of course, since these are maps derived not from individual neurons but from volumes of about half a

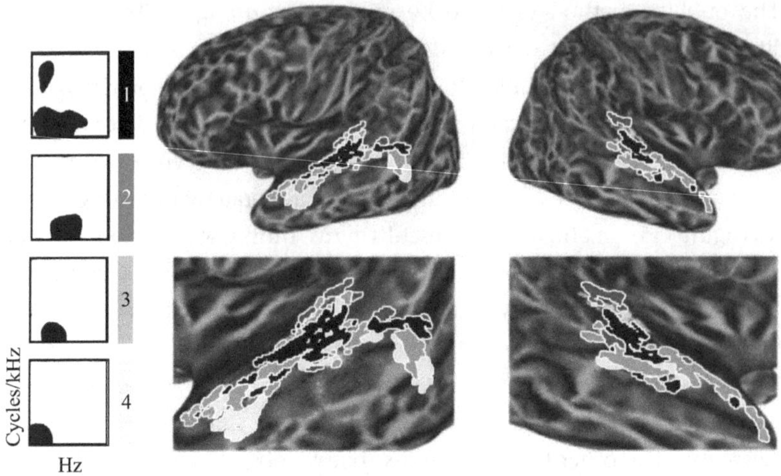

Figure 6.5
Speech-driven spectrotemporal modulation tuning map derived from fMRI. Left panels are example spectrotemporal modulation tuning plots, and the grayscale bars labeled 1–4 correspond to the grayscale shading on the brain maps showing the location of each spectrotemporal modulation tuning type. Note that tuning involves a broader range of spectral and temporal frequencies (black-1) in and around core auditory areas and a more posterior zone, while more lateral and ventral areas are tuned more focally to low spectrotemporal modulations (light gray-3 and white-4) relevant for speech. The fastest temporal modulation tuning (dark gray-2) is mostly confined to the planum temporale region. Modified with permission from Venezia et al. (2019).

centimeter square, each tuning preference may represent a conglomerate of finer-grained tuning properties. Finally, just because we can reliably measure an STRF in a given area does not necessarily mean that we are measuring a response to acoustic signals. For example, an area that responds to morphosyntactic features may be associated with a low-spectral/low-temporal modulation tuning function because that part of spectrotemporal space is needed for intelligibility, which enables morphosyntactic processing. The pattern 4 areas in the superior temporal sulcus (STS) and even the middle temporal gyrus may represent this kind of higher-level processing aliasing in our measurements (see part III).

Other approaches to measuring the tuning properties of auditory signals reveal some further details about how speech signals are coded in the STG. Using direct recordings from the human cortex via electrode grids, Nema Mesgarani, Edward Chang, and colleagues found that individual electrodes placed on the middle and posterior STG exhibited selective responses to classes of phonetic features. These included plosives, fricatives, place of articulation, front vowels, back vowels, and more. In fact, they were able to characterize "the STG representation of the entire American English phonetic inventory" (Mesgarani et al., 2014, p. 1009). A subsequent study from this team further identified a distinction between the middle and posterior STG: "The posterior STG is tuned for temporally fast varying speech sounds that have relatively constant energy across the frequency axis (low spectral modulation) while the anterior STG is tuned for temporally slow varying speech sounds that have a high degree of spectral variation across the frequency axis (high spectral modulation)" (Hullett et al., 2016, p. 2014). Hints of this pattern can also be seen in the bubbles-based spectrotemporal tuning map, but it also suggests that the functional organization of the superior temporal region is rather more nuanced.

I hope that the foregoing discussion helped to clarify my earlier statement that we have a basic understanding of the early stages of acoustic analysis up to the level of primary auditory cortex, but beyond that, we have many clues and few details. What have we learned? Most auditory neuroscientists would agree, I think, that primary auditory cortex functions as a kind of spectrotemporal modulation filter bank. Most would also agree that various regions of the STG code spectrotemporal features more specifically tuned to bits of speech, phonemic-level speech codes are distributed in this region, there is an interactive hierarchy of speech-related coding (Gwilliams et al., 2024), and there are differences between more anterior versus posterior regions.

Many questions remain, however. A major one in my mind concerns the anterior-posterior differences. Some view it as a fundamental distinction between ventral-comprehension (anterior) and dorsal-production (posterior) streams (Rauschecker & Scott, 2009). Others argue for finer-grained spectral (anterior) versus temporal (posterior) coding, as we just saw. A more recent idea is anchored by the observation of onset-only responses, which have been reported in subregions of the posterior STG (Forseth et al., 2020; Hamilton, Edwards, & Chang, 2018; Hamilton et al., 2021). These are areas that respond at the onset of speech, and with a very short latency, and then turn off for the rest of the stimulus duration, as recorded intracranially. Liberty Hamilton and colleagues (2021) have suggested that this is evidence for a parallel posterior superior temporal processing stream for speech and other signals that does not flow through the auditory core.

I don't think that any of these proposals has it quite right by itself, although each may have captured part of the story. Regarding the ventral/dorsal hypothesis, we'll see in chapter 8 that anterior temporal areas comprise one part of a distributed semantic network, which is obviously critical for understanding speech. I suspect that anterior temporal intelligibility effects are somehow related to their connection with meaning systems (possibly combinatorial semantics or integration of emotional prosody with meaning) rather than to speech analysis or word recognition per se. Further, we'll see in chapter 10 that posterior temporal areas play a critical role in speech production, leading some to align the region with the dorsal stream, but this does not preclude involvement in receptive speech (a ventral stream function) as well. As for the spectrotemporal modulation hypothesis, I think there's something to it. For example, we have already learned that there is a region lateral to Heschl's gyrus that responds relatively selectively to hearing vocal sounds (Belin, Zatorre, & Ahad, 2002; Belin et al., 2000), which involve finer spectral than temporal modulation cues, and the bubbles fMRI study found pitch-related coding in a similar area (Venezia et al., 2019). But I don't think it's that simple because pitch-related responses have also been observed in more posterior areas (figure 6.5, black, pattern 1; see Hamilton et al., 2021). Perhaps these are related to higher-level prosodic cues that are relevant to lexical or morphosyntactic processes, which also implicate more posterior temporal regions (see chapters 7 and 9). As for the parallel pathways hypothesis, I think it is reasonable, but I also think that the existence of rapid onset response properties in the posterior superior temporal lobe is better explained in terms of auditory stream segregation, which could be a parallel processing stream in its own right and which has previously been shown by our lab and others' to implicate auditory regions posterior to Heschl's gyrus (Smith et al., 2010; Zatorre et al., 2002). It makes sense that such a system turns on rapidly when a sound is presented in order to sort out the source of the signal and then turns off once its job is done.

Another open question concerns what information the various hierarchical levels code. The direct cortical recording study that I just discussed points to codes for classes of phonetic features, but not specific phonemes. As I pointed out earlier, there are debates regarding whether phonemes are coded as units of analysis in speech perception as is typically assumed. I agree with those who say that the phoneme is not the basic unit of analysis in receptive speech, but debate remains (Hickok, 2014a, 2014c; Kazanina, Bowers, &

Idsardi, 2018). And there are important acoustic features of speech that have been relatively neglected to date, such as prosody, both linguistic and emotional, and tone in tone languages.

Rhythm is another feature that I have already highlighted as a feature that is signaled acoustically in the form of the amplitude envelope of speech. This is a feature that has attracted a ton of attention in speech research, although mostly from a dynamic rather than an architectural perspective. Given the prominence of this line or work in the neuroscience of speech, it is worth a few pages of discussion here.

The Rhythm of Perception?

The excitement over speech rhythms is difficult to overstate. I might have done so, however, in a 2020 Twitter (now X) post when I wrote, "Ok. I'm gonna come out and say it. I think neural oscillations are the 2010s equivalent of mirror neurons. Real, but not doing all the work people think they are. I'm willing to be convinced otherwise but the more I look the more confident I get that there's not much there."

I'll let you be the judge of whether I crossed the line, but here are some of the exuberant scientific paper titles that ultimately led to my post:

- *Neural Entrainment Determines the Words We Hear* (Kosem et al., 2018)
- *Speech Intelligibility Predicted from Neural Entrainment of the Speech Envelope* (Vanthornhout et al., 2018)
- *Out-of-Synchrony Speech Entrainment in Developmental Dyslexia* (Molinaro et al., 2016)
- *Frequency Modulation Entrains Slow Neural Oscillations and Optimizes Human Listening Behavior* (Henry & Obleser, 2012)
- *Neural Oscillations Carry Speech Rhythm through to Comprehension* (Peelle & Davis, 2012)

There are some rather lofty claims here. Before evaluating them—in case you're not yet intimate with the field of neural oscillations and entrainment—I'll provide a few paragraphs of background.

In 1890, American psychologist William James (1842–1910) famously likened our conscious experience to the flow of a stream: "Consciousness, then, does not appear to itself chopped up in bits. . . . It is nothing jointed; it flows. A 'river' or a 'stream' are the metaphors by which it is most naturally described. *In talking of it hereafter, let's call it the stream of thought, consciousness, or subjective life*" (James, 1890, p. 239, emphasis in the original). While there is no disputing James's metaphor for our subjective experience, recent research has suggested that it might be an illusion.

The first hint of a rhythm rather than a stream of perception came more than a decade before James penned his famous metaphor. In 1875, Richard Caton (1842–1926) reported the remarkable discovery that the brain's cortex emitted electrical current, even at rest, and these currents were related to function (Brazier, 1963). It was already well established that electrical currents flowed through peripheral nerves—Luigi Galvani's (1737–1798) work on the frog in the 1780s pretty much confirmed the theory of "animal electricity" (Piccolino, 1998, 2006)—and it was known that electrical stimulation of the brain could produce movement (Fritsch & Hitzig, 2009/1870). But Caton, working mostly with the rabbit, was the first to document intrinsic electrical signals emanating from the brain's gray matter. Importantly, for our story, Caton noted an interesting property of these currents—that they are "usually in constant fluctuation; the oscillation of the index being generally small" (Caton, 1877, reprinted in 1971, p. 23). In the 1920s, Hans Berger (1873–1941) discovered electroencephalography (EEG) in human participants, confirmed the oscillatory nature of "brain waves," and observed that at least two distinct rhythms could be discerned: slower alpha (about 10 Hz) and faster beta (about 25 Hz) waves (Millett, 2001). Subsequent research catalogued a whole spectrum of rhythms, including the very slow delta (.5–2 Hz) and theta (4–8 Hz) waves and the very fast gamma waves (30–100+ Hz), although it's debatable whether gamma activity is oscillatory.

Despite clear evidence of neural rhythms, this research did not shift scientific thinking about the streamlike nature of perception or even the relevance of the rhythmic pattern of neural processing. Most EEG research in the twentieth century was content to document correlations between different brain rhythms with mental states (alert, drowsy, sleeping, dreaming, comatose, seizure) or to document the ability of sensory stimulation or motor action to generate *evoked response potentials (ERPs)*, also called *event-related potentials*, embedded in the electrical rhythms (Pagni, Naddeo, & Mascari, 1988). But in the last decades of the twentieth century, some clues started accumulating, which hinted that the rhythms themselves mattered to cognition.

The slower theta rhythms (about 4–8 Hz) mentioned here were discovered in the human EEG signal not long after Berger's initial measurements (Mundy-Castle, 1951) and studies in animals (O'Keefe & Recce, 1993; Winson,

1978) and later humans (Kahana et al., 1999) had found a prominent theta rhythm in the hippocampus that correlated with spatial navigation behavior. Research on higher-frequency gamma energy (Hughes, 1964; Jia & Kohn, 2011) led to new hypotheses regarding the role of synchrony within and between brain networks in binding information together, enabling top-down processing and prediction, and even consciousness (Crick & Koch, 1990; Engel, Fries, & Singer, 2001; Singer, 1993).

Recent research has yielded impressive demonstrations of the perceptual reflection of the brain's rhythmic nature. You may be familiar with the *wagon wheel illusion,* in which the spokes on a wheel are sometimes perceived to reverse their rotation direction. This illusion is readily induced with a strobe light (or movie) if the rotation of the wheel (or frame rate) is such that each flash (or frame) captures the spoke location slightly behind the location captured in the previous exposure, which leads to the perception of reverse motion. The effect is a consequence of sampling the scene in discrete frames. What's telling is that the illusion can occur during live observation of a rotating wheel in full daylight. I recall noticing this effect in automobile wheels while staring out of the back seat of a car window as a child. Vision scientists have since quantified the effect in the lab (Purves, Paydarfar, & Andrews, 1996).

Even the ability to detect an event, like the presence of a flash or a tone, can oscillate over time, cycling several times a second. For example, Haleh Farahbod, Kourosh Saberi, and I presented listeners with white noise that was amplitude modulated (AM) at 3 Hz and then transitioned into unmodulated noise after the 3-second AM entrainment period. We then measured the participants' ability to detect a faint tone in the unmodulated noise at various time points after the entrainment phase ended. We found that tone detection accuracy oscillated at the entrainment frequency by up to 25 percent (Farahbod, Saberi, & Hickok, 2020; Hickok, Farahbod, & Saberi, 2015; Saberi & Hickok, 2023). Other research linked such effects directly to brain rhythms by showing that perceptual oscillations are correlated with the phase of neural oscillations (Landau et al., 2015; Ten Oever et al., 2017; van Dijk et al., 2008). All of this suggests that the brain doesn't sense the world in a continuous stream but instead samples it in more or less discrete time chunks (Schroeder & Lakatos, 2009).

Importantly, the sampling frequency and phase of neural and perceptual oscillations are not fixed like a metronome clicking off beats. They can be synced to, or indeed driven by, what's going on in the perceptual environment. We saw this in variants of our perceptual rhythmic entrainment experiments: by changing the AM frequency during the entrainment phase, we could change the frequency and phase of the perceptual oscillation, at least between 2–5 Hz, which is relevant for speech. Entrainment of neural oscillations to rhythmic signals is also well established in humans and nonhuman animals and has been demonstrated using a variety of measurement tools (Doelling et al., 2014; Henry & Obleser, 2012; Keitel, Quigley, & Ruhnau, 2014; Lakatos et al., 2013; Peelle & Davis, 2012), including in our own direct recordings in human auditory cortex using the same AM noise paradigm that I described previously (Forseth et al., 2020).

What do all these oscillation and entrainment facts mean? Let's review and assemble the pieces. First, neural networks oscillate, at least in some measurements. Second, perception also can oscillate and does so in sync with neural oscillations. Third, neural oscillations can be entrained, within limits, by the rhythmic patterns in the sensory environment. So it appears that the brain is tuning into the frequency and phase of rhythmic information in the environment and therefore can optimize human listening (Giraud & Poeppel, 2012; Lakatos et al., 2008; Lakatos et al., 2005; Ten Oever et al., 2014). At least that's the hypothesis. Speech, it turns out, has been a productive testing ground for this idea.

The Nature of Speech Rhythms

Earlier in this chapter, I pointed out that the acoustic speech signal contains cues to the rhythm of speech. Here, I'll motivate its perceptual prominence, which in one sense is unquestionable. Poets have long recognized the rhythm of language and honed their craft by enhancing and manipulating it. Edgar Allan Poe defined "the poetry of words as the rhythmical creation of beauty" (quoted from Poe, 2000) and Virginia Woolf, commenting on writing style, said simply, "It is all rhythm. Once you get that, you can't use the wrong words" (Woolf, 1984, p. 107). The very existence of poetry as a popular art form, and its further elaboration in music, is evidence that humans adeptly perceive and indeed enjoy the rhythms of language.

In a decidedly less aesthetic analysis of natural speech, scientists have quantified linguistic rhythms by computing the dominant envelope modulation frequency. That frequency is a remarkably consistent 2–8 Hz across the world's languages (Poeppel & Assaneo, 2020), which lines up with the syllable rate and is in a frequency range that is well represented in human auditory cortex, as we learned earlier. This is not a coincidental collection of facts. This rate appears to stem from the natural frequency of the open-close cycle of the jaw, which falls in this range not only during speech, but also in prelingual infant babbling (de Boysson-Bardies et al., 1981) and even orofacial gestures of nonhuman primates (Ghazanfar et al., 2012). Syllables are yoked to this open-close cycle, with the vowel nucleus corresponding roughly to the most open vocal tract position bookended by the most closed positions at syllable onsets and codas. The syllable "from," for example, starts with a closed position to form the *f* sound, opens a bit for the *r*, then opens more for the vowel *o* before closing again to make the *m*. Looking back at the waveform figure 6.1, you can see how this close-open-close cycle is reflected in the acoustic envelope, forming an increasing and then decreasing lobe of energy. In this way, the open-close cycle of the jaw is the biomechanical frame for syllables (MacNeilage, 1998), which generates the 2–8-Hz rhythm of amplitude modulations and which the brain appears to be well tuned to detect.

While the speech rhythm has an *average* sweet spot, the actual rhythm of any given utterance is more variable. Consider the sentence *Juan wrote a draft and then he edited it.* Clap along with the syllables as you read it and note how much variation there is in the rhythm. Can neural oscillations entrain to such a variable rhythm? (Recall that we are exploring the hypothesis that neural oscillations entrain to sensory rhythms to maximize neural sensitivity; if the oscillations can't align to the actual, variable rhythm, this would undermine the claim.) Answer: yes. Recordings made from outside the head with magnetoencephalography (MEG) have shown in several studies that neural oscillations track the actual amplitude envelope of speech (Ahissar et al., 2001; Luo & Poeppel, 2007; Peelle & Davis, 2012).

Recordings made directly in the human auditory cortex confirm the envelope-tracking ability of neural signals. In the same report that described neural entrainment to a regular AM signal, Kiefer Forseth, Patrick Rollo, Nitin Tandon, and I reported neural oscillation entrainment to the amplitude envelope of spoken sentences (Forseth et al., 2020). Two further observations from this study suggest that neural entrainment can be a fairly low-level, stimulus-driven effect, although other research suggests that it can be influenced by top-down information too (Broderick, Anderson, & Lalor, 2019). First, as extracranial recordings had shown previously (Howard & Poeppel,

2010), entrainment in our study was observed not only during listening to normal speech, but also time-reversed speech in which the linguistic content is unintelligible. And second, entrainment effects were found in early auditory areas in and around Heschl's gryus, but neither in higher-order regions on the lateral surface of the superior temporal lobe nor in the more posterior zone in the planum temporale, which exhibited very different kinds of responses.

OK, so entrainment to speech is a real thing. It happens early and often. But does it actually *do* anything, as many have claimed? Let's consider the first study in the list of titles that I bulleted earlier, "Neural Entrainment Determines the Words We Hear" (Kosem et al., 2018), which makes a bold claim. The study itself was quite clever. Researchers presented listeners with sentences at two rates, fast and slow, thus changing the tempo of the speech rhythm. At the end of each sentence was a word containing a long versus short vowel ambiguity, similar to a sound halfway between "peel" and "pill." (The study was conducted in the Netherlands with Dutch speakers; an actual example is the syllable *t[vowel]k,* where the vowel was ambiguous between *tak*, which means "branch," and *taak*, which means "task.") These ambiguous words were identical in both the fast and slow sentence conditions. The idea of the experimental design was that a sentence context with a faster tempo would entrain neural oscillations to that rate, and then bias perception of the ambiguous vowel for the long tempo and vice versa for the slower tempo. Behaviorally, this is exactly what happened. Listeners reported hearing *tak* more frequently after the slow sentence condition and *taak* more often after the fast sentence condition. Moreover, the behavioral bias was correlated with the neural oscillation entrainment strength, as measured using MEG. This seems like pretty solid evidence that neural oscillation entrainment has a direct effect on subsequently presented speech.

However, a closer look reveals that the tempo manipulation affected the perception of *tak* or *taak* by only 5 percent on average. In the slow context, *taak* was reported about 46 percent of the time on average, which nudged up to 51 percent in the fast context. This might be an impressive victory margin in a political election, but it falls far short of determining "the words we hear." The details of the correlation with entrainment strength only adds further squints of concern. While the effect is statistically significant, it is underwhelming: variation in entrainment strength accounted for only 29 percent of the variation in behavior. For me, this lets the air out of the lofty claim promised in the title. Bringing it back down to Earth, what we can conclude is that rather than determining the words we hear, neural entrainment accounts for 29 percent of the variance of a 5 percent change in what we *report* we hear when listening to ambiguous words.

Underwhelming effect sizes are not unique to this study. Broader surveys of the field indicate effect sizes no better than 10–20 percent (VanRullen, 2016). That's nothing to sneeze at, but keep in mind that one gets an effect at all only if the stimuli under investigation are ambiguous or barely detectible. This was true in my own study that I described earlier (Hickok et al., 2015). We discovered a roughly 20 percent perceptual oscillation in our tone in noise detection experiment, but only when the tone was weakly detectible in the first place. So, while I was quite excited about the "rhythm of perception" at first, now I tend to think of it as a real, but relatively minor, background fluctuation.

It's not just small effect sizes that lead me to this view. The fact that neural rhythms entrain to the actual envelope of the speech signal is consistent with an alternative hypothesis—namely, that the neural rhythm is nothing more than the evoked response to a rhythmic signal (Haegens & Zion Golumbic, 2018). Imagine a neuron that fires every time you flash a light. If you flash a light rhythmically, the neuron will track the rhythm, not because it cares about the rhythm per se but simply because it is responding to the stimulus whenever it occurs. Now imagine a large population of neurons doing the same thing, and voilà, you have what looks like neural oscillation entrainment to the rhythm. Translating this evoked response account back into speech, the reason why neural oscillations appear to track the envelope is not that they are entraining to a rhythm per se, but simply responding to the bursts of acoustic energy that happen to occur rhythmically.

The Dutch *tak-taak* study considered earlier was intended to speak precisely to this debate. If the neural oscillation persisted and had a perceptual effect *after* the speech rhythm ended, then this should indicate that it is not just an evoked response, which would predict that the preceding context should have no effect on the response to a subsequent word. This is solid logic, but there's a loophole opened by the well-established fact that attention can modulate neural oscillations (Haegens & Zion Golumbic, 2018; Lakatos et al., 2008). Returning to the light flash example, suppose that we flashed that light every second for 10 seconds and then stopped. Since we, as perceivers, can detect rhythms and generate predictions based on them, which may be accomplished by something quite different than neural oscillations (Breska & Deouell, 2017), we might very well expect an eleventh flash one second after the tenth, and this might drive activity from the top-down attentional signal in the very neurons that had responded to the bottom-up stimulus. This would make it look like the entrainment of the neural rhythm persisted after the stimulus ended, when in fact we are looking at two mechanisms for evoking a neural response, one bottom up and one top down.

There is evidence for this hypothesis. In a follow-up to our perceptual oscillation study—the one that presented amplitude-modulated noise as the "entrainment" stimulus—we assessed the role of attention and found it to be the major driver (Farahbod et al., 2020). Moreover, in the direct cortical recording study of the same stimuli, the "high gamma" signal—the response most closely tied to local neural activity—showed no poststimulus continuation or oscillation; we *did* see a poststimulus continuation in the low-frequency energy bands of neural recordings, but this is more likely associated with top-down attentional effects. So, unfortunately, poststimulus persistence of a neural rhythm is not solid evidence against the evoked response account.

Suffice it to say, this is an interesting and contentious topic with very little consensus on whether oscillations themselves play a critical role in speech processing or whether they are something closer to neural exhaust fumes (Obleser, Herrmann, & Henry, 2012). We could spend another chapter, or even a book-length amount of time, on the issue—indeed, some authors have (Buzsáki, 2006; Murphy, 2020)—but my reading of the literature is that we would have to conclude at the end that the jury is still out. So, rather than going further down the oscillation rabbit hole, I'm going to suggest that the interested reader look at these other sources. But before moving on, indulge me one final comment.

Even if oscillations turn out to add some computational heft to neural processing, oscillations alone are going to be only a fraction of the story. Consider the acknowledged fact that oscillations are ubiquitous in neural systems from single cells to large-scale networks and from memory systems in the hippocampus to language networks in the frontal, parietal, and temporal lobes. To say that oscillations are the basis of this or that and leave it there is not far from saying that neurons are the basis of this or that and leave it there. Well, yeah, neurons are the basis of this and that, but there's a whole lot more to the story. A core part of it is the arrangements, the *architectures* that neural populations live within, and their patterns of communication, the connections, within those architectures.

The same is true of oscillations. It's not that the whole brain oscillates at various frequencies, and that's why we remember or see or talk. These oscillations are taking place in the context of some neural architecture, which surely constrains how the network oscillates (Buzsáki, 2006). For example, one study reported that the scale of the architecture was related to the frequency of the oscillation: small, local networks oscillated rapidly, long-range networks oscillated more slowly, and medium-scale networks oscillated at an intermediate rate (von Stein & Sarnthein, 2000). In terms of language systems, then, we are going to need to talk about oscillations at various network scales, not just how some summed EEG oscillation entrains to the overall speech envelope. Already, we know that when presented with a rhythmic acoustic stimulus, some parts of the auditory/language system in the temporal lobe entrains and other parts simply show a response at the onset of the stimulus (Forseth et al., 2020; Hamilton et al., 2018).

A final note on speech rhythms: all this debate over the status and usefulness of neural oscillations does not detract from the importance of the beat of language, which provides important cues to syllable and phrase boundaries, topic focus, and other linguistic features, as we reviewed previously. The brain certainly uses these leaky funnel cues. The question is how. Neural oscillations are a candidate mechanism for utilizing some of these cues, but many questions and concerns remain.

Left Brain, Right Brain. Again.

I would be remiss if I didn't revisit the topic of hemispheric asymmetries for speech perception in the context of this chapter's finer-grained discussion of how the brain analyzes speech signals. Do any of these finer-grained analysis tools, such as spectrotemporal modulation tuning and neural entrainment patterns, reveal asymmetries?

Yes, they do. Using fMRI, researchers have manipulated the spectral versus temporal complexity of sounds and found a leftward bias for temporal complexity and a rightward bias for spectral complexity (Albouy et al., 2020; Schonwiesner, Rubsamen, & von Cramon, 2005). This is consistent with the claim encountered earlier that the left auditory system is more specialized for temporal processing, while the right excels at spectral processing (Zatorre et al., 2002). Other researchers have used MEG to measure neural entrainment and report, paradoxically for some theories, that the *right* hemisphere shows stronger entrainment to speech (Assaneo et al., 2019; Flinker et al., 2019; Lam et al., 2018).

However, there are notable caveats for lateralization models and the data behind them. Let's consider the spectral versus temporal perspective first. I find the published data fairly compelling in demonstrating the existence of some kind of bias under some stimulus and task conditions. Further, I find the idea of differing spectral and temporal biases attractive because, as Robert Zatorre and colleagues have pointed out (Zatorre et al., 2002), there is a fundamental trade-off between spectral and temporal resolution: high-spectral resolution requires sampling a longer (i.e., lower-resolution) temporal window and vice versa, an acoustic version of Werner Heisenberg's *uncertainty principle* from quantum physics. David Poeppel's asymmetric sampling in time hypothesis is also consistent with this, in that shorter temporal windows would correspond to a high-temporal/low-spatial resolution analysis and longer temporal windows would correspond to low-temporal/high-spatial resolution. A nice convergence of models. But while divvying up these computational leanings by hemisphere scratches our chronic itch for left-right asymmetry, it doesn't *have* to work that way. In fact, as we learned earlier, different levels of the auditory system and regions of auditory cortex vary quite a lot in the spectral and temporal modulation preferences, such that even within a hemisphere, they surely cover the range of spectrotemporal trade-off settings needed to analyze just about any kind of signal.

Perhaps not surprisingly, then, between-hemisphere spectral versus temporal biases tend to be very subtle indeed. A recent report published by Zatorre's group in the journal *Science* compared the brain's response to sung speech that was degraded either in terms of spectral or temporal modulations (Albouy et al., 2020). Interestingly, when study participants listened to these stimuli and had to extract speech versus melodic information from them, spectral degradation led to more difficulty with melody recognition and temporal degradation led to more difficulty with speech content recognition. This is a very clear behavioral distinction. Then the team presented these same stimuli to listeners who were undergoing fMRI scanning. Do the left and right hemispheres respond differently? Not in a standard analysis. The authors report "no significant difference in the hemispheric response to either dimension." (Albouy et al., 2020, p. 1044). However, a more nuanced analysis asked whether there were any regions that could decode the speech versus the melody parts of the signal. Two regions were found—one in the left hemisphere, one in the right, both just anterior to the primary auditory cortex. The left hemisphere was reported to decode speech better than the right, and the right hemisphere was reported to decode melodies better than the left. How much better? About 5 percent on average, a modest effect.

Jonathan Venezia's bubbles study, which more directly measured the left-right distribution of spectrotemporal modulation tunings, came to a different conclusion, however (Venezia et al., 2019). In that study, we mapped tunings in both the left and right hemispheres and looked both for the existence of a spectral-temporal trade-off, as argued by Zatorre, and for left-right asymmetries in zones that leaned toward higher spectral versus temporal resolution. Consistent with Zatorre's prediction, we found that a spectral-temporal trade-off does exist. Auditory cortical zones that were tuned to faster temporal modulations tended to have coarser spectral tunings and vice versa. This pattern was strongest in and around primary areas. However, we did not find the expected left-right asymmetry. In fact, if anything, the right hemisphere contained a preponderance of fast temporal modulation–tuned networks.

Turning to the envelope entrainment asymmetries, a recent reevaluation of these effects by Poeppel's group clarifies their sources, but at the same time, it muddies the waters regarding the significance of proposed inherent spectrotemporal asymmetries in auditory cortex (Assaneo et al., 2019). These researchers showed that the right-hemisphere entrainment dominance disappears and can even be reversed by top-down influences such as the meaning contained in the acoustic signal (see also Peelle, Gross, & Davis, 2013). They propose that entrainment effects are driven by multiple sources, including the intrinsic signal processing biases in the auditory system as originally proposed, as well as any number of top-down, extra-auditory sources (Assaneo et al., 2019; Rimmele et al., 2018). When impoverished acoustic stimuli are presented, like modulated white noise or other meaningless sounds, entrainment reflects only the intrinsic biases and reveals right-hemisphere dominance. However, when top-down information is present, it can drive entrainment and override or swamp the intrinsic biases. This seems reasonable. However, one now wonders how much work the intrinsic asymmetries are doing. If just adding some meaning flops the dominance pattern by presumably engaging the left-hemisphere system more strongly, what good are those supposedly right-dominant preferences in the first place? Maybe they don't exist at all. Maybe what looks like an intrinsic right dominance is just some top-down effect involved in processing meaningless noises, like the mental effort required to quell our annoyance with listening to meaningless sounds for long periods of time.

I'd like to push Poeppel and colleagues' idea one step further with the following two-part proposal.

Part 1: The spectrotemporal analysis capabilities of the auditory cortex are the same in the left and right hemispheres. Put differently, in contrast to Zatorre and Poeppel, I do not think that there are spectral versus temporal asymmetries or short versus long temporal integration window asymmetries in auditory cortex in the two hemispheres.

Part 2: Whatever asymmetries that fMRI or MEG studies document are *exclusively* the result of top-down selection of subnetworks of the overall system. In this view, asymmetry is not a property of auditory networks, but the higher-level networks that are ultimately engaged by the stimuli or the task. If these higher-level networks are asymmetric to some degree, then top-down feedback to the auditory system will be asymmetric, which will show up in our measurements of the lower-level auditory system.

Part II Summary: The Analysis of Speech

We've covered a lot of ground in the last several chapters, from the remarkable discoveries of early neurologists to functional imaging–generated brain maps of speech perception to spectrotemporal filter banks and oscillation entrainment. Let's take a look at the big picture.

Our overall aim in part II of this book has been to understand how the brain transforms acoustic wave patterns into phonological-level information that can allow access to higher levels of linguistic information. We have a moderately clear, if low-resolution, picture of where this is happening in the brain. The relevant zone is the superior temporal gyrus (STG), both left and right. This region of the brain is organized hierarchically, but probably with distinguishable parallel streams embedded within the hierarchy. Primary auditory areas on the dorsal plane of the temporal lobes are the lowest level of the hierarchy and function as a spectrotemporal filter bank. These filters cover a wide range of the acoustic spectrotemporal space that is relevant for humans. The later stages of the hierarchy are found closer to the crown of the gyrus, the part that is visible from the lateral view of the brain and extends into underside of the gyrus in the superior temporal sulcus (STS). Here, we find a more robust tuning to lower spectral and temporal modulation frequencies and codes for various phonetic feature classes.

The middle-to-posterior portion of the lateral STG extending into the STS bilaterally shows consistent responses to speech in functional activation studies and correlates with perceptual judgments of speech intelligibility. Activation in this region can be decoded reliably to recover the acoustic speech signal to a reasonable approximation. Damage to the superior temporal lobe bilaterally causes severe speech perception deficits (word deafness). Unilateral damage to the superior temporal lobe does not disrupt receptive speech at the phonological level in most people, but it is associated with decreased performance when degraded speech is presented. For these reasons, we termed this region the *auditory phonological area (APA)*. Damage outside the superior temporal lobe does not disrupt phonological-level speech perception when measured using ecologically valid tasks, but it can disrupt speech perception on tasks that rely on phonological working memory or conscious attention to segmental information.

This STG/STS system interfaces with two broad networks according to the dual stream model that David Poeppel and I developed, building on the work of Carl Wernicke and others. One stream continues ventrally in the temporal lobe and connects phonological representations with lexical, syntactic, and semantic networks. This is the primary network for language comprehension. The other courses into more dorsal networks, linking phonological representations in the temporal lobe with parieto-frontal, motor-related networks. We will tackle the details of the production network in part IV of this book. For the next journey in part III, though, let's continue down the increasingly murky ventral stream and see how much of the sound-to-meaning network we can resolve.

III From Words to Sentences

7 Word Search

In the beginning was the Word.
—John 1:1, *King James Bible*

I know nothing in the world that has as much power as a word.
—Emily Dickinson, *Letters*

Words are, in my not-so-humble opinion, our most inexhaustible source of magic.
—Dumbledore, in J. K. Rowling's *Harry Potter and the Deathly Hallows*

It's not hard to find sage quotes about words. It's much harder to define them. The fifth edition of the *American Heritage Dictionary* tried: "A sound or a combination of sounds, or its representation in writing or printing, that symbolizes and communicates a meaning and may consist of a single morpheme or of a combination of morphemes." Yet we frequently utter communicative sounds like the indignant "psh," the contemplative "hmm," the pejorative "shh," or the befuddled "huh?" Are those words? Most would say no, but they seem to satisfy that dictionary definition. Or consider "of" as in *I've been writing a lot of words lately*? Clearly, it's a word, but what meaning does it communicate, especially given that we can do without it in *I've been writing words a lot lately*? How about the "-ed" in "walk*ed*"? It's not a word on its own, but it communicates a clear meaning: something that happened in the past. Are "dunno" and "I don't know" the same word? If yes, what about when we just hum the clearly interpretable melody of "I don't know" in response to a direct question? Is that a word?

The *Stanford Encyclopedia of Philosophy* (https://plato.stanford.edu) tried harder. It devoted over 17,000 words—whatever they are—to the topic of words and word meaning. Foreshadowing the complexity of the matter, the entry starts with an understated heads-up to the reader: "The notions of *word* and *word meaning* are hard to pin down," and then proceeds to review dozens of ideas, problems, and controversies, citing over 200 references.

The Oxford Handbook of the Word (Taylor, 2015) also gave it a go, dedicating forty-two chapters and over 800 pages to the understanding of "the word." The volume starts with an introductory sentence by editor John Taylor, "Words are the most basic of linguistic units," and ends with a chapter by Alison Wray titled, "Why Are We So Sure We Know What a Word Is?" (Wray, 2015). In her chapter, she starts with the problem that "defining the word is far from easy" and argues "that the concept of the word looks vague because it is inherently so" (p. 725). As proof of Wray's assessment, consider Martin Haspelmath's recent attempt at "defining the word" (the title of his paper). After noting that the literature contains "no clear definition of *word,*" he proposes one: "A word is (i) a free morph, or (ii) a clitic, or (iii) a root or a compound possibly augmented by nonrequired affixes and augmented by required affixes if there are any" (Haspelmath, 2023, p. 285). This level of complexity is more than enough to ruin a good quote: "I know nothing in the world that has as much power as a [free morph, or a clitic, or a root or a compound possibly augmented by nonrequired affixes and augmented by required affixes if there are any]."

The reason why words are cumbersome to define scientifically is that they are not the poetic monoliths of our imagination, but rather assemblies of smaller bits, *morphemes*. Morphemes can exist in bound forms that cannot appear on their own (like *-ed*, *-s*, or *un-*), unbound forms, which can stand by themselves (like "walk" and "talk"), and can be combined to make complex forms (like "walked" or "antiauthoritarianism"). It used to be thought that the rules or processes for combining morphemes to make words were separate from the syntactic rules or processes for building phrases. But an alternative view is gaining steam—namely, that morphological operations are nothing more than syntactic operations applied at the level of words (Krauska & Lau, 2023; Marantz, 2013, 2016). On this view, the language system is composed of three main parts: a phonological system, a semantic system, and an in-between system that computes the mapping from sound to meaning or meaning to sound. The in-between system is sometimes termed *morphosyntax*, which is composed of morphemes and a hierarchical syntactic system that link the levels.

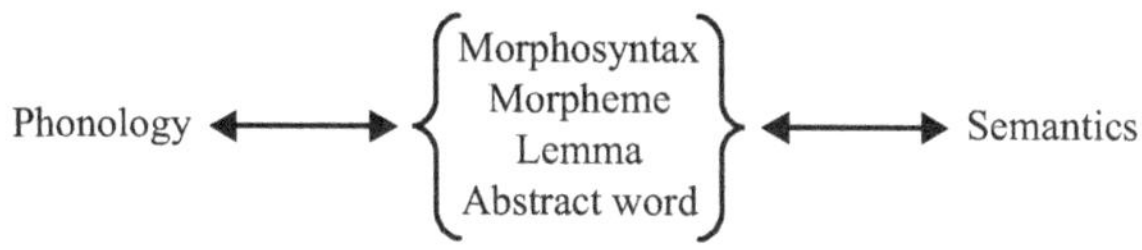

Figure 7.1
Three-level architecture of the language system. The term used for the middle level varies, as do the details of the processes and representations associated with it.

A similar three-level picture has emerged from the psycholinguistic tradition, with the phonological and semantic systems mediated by an in-between realm called the *lemma* level (Levelt, Roelofs, & Meyer, 1999).[1] Like the morphosyntactic level in linguistics, lemmas link phonological and conceptual representations and are assumed to contain packets of syntactic information, including grammatical category (noun, verb, etc.) and syntactic frames (subject, object, etc.). Some psycholinguistic theorists, particularly those studying picture-naming ability, simply call this in-between layer the *word* level (Foygel & Dell, 2000; e.g., see their Figure 1), where the units correspond to abstract, wordlike representations, meaning that they don't themselves contain meaning or phonological form but rather link to both.

While the differences in terminology can be confusing, there is general agreement on a three-part architecture, which is the important point for us at this stage (figure 7.1). There are important differences in terms of how researchers conceptualize *what* is represented in the middle level and *how* they link to other levels (Jackendoff & Audring, 2016; Krauska & Lau, 2023). My own view currently leans toward the morphosyntactic perspective, where the middle level aligns with the *process* of computing a hierarchical structure for words and sentences, which is required for the mapping between sound and meaning. But we don't need to concern ourselves with these details here. Our focus in this chapter is to determine whether the middle level of the functional architecture of language—whatever it is, precisely—is resolvable in terms of neural architecture. And as the chapter title implies, we are going to focus on word-level processing. We will scale up to sentences in chapter 9, where we will find some overlap in the neural architecture of word and sentence processing, as the morphosyntactic perspective predicts.

Before moving on, some terminological housekeeping is in order. Let's start by considering where our common notion of the *word* comes from. I think Ray Jackendoff and Jenny Audring's definition captures the source, nicely: "A word such as *cat* consists of a piece of semantic structure (the meaning of the word), a piece of phonological structure (/kæt/), and the syntactic category Noun" (Jackendoff & Audring, 2020, p. 3). In other words, what we think of as a *word* in the colloquial sense is a linked triad of information, spanning its phonology, its "middle word" information, and its semantics. To illustrate, consider that phonological forms without an attached meaning in a particular language (*glander, frink,* or *kolexum*) are not considered words in that language; they are not in the dictionary and can't be used in Scrabble or Wordle. Likewise, a concept that cannot be named by a word in one language is not considered a word even if that same concept can be named by a word in another language. There is no word in English for the act of tenderly running your fingers through someone's hair, but there is such a word in Brazilian Portuguese: *cafuné*. It seems that to gain colloquial word status, the thing has to have both a meaning and a (nonphrasal level) phonological form. If language scientists tell us that there is something between phonological forms and meanings, then our common conception of the *word* must map onto the phonology, the semantics, and the system in between, whatever that may be.

I will continue to use the term *word* in the colloquial sense. so interpret it as you would in a nontechnical sense. When I refer to the in-between level, I'll use the term *lemma* or *abstract word* or my own neologism, *middle-word,* when the discussion is about word-level processing. If we are talking about phrasal- or sentence-level processes, I'll use *morphosyntax,* or just *syntax*. The term *lexical* is also sometimes used in the word-related literature. Although different authors may have different intended referents in mind, I will restrict my use of the term *lexical* as being basically synonymous with the colloquial interpretation of *word*—that is, referring to the phonological-lemma-semantic triad. This aligns with the common conception of a *lexicon* as a dictionarylike mental store of words, including their phonology, grammatical category, and meaning.

With the housekeeping done, we can now turn our attention to identifying neural systems involved in lemma-level processing. There is a large literature on lemmas that includes important work on their neural basis (Dell & O'Seaghdha, 1992; Indefrey, 2011; Indefrey & Levelt, 2004; Roelofs, Meyer, & Levelt, 1998; Vigliocco, Antonini, & Garrett, 1998). However, almost all this research concerns speech production, not comprehension, our primary focus here. Nonetheless, it would be surprising, computationally, if a lemma level were needed for mapping in one direction and not the other, so I'll draw on evidence from all sources, not just comprehension. Let's start the abstract word search with a classical aphasia syndrome.

Transcortical Sensory Aphasia as a Lemma Impairment?

On the night of May 16, 1884, Mr. Schwarz became restless, talked in his sleep, and wandered about. At 4 a.m., something was clearly wrong. Mr. Schwarz's speech was "mutilated," and he didn't, or couldn't, answer questions. He appeared to be trying hard to communicate, but he just kept repeating the same nonsense expressions. His wife eventually gathered that he was trying to express the fact that he was suffering severe pain in his forehead and neck. He deteriorated further and was taken to the hospital, where he was evaluated by Dr. Ludwig Lichtheim (1885, p. 454, emphasis added):

> Patient is a rather fat, pale, man. Features flabby; expression dull. On admission, he gave the impression of a subject with psychical deficiencies. He talked a good deal, but repeated the same phrases. He points to his head and says, "Oh how stupid I am, I cannot," &c. It is difficult to make anything out of him, for he understands little that is said to him; but *it is easy to ascertain that he is in possession of his intelligence*, and that the apparent psychical deficiency consists only in his difficulty in understanding others, and in making himself understood. His actions leave no doubt on this point.

Language evaluation revealed significantly impaired comprehension. "If one stands behind him and talks to him, he turns round and asks: 'Do you speak to me?' The simplest requests . . . are answered by: 'I don't know what one wants.'" Speech production was fluent but paraphasic: "He talks a good deal in a flowing manner: he is seldom short for a word; occasionally uses a wrong or a mutilated one."

Object naming was significantly impaired: "He is in great difficulty when he has to name objects shown to him; he finds the names with the greatest difficulty, and assists himself with descriptions. Instead of 'wine,' he says 'that is strong;' for 'water,' 'that is weak,' &c" (p. 454). This ability to describe attributes of the objects he couldn't name no doubt contributed to Lichtheim's conclusion that Schwartz was in possession of his intelligence.

So far, these symptoms are consistent with Wernicke's sensory aphasia: impaired comprehension with fluent but paraphasic speech production. Lichtheim tested another language ability, however, which indicated something different. Lichtheim reported that Schwarz could "repeat correctly all that was spoken before him; but he apparently did not understand what he did repeat, as for instance the words, 'My name is Peter Schwarz, and I am already 4 years old,' which did not draw any signs of denegation on his part" (pp. 454–456).

Schwarz was the first reported case of *transcortical sensory aphasia*. Lichtheim theorized that this syndrome resulted from an interruption of the white-matter pathways linking Wernicke's auditory word center with widely distributed conceptual representations (see figure 3.1). Lichtheim's logic for his diagnosis was flawless, assuming that his premises were correct: preserved intelligence indicates that Schwarz's conceptual network was intact and preserved repetition ability indicated that his auditory word center (the phonological network) was intact. If the connection between sounds and concepts were impaired, this would explain the poor comprehension because there is no way to properly link sounds with meaning, but repetition would be possible because auditory and motor word centers were intact, along with the connection between them. In our terms, Lichtheim seems to have discovered a pure case of lemma impairment, with damage to the middle-word system.

But alas, the interpretation of transcortical sensory aphasia is far murkier than the tidy picture presented by Lichtheim. The core of the problem for us in our effort to uncover the neural basis of lemma representations is that it is unclear whether more general semantic deficits are the source of the comprehension problem for people with transcortical sensory aphasia. It is quite common to find nonverbal semantic deficits alongside word comprehension deficits (Jefferies & Lambon Ralph, 2006), leading some theorists to conclude that transcortical sensory aphasia is a type of semantic rather than linguistic impairment (Reilly & Martin, 2018). The typical pattern of brain damage behind the syndrome is consistent with this conclusion because it tends to involve the temporal-parietal-occipital junction regions (Berthier, 1999), which have also been implicated in aspects of semantic processing (Binder & Desai, 2011), as we'll learn in chapter 8.

The murk only thickens as we dig deeper. Around the edges of the core syndrome—impaired comprehension, fluent but error-prone output, good repetition—we find variable performance on other types of tasks, like picture naming, reading, and writing, which further complicates our interpretation of what's going on (Berthier, 1999). We also find variable lesion locations. Some reports implicate the underbelly of the brain, the ventral occipital-temporal junction (Kertesz, Sheppar, & MacKenzie, 1982), while others point to a more lateral temporal-parietal junction zone involving the posterior middle temporal gyrus, angular gyrus, and supramarginal gyrus (Berthier, 1999; Damasio, 1991; Jefferies & Lambon Ralph, 2006; Kertesz et al., 1982; also see figure 5.1). Finally, transcortical sensory aphasia tends to be a transient syndrome, more often appearing during the acute phase of a stroke and resolving into milder deficits or sometimes none at all after recovery (Berthier, 1999). It is hard to draw any firm conclusions regarding lemma-level deficits from this soup.

A ray of clarity was provided by Dana Boatman and colleagues (2000), who were able to induce the core symptoms of transcortical sensory aphasia in six people undergoing evaluation for surgical treatment of epilepsy. All six had electrode grids implanted on the surface of their left mid-to-posterior temporal lobes. The researchers delivered electrical stimulation between adjacent pairs of electrodes spaced 1 cm apart, thus disrupting the function of the underlying tissue. Stimulation of a range of zones in the posterior lateral temporal lobe was found to produce the symptom pattern of transcortical sensory aphasia: poor comprehension (measured by the inability to follow verbal commands involving plastic tokens of different colors and shapes), spared word and nonword repetition (93 percent or better), and fluent but paraphasic speech output. As figure 7.2 shows, the locations that induced this symptom complex varied, but there is a substantial concentration around the posterior middle temporal gyrus, which is consistent with at least some of the stroke-induced cases (Damasio, 1991). Semantic abilities were not assessed, unfortunately, so we can't be sure whether the stimulation disrupted lemma, as opposed to even higher-level semantic processes, which some studies have also linked to the posterior middle temporal gyrus region (Binder & Desai, 2011; Hart & Gordon, 1990).

Indeed, distinguishing networks involved in lemma-level processes is inherently problematic, given that this level is hypothesized to be the intermediary between phonology and semantics. What that means is that lemma networks, when involved in a task, will almost always be coactive with phonological and semantic networks. (I certainly can't think of a task that would activate an abstract word with *no* phonological or semantic information coming along for the ride.) As an example of the problem, consider the bear-pear-moose-grapes word comprehension task discussed in chapters 3 and 5. We used this task to address the question of whether word comprehension problems in aphasia primarily have a phonological source as predicted by classical theories. We found that many patients make at least as many, if not more, semantic errors on these tasks, indicating that much of the deficit

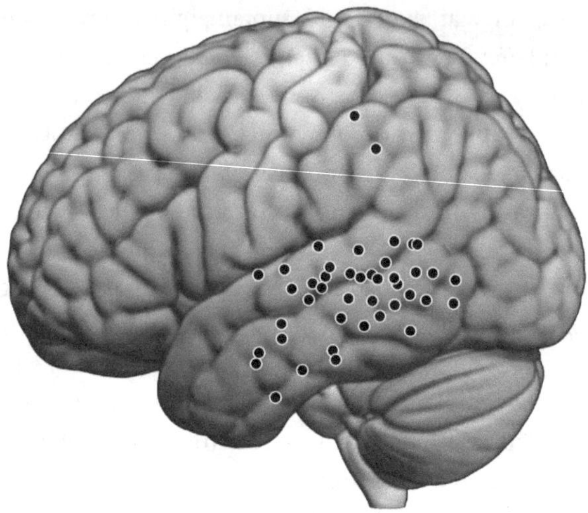

Figure 7.2
Distribution of electrode locations that, when stimulated, induced the symptoms of transcortical sensory aphasia. Redrawn from Boatman et al. (2000).

is postphonemic—but how postphonemic? Are these lemma problems or semantic problems? We can't tell because both lemma and semantic problems could lead to semantic errors.

In theory, we could parcel out lemma from semantic problems by separately assessing semantics using a nonverbal task, such as an all-pictorial task that involves arrays of pictures (say, a camel, an oak tree, and a cactus) and deciding which are semantically related. If we found word comprehension problems with semantic errors in the face of preserved nonverbal semantic abilities, this could point to a lemma deficit. The problem, however, is threefold. First, as we learned earlier, single-word comprehension deficits on the bear-pear-moose-grapes task are rare due to the more bilateral capacity for word comprehension. Second, when postphonemic word comprehension deficits do exist, nonverbal semantic deficits are often found as well (Hart & Gordon, 1990; Reilly & Martin, 2018).[2] Third, most studies simply do not assess for both a postphonemic source of word comprehension deficits *and* nonverbal semantic ability, which is what we need to pin down a lemma-level impairment. There have been some case reports (e.g., Warrington & Crutch, 2004), but these are rare enough to raise questions about the generalizability of the results.

Despite these difficulties, I think we can identify lemma-level processes for both comprehension and production. Let me build a case for their existence and localization in the posterior middle temporal gyrus, as suggested by the cortical stimulation mapping study of Boatman and colleagues (2000).

The Computational Argument

Getting from a phonological pattern to a semantic representation is a transformation, a translation between one feature space inhabited by abstracted bits of sound, such as */kæt/*, to another feature space filled with bits of meaning, such as animal, mammal, carnivore, purrs, furry, pet, and domesticated. Both spaces contain features that are shared, more or less, with other inhabitants of their space (*/kæp/*, */tæk/*, */sæt/*; animal, mammal, omnivore, barks, furry, pet, domesticated). This makes the translation between these spaces computationally complicated—nonlinear in mathematical terms. This is true even ignoring the additional complications added by morphosyntax. In neural network models, the nonlinearity problem is solved by adding an intermediate level, or even multiple intermediate levels in the case of deep neural networks. For example, the highly successful neural network model of word production developed by Gary Dell and colleagues simulates the translation from semantic to phonological features using a semantic layer, an intermediate layer that they call the word level, and a phonological layer (Dell et al., 1997; Foygel & Dell, 2000). Without the intermediate level, the transformation from meaning to sound would be computationally impossible (Dell et al., 1997). The reverse mapping from phonology to meaning in receptive speech is no less complicated, which means that some intermediate level must exist on computational grounds alone.

The Neurolinguistic Argument

Alex Marantz, Liina Pylkkänen, and colleagues have identified a particular component in the magnetoencephalography (MEG)–measured neural response to words, the M350, which shows sensitivity to morphological word features and can be dissociated from phonological form and semantics (Pylkkanen, Stringfellow, & Marantz, 2002; Stockall & Marantz, 2006). For example, the M350 is sensitive to morphological relatedness of prime-target pairs ("teach"-"taught") but not phonological and semantic relatedness ("boil"-"broil"). Interestingly, the neural source of the M350 appears to involve the posterior middle temporal gyrus.

The Phonological-Semantic Boundary Argument

We saw how the in-between space that lemmas inhabit makes it difficult to parcel them out from phonological and semantic networks. In functional imaging studies, this means that there will be substantial coactivation between the lemma level and the phonological and semantic levels that it translates between. But suppose that we put that coactivation property to use for providing a clue to the location of the lemma network. Here's the logic. We know that phonological and semantic networks are separable. For example, people with word deafness can't perceive or repeat speech but seem to understand the world quite well through other sensory modalities. Conversely, people with transcortical sensory aphasia or the semantic variant of primary progressive aphasia (discussed in chapter 8) show various degrees of the reverse pattern: preserved phonology with impaired semantic processing. Since lemmas interact with both of these systems, the network that supports lemmas should exhibit functional coinvolvement with both.

By analogy, consider the following, admittedly strange, hypothetical scenario involving a secret summit between two world leaders, President Phon and Prime Minister Sem, who don't speak the same language. It is understood that the mysterious Dr. Lem is slated to translate, but his identity is not known to anyone outside the meeting. While these are private talks between the two leaders, during breaks in the discussion, both the president and the prime minister retreat to separate annex rooms to strategize with their entourages. It is not known where Dr. Lem goes during these breaks. A spy group wants to know more, so they covertly implant each person at the summit with a top-secret neural monitoring system. Dr. Lem is there too, as outsiders assume, and receives one of the implants, although his identity is unknown to the spies. The top-secret device signals each time a person is communicating linguistically, whether talking or listening. It provides no information, however, regarding where in the building the signal is coming from. Now, the spies know which signals are coming from the two world leaders, but they don't know anything about who is generating the other signals. We are brought in to help in the analysis. Our assignment: figure out which signal is coming from Dr. Lem.

We can do it easily by looking at coactivity patterns. When the president is communicating with her team, all their signals will be synchronized, turning on whenever someone is talking and off when there is a break in the conversation. Ditto for the prime minister's team. And during face-to-face talks between the leaders, *their* two signals will be coordinated, along with a third, that of Dr. Lem, who is in the room coordinating the conversation. Simply put, to find Dr. Lem, all we have to do is look for a signal that is coactive with both leaders together, as opposed to coactive with each leader separately.

We can apply this logic to functional imaging activations for phonological and semantic tasks. Given the dissociability of phonological and semantic abilities, we can expect that at least some phonological tasks will drive phonological networks much more than semantic networks (e.g., nonword repetition or phoneme detection) and vice versa. This is analogous to the two leaders separately talking to their teams. But many tasks will drive both networks, such as listening to words or naming. This is analogous to the two leaders talking face to face. Given this, the overall pattern of activation for phonological and semantic tasks in general should be some regions that are specific to the task type, reflecting their separate networks, and regions of coactivation reflecting their interaction. Following our logic, these shared regions of coactivation should contain the lemma system.

Notice that I said "contain," not "define the lemma system." The reason for this choice of words is that unlike in our political summit analogy, real brain activations are messy and hard to interpret. For example, some of the regions of shared activity could simply reflect coactivation of a phonological or semantic network by a task that targeted the other. That is, it could be that all phonological tasks activate phonological area *P* (because it is a phonological area) and some semantic tasks also activate phonological area *P* (because they call words to mind), thus yielding a region of overlap in phonological space, not lemma space. So this overlap analysis will not provide a conclusive case for a lemma network, but it will, I suggest, constrain our search and, when combined with other data sources, will be part of a stronger case.

Alex Teghipco and I applied this logic in a large-scale meta-analysis of hundreds of functional imaging studies using the keywords "phonological" and "semantic." This kind of meta-analysis involves first tagging terms that occur frequently in each report and then compiling the reported activation patterns for all the reports that frequently used a given term compared with the activation patterns from reports that didn't use that term frequently. It

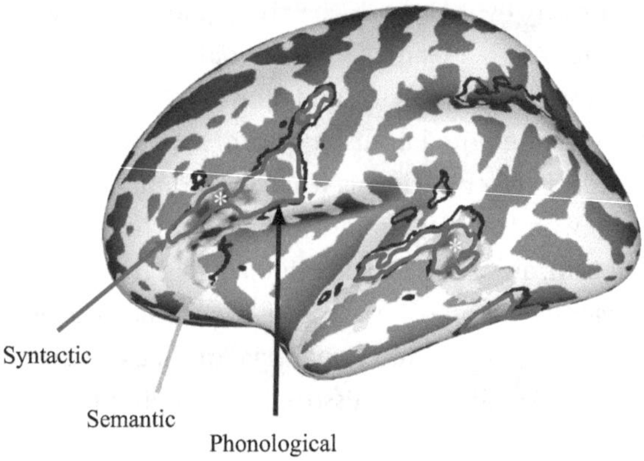

Figure 7.3
Overlap of regions involved in "phonological" (black outline), "semantic" (grayscale shading), and "syntactic" (gray outline) processing in functional imaging studies as indexed via an automated meta-analysis on Neurosynth.org. Asterisks indicate notable regions of overlap of all three domains. Courtesy of Alex Teghipco.

works surprisingly well for basic network mapping. You can try it for yourself at Neurosynth.org.

As we predicted, the maps associated with "phonological" and "semantic" were different, but they also overlapped, as figure 7.3 shows. Of particular interest to me is the region of overlap in the posterior superior temporal sulcus (STS)/middle temporal gyrus (indicated by the asterisk), the same zone that, when electrically stimulated, was reported to cause transcortical sensory aphasia. This region also exhibited overlap with syntactic processing consistent with an alignment between the lemma/morphosyntactic level and syntax, a fact that I'll return to shortly. (I'm ignoring the frontal overlapping regions, but we'll return to this in chapter 9.) The existence of overlap in the posterior temporal lobe doesn't prove anything—the data so far are consistent with that region being either a lemma zone or a semantic zone—but at the very least, it suggests that the stimulation study was not a fluke and the posterior middle temporal area is a strong candidate for a lemma-related zone.

An advantage of using large-scale, data-driven meta-analyses is that we can take advantage of the results of many studies, hopefully avoiding spurious results from individual studies. However, by mapping neural regions associated with general terms, one may rightfully wonder exactly what we are mapping. For this reason, it is helpful to look at more targeted individual studies as well to see if we can find corroborating evidence. I'll describe two studies using different paradigms, which together nicely frame the posterior middle temporal gyrus as the phonological-semantic interface zone.

The first experiment used a paradigm based on classic work by Loraine Tyler and William Marslen-Wilson on the *cohort model* of spoken word recognition (Marslen-Wilson & Tyler, 1980). The basic idea is that when we are listening to a spoken word, such as "alligator," we don't hear it all at once, but bit by bit from start to end. For many words, the early bits are consistent with a cohort of candidates. For example, after hearing the beginning /æl/ bit of "alligator," there is no way to know whether it will continue as "alligator," "alabaster," "alcohol," "altitude," or something else. Tyler and Marslen-Wilson argued that when faced with this temporary ambiguity, the brain proactively activates the entire cohort consistent with what has been heard up to that point rather than waiting for more information before doing something. Then, as more information streams in, the cohort is winnowed down until there is only one candidate consistent with the phonological evidence. This is the *uniqueness point* of a word enabling recognition, which may occur well before the full word is even heard (e.g., *allig . . .*).

Crucially, for our discussion, this bottom-up process can be augmented by top-down semantic information, which can speed the process. For example, listeners are quicker to recognize the word "alligator" in a context like *I'm not sure if that's a crocodile or an æl . . .* , compared to a context that doesn't prime the target word. Thus, phonological ambiguity interacts with semantics during word recognition. One functional magnetic resonance imaging (fMRI) study, headed by Jie Zhuang in Tyler and Marslen-Wilson's group, looked at the neural correlates of this interaction and reported an effect involving the superior and middle temporal gyri (Zhuang et al., 2011).

Another fMRI study by cognitive neuroscientist Jennifer Rodd and colleagues mapped the brain regions that were sensitive to lexical-semantic ambiguity rather than phonological ambiguity in sentences such as *The shell was fired toward the tank*. Although the team reported that most study participants didn't even notice that "shell," "fired," and "tank" are all ambiguous, their brains did. Regions in both the frontal lobe (the inferior frontal gyrus (IFG), Broca's area) and the posterior temporal lobe (the middle temporal and inferior temporal gyris) activated more strongly to the ambiguous sentences compared to unambiguous control sentences.

These two experiments studied word ambiguity from complementary angles. The first took advantage of phonological-lexical ambiguities, while the second took advantage of lexical-semantic ambiguities. The pattern of brain activation in these two studies formed a kind of tidy Venn diagram, with the overlap centered on the middle temporal gyrus, presumably reflecting their shared lemma involvement. The usual cautionary comments hold, of course. These studies alone don't build an airtight case for the middle temporal gyrus as the neural warehouse for lemmas, but they do build further support.

The Syntactic Argument

We can strengthen our case further by adding another network to the party. Lemmas are hypothesized to translate not only between phonology and semantics, but code aspects of syntax. More forcefully, assuming the morphosyntactic view of middle-word, that level of processing *is* syntax. Applying our coactivation logic, then, we should expect a lemma network to exhibit three-way coactivation between phonology, semantics, and syntax. To test this prediction, Alex Teghipco and I added the keyword "syntactic" to our meta-analysis and found the expected three-way overlap in the posterior middle temporal gyrus (figure 7.3).

As a reality check, we looked to see whether processing nonwords, which by definition don't have a lemma-level representation, activated the same posterior middle temporal gyrus region. They did not, suggesting that we were not merely tapping into phonological networks. A similar conclusion was reached in an fMRI study that directly contrasted the perception of spoken words and nonwords in the same participants: the posterior middle temporal gyrus showed sensitivity only to the words (Gagnepain et al., 2008). The case for this region as a lemma network grows stronger.

A skeptic (like me) might agree that we have reasonably ruled out the possibility that this region of overlap is a phonological area, but we may counter, correctly, that we haven't ruled out the possibility that it is a semantic area. Indeed, it's hard to come up with a task that taps only into semantics without a strong possibility of also involving words, since one of the powers of language is to translate any old thought into speech. We might counter this concern by noting that there are semantic zones that do not show up as overlapping with phonology and syntax: if the overlapping zone really is just the semantic network, why isn't the whole semantic network involved? But our skeptic might counter the counterargument by pointing out that different parts of the semantic network are doing different things (Binder & Desai, 2011; also see chapter 8), and maybe the posterior middle temporal gyrus in the semantic network is just the part that is most closely related to meanings called up by words, so it tends to show up more robustly in our meta-analysis. While this line of thinking strikes me as a bit handwavy, it is a crack in the argument and thus deserves some spackle. For that, we will have to turn to another source of data.

The Neuropsychological Argument

One way to patch the crack in our case is to show that semantic deficits caused by brain damage do not typically involve our proposed posterior middle temporal gyrus lemma zone. The logic is this: if the posterior middle temporal gyrus zone is a semantic area, then damage to that zone should produce some rather profound semantic deficits for basic concepts, like those associated with words.

Right off the bat, though, this prediction seems to fall short of the facts. Consider Wernicke's aphasia, which typically involves lesions covering the posterior middle temporal gyrus. While it's true that many of their errors in understanding appear to reflect semantic confusions ("moose" for "bear"), there is a characteristically aphasic inconsistency in their performance on a given item. On one attempt, they may commit a semantically related flub, and yet on other trials or other tasks probing the same knowledge, they may get it perfectly correct (Jefferies & Lambon Ralph, 2006). This hit-or-miss performance is typical of stroke aphasia (Freed, Marshall, & Chuhlantseff, 1996) and suggests a failure of *access* to semantics rather than a degeneration of semantic knowledge per se. If you've lost a document in the cloud, it doesn't matter how many times you try to access it or through which channel; it's gone; you'll fail every time you attempt to retrieve it. On the other hand, if the document is there and you fail to access it one time because of an unstable Internet connection, say, a second or third try may do the trick. Thus, the variability of item-level performance in aphasia suggests an access problem rather than a content problem.

In contrast to poststroke aphasia, more stable item-level error patterns *are* found in semantic dementia, also called *semantic variant primary progressive aphasia* (Jefferies & Lambon Ralph, 2006), suggesting a deficit involving semantic content. People with semantic dementia also tend to have better syntactic ability than individuals with Wernicke's aphasia (Ogar et al., 2011). This is evident both on sentence comprehension tasks, where people with Wernicke's aphasia perform significantly worse, and in elicited speech production. Consider the following speech samples reported by Ogar et al. (2011, p. 30):

> Wernicke's aphasia: "And the /kwer ledi/ or leather here it's /leow/ fire, farm, fur, something feet and um, dime farm here. And the farm any the farm here."
>
> Semantic dementia: "Looks like there's a husband and wife. She's reading and she's drinking and then there's another young man who has a . . . I want to say 'flag', but it's not a flag, but he's holding something up in the air (referring to kite)."

Crucially for the argument that we are constructing, the neural dysfunction in semantic dementia does *not* typically involve the posterior middle temporal gyrus and instead predominantly implicates more anterior temporal lobe areas (Hodges & Patterson, 2007; Mesulam et al., 2015; Ogar et al., 2011). In contrast, receptive syntactic deficits and syntactic misfires in speech production are associated with posterior temporal lesions (Fahey et al., 2024; Matchin et al., 2020, 2023), as we will see in chapter 9. This suggests that when Boatman and colleagues (2000) induced transcortical sensory aphasia by stimulating the posterior middle temporal gyrus and its surrounds, they were interfering with *access* to semantic networks, not with semantic knowledge generally. Just to connect the dots, what kind of system that, when disrupted, can destabilize access to semantics while leaving both phonological ability (preserved repetition) and semantic knowledge intact? The answer is: the in-between lemma/morphosyntactic system.

Although the argument sketched here supports the view that the posterior middle temporal gyrus supports lemma processing, it is not crack free. Perhaps, for example, the hit-or-miss performance in stroke aphasia involving the posterior middle temporal gyrus is due to the unilateral nature of lesions typical of stroke aphasia; if the lesions were bilateral in that zone, a more consistent semantic deficit might be found.

The Speech Production Argument

Most research on the lemma level has come from speech production studies, but I think most language scientists would agree that a lemma system exists for receptive speech as well (especially if one buys the argument that lemma = morphosyntax). And there are a handful of studies supporting the existence of lemmas in receptive word processing (Stockall & Marantz, 2006; Tsuboi, Francis, & Jameson, 2021). Assuming that the middle-word system is shared for expressive and receptive language, attempts to localize lemmas in production will be relevant for comprehension. I'll highlight two relevant studies that place the lemma system in the middle temporal gyrus.

The first is a seminal meta-analysis of functional imaging studies by Peter Indefrey and Willem (Pim) Levelt of the Max Planck Institute for Psycholinguistics, which David Poeppel and I had the privilege of guest editing for a special issue of the journal *Cognition* on the functional anatomy of language. Indefrey and Levelt analyzed the spatial distribution and timing of activations resulting from picture naming, word generation, word reading, nonword reading, word listening, and nonword listening from dozens of published studies with the aim of localizing the networks involved in various stages of speech production, including lemma selection. They identified a single area that satisfied their criteria for lemma engagement: the middle section of the middle temporal gyrus (Indefrey, 2011; Indefrey & Levelt, 2004).

The second is a compelling neuropsychological study was carried out by Juliana Baldo, Nina Dronkers, and colleagues, who studied the picture-naming abilities of ninety-six stroke patients (Baldo et al., 2013). They were specifically interested in isolating the middle process between visual recognition and articulation. Picture-naming ability was correlated with a large network that included the superior and middle temporal gyri and portions of the parietal lobe. After factoring out speech fluency and controlling for low-level articulatory problems and semantic knowledge of the items that they failed to name, damage to the posterior middle temporal gyrus remained as a location causally associated with deficits in picture naming.

Speech production studies, therefore, also point to the posterior middle temporal gyrus as a key node in lemma-level processes.

Toward a Multilateral Agreement

There is a range of additional studies that are consistent with the model that we are developing here but themselves are ambiguous regarding whether the effects are semantic or reflect lemmas. Neurolinguist Ellen Lau and colleagues surveyed much of the relevant literature, focusing on the electrophysiologically measured N400 effect and fMRI-measured lexical-semantic priming effects (Lau, Phillips, & Poeppel, 2008). *N400* refers to a negative-going brain wave that occurs about 400 msec poststimulus. The size of this wave is increased in response to words that are semantically incongruous with the preceding sentence. The classic example from the original study by Marta Kutas and Steven Hillyard (1980) is "I take coffee with cream and socks," in which "socks" yields a sizable N400 effect compared to "I take coffee with cream and sugar." Lexical-semantic priming is when one's response to a word such as "nurse" is facilitated by previous exposure to a related word, such as "doctor" (Meyer & Schvaneveldt, 1971). The fMRI reflection of this is a decrease in activation to a stimulus when it is primed.

The analysis by Lau and colleagues of the literature on the neural basis of these lexical-semantic effects pinpointed the posterior middle temporal gyrus as the source. Their interpretation of this result aligns neatly with our model: "Amodal lexical representations, linked to distributed semantic representations throughout the cortex, are stored and activated in the posterior middle temporal cortex and in parts of the neighbouring STS and IT [inferior temporal]" (Lau et al., 2008, p. 929). I submit that this conclusion steps just a bit past the data they reviewed, as it's hard to rule out that these effects are occurring in semantic rather than lemma networks, However, again, combined with the evidence reviewed in this chapter, I think that the conclusion is right.

Summary

In chapter 6, we traced the speech perception network from the earliest stages of spectrotemporal modulation filters in primary auditory cortex to representations of learned phonological sequences in the auditory phonological area (APA) of the lateral superior temporal gyrus (STG) and the STS. In this chapter, we built a case for the posterior middle temporal gyrus as the core region involved in processing abstract words or lemmas, which appears to overlap with regions implicated in morphosyntax. In chapter 9, we will consider the role of the posterior middle temporal gyrus and other regions for syntactic ability, which will provide further neuroscientific support for the idea that the middle-word level is actually a core system for hierarchical syntactic computation.

Even after Poeppel and I surveyed the literature three times regarding the neural basis of what we called the "sound-conceptual interface" (Hickok &

Poeppel, 2000), the "sound-meaning interface" (Hickok & Poeppel, 2004), and the "lexical interface" (Hickok & Poeppel, 2007)—purposefully theory-neutral terms for the in-between system—I've felt that this was the weakest part of the neuroanatomy of the dual stream model. As I looked again at this literature for this chapter, it still feels hard to pin down. Part of this has to do with different terms and conceptions regarding what the middle-word realm is, and part of it comes from the difficulty of isolating it neuroanatomically. At the same time, I'm struck by the remarkable consistency in how researchers in multiple subdisciplines have concluded that there *is* an in-between level of some kind, that it is neither phonological nor semantic, that, it has morphosyntactic features, and that, the posterior middle temporal gyrus is involved. This gives me confidence that we are on the right track.

Now that we have looked at the auditory, phonological, and middle-word levels of the speech comprehension process, it makes sense to consider what we know about the end stage of the mapping between sound and meaning. So we will turn next to the recent progress in understanding the organization of conceptual knowledge in the brain. To some readers, it might make more sense at this stage to pursue the relation between the lemma/morphosyntactic system and the literature on the neural basis of syntax. If you agree with that, feel free to invert the order of the next two chapters. My organizational logic, though, is that by moving to semantics next, we can stick roughly to the word level of processing and then tackle the complexities of the syntactic literature on a foundational understanding of the mapping between sound and meaning.

8 Where Do You Know What You Know?

> Mr M, a patient with semantic dementia . . . was being driven through the countryside to visit a friend and was able to remind his wife where to turn along the not-recently-travelled route. Then, pointing at the sheep in the field, he asked her "What are those things?" Prior to the onset of symptoms in his late 40s, this man had normal semantic memory. What has gone wrong in his brain to produce this dramatic and selective erosion of conceptual knowledge?
>
> —Karalyn Patterson, Peter Nestor, and Timothy Rogers, 2007

Knowledge about the world is much of what language communicates. It is the end point of the transformation between sound and thought, and the starting point for the transformation between thought and sound. Philosophers have debated the nature of knowledge for millennia and have an entire subfield, *epistemology*, dedicated to its study. In this chapter, though, we are concerned with a simpler question, one that cognitive neuroscientist Karalyn Patterson and colleagues posed in the title of their influential 2007 review article (Patterson, Nestor, & Rogers, 2007): "Where Do You Know What You Know?"

You may wonder why we should care about this question. Surely, figuring out *where* something happens doesn't help in understanding *how* it happens or *what* it contains (or does it?). Moreover, this is a book about language, not semantic knowledge. But there are at least two reasons why we, as students of language, should care.

One is that an answer to this question has the potential to make significant inroads into an age-old debate regarding the relation between language and thought, which has been featured repeatedly in neurological thinking about language disorders. Carl Wernicke, for example, argued for a clear distinction, stating that "it is necessary to regard the concept and the word as different" (1874/1969, p. 63). An influential contemporary, the English neurologist John Hughlings Jackson, agreed with Wernicke that words could be distinguished from concepts—the word "bell" is not the same thing as the idea of a *bell*—he nonetheless believed words to be a core part of our intellectual capacity, hedging ever so slightly: "Words are required for thinking, for most of our thinking at least" (1878, p. 323). Adolf Kussmaul (1822–1902), whom we will meet again in chapter 9, succinctly summarized the state of the field in his 1877 volume: "It is still a vexed question whether intelligent thought is bound up with words (Condillac, Max Mueller, Bastian, and others), or is quite independent of words (Lock, Helmholtz, Maudsley, Finkelnburg, and others)" (Kussmaul, 1877, p. 597). In modern times, the debate continues in more circumscribed forms. For example, there is the claim that syntax and semantics are not separable, as championed by Evelina Fedorenko and colleagues (2020, p. 1): "Contra many current proposals of the neural architecture of language, syntactic/combinatorial processing is not separable from lexico-semantic processing at the level of brain regions-or even voxel subsets-within the language network."

The two positions make different predictions. If language and thought are distinct, language and intelligence should dissociate following neurological disruption and have distinguishable neural substrates; if they are bound up together, language and intelligence should hang together behaviorally and neuroanatomically. So whether concepts and words map onto the same or different neural networks is a *where* question that can help move us closer to settling this debate.

The second reason why we should care about the *where* question is that its answer will provide a neuroanatomical bracket around a core part of the language network. In the previous chapters, we have reviewed a wide range of evidence arguing for the existence of an auditory phonological area (APA) in the superior temporal lobe and a middle-word level in the middle temporal gyrus. If we can nail down key regions involved in conceptual knowledge, we will have fixed the end point of the mapping from sound to meaning and the starting point of the mapping from meaning to sound. Because of the tendency to minimize distance in neural wiring patterns (see the appendix of this book), knowing where phonological and semantic networks live in the brain will potentially constrain our models of where the in-between, morphosyntactic networks live.

So let's survey what we know about where we know what we know and how it relates to language. I'll start, as usual, with a little backstory.

Intelligence and Aphasia

Despite Wernicke's claims to the contrary, by the early part of the twentieth century, the dominant view was that "intelligence is always lamed" in aphasia, to use the words of French neurologist Armand Trousseau (1801–1867) (Trousseau, 1865, quoted in Benton, 1991, p. 12), who also coined the term *aphasia*. This view was put to the test in the 1930s by Theodore Weisenburg (1876–1934) and Katharine McBride (1904–1976) (Risser, 2018; Weisenburg & McBride, 1935). Weisenburg, a neurologist, had secured funding for a large-scale, systematic study of aphasia. The lack of a standardized test battery was a major problem in aphasia assessment and a likely source of varying opinions on the topic. Henry Head's (1861–1940) influential two-volume 1926 work, *Aphasia and Kindred Disorders of Speech*, pioneered the idea of a standard set of assessment tools, but Weisenburg identified a need for improvement. As editor-in-chief of a major neurological journal and president of the American Neurological Association, Weisenburg was a well-established and respected leader in neurology, but he was no expert in psychological assessment and realized that he needed help. He recruited McBride, then a graduate student, who had impressed her mentors sufficiently to earn their strong endorsement for the role in Weisenburg's project despite having not yet started her dissertation proposal.

McBride crafted an impressive assessment battery that would be familiar to clinicians today, comprising a range of language tasks involving spontaneous speaking, automatized sequences, naming, repetition, comprehension, reading, and writing, to name a few. She also used a range of nonverbal tasks drawn from some of the early IQ assessment tests. It took an average of nineteen hours to administer the whole thing to each patient! Beyond the assessment tool itself, a major advance of the team's research was the inclusion of a control group to provide a performance baseline from neurologically healthy individuals. These controls were recruited from surgical and orthopedic wards of the same hospitals that provided the neurological sample to account for the effects of hospitalization itself. The final study included an impressive 234 individuals: 60 people with aphasia, 38 people with unilateral lesions but no aphasia, 25 cases with miscellaneous neurological conditions, and 85 neurological healthy controls; another 26 individuals were examined but not included in the final study. At the time, it was the most comprehensive study of aphasia ever undertaken.

Weisenburg and McBride appear to have started their study decidedly sympathetic to the views of Hughlings Jackson and his ideological inheritors such as Head. Indeed, the duo devote nine pages to a summary of Hughlings Jackson's views compared to not even three pages to Broca, Bastian, and Wernicke combined. Among Hughlings Jackson's contributions that are lauded by Weisenburg and McBride is his "appreciation of the complex relations of language and thinking" (1935, p. 19), which further tips their hand toward a codependency viewpoint. At the same time, Weisenburg and McBride remain pragmatic and data-driven in their approach to the issues of the day. An extended quote on this point in relation to the intelligence and aphasia debate is worth our time:

> The question of a general intellectual defect in cortical sensory aphasia has been argued pro and con for many years and has been particularly difficult to settle because the receptive disturbances masked whatever intellectual changes might exist. There are still investigators who believe that the receptive disturbances are actually responsible for the apparent intellectual defect, and others who think the opposite relationship is the true one. . . . Until "non-language" tests were introduced, no one could determine the relative efficiency of behavior in verbal and non-verbal problems. Non-language tests have not yet been used in sufficient variety, nor have they been employed for enough cases. (Weisenburg & McBride, 1935, pp. 75–76)

Weisenburg and McBride's study promised an answer. What did they find? After noting that nonlinguistic deficits are often present in aphasia, they conclude that "it is evident that purposeful and effective thinking may be carried through when language is extremely inadequate or confused, and must therefore depend largely on nonlinguistic symbols such as visual or kinaesthetic images" (Weisenburg & McBride, 1935, p. 426).

This was the first clear large-scale demonstration of the separability of language and thought. Others followed. One study evaluated 111 people with aphasia using a modern aphasia assessment, the Western Aphasia Battery, and the Raven's Progressive Matrices task (Kertesz & McCabe, 1975), which involves a matrix of visual patterns with the last cell left blank for the participants to fill in according to the pattern (figure 8.1 provides an example). It has become a standard tool in assessing nonverbal intelligence. Like Weisenburg and McBride, the study reported that people with aphasia can be impaired

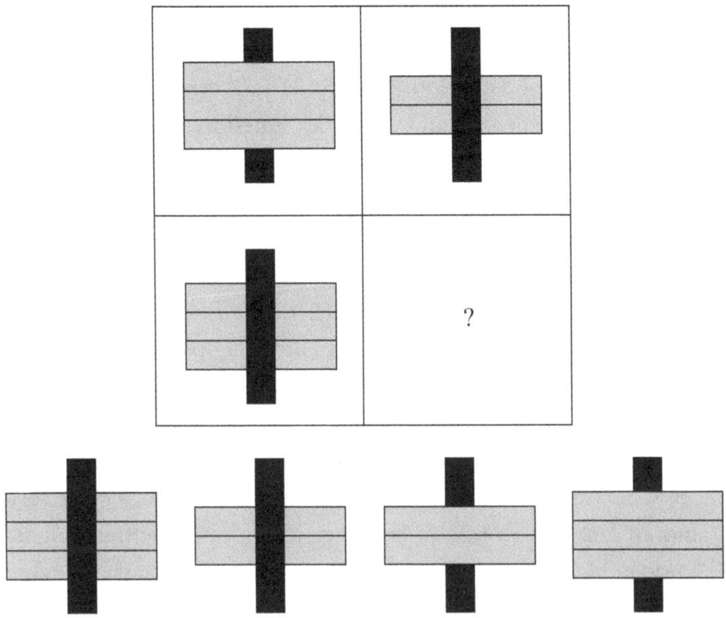

Figure 8.1
Example of a nonverbal task item that might benefit from verbal recoding. (The correct answer is the third pattern from the left.)

on the nonverbal task, but the degree of impairment did not correlate with the severity of aphasia, and, importantly, a sizable fraction of the patients (42 percent) scored within normal limits on the Raven's task. A more recent study published in 2020 tested 200 people with aphasia and found the same pattern using a correlation analysis (Gonzalez, Rojas, & Ardila, 2020). Aphasia and nonverbal intelligence were found to be statistically significantly correlated, which is consistent with previous findings that language and nonlanguage deficits can cooccur. But, again, the correlation values themselves reveal a rather weak relation. In particular, the overall "aphasia quotient" score accounted for only 11.5 percent of the variance on the Raven's Matrices task. This result is more consistent with the correlation being driven by an incidental disruption of separate systems rather than damage to a single system that supports both linguistic and nonlinguistic ability.

Other studies, which evaluated performance on a broader array of nonverbal tasks, reported similar findings: aphasia and nonverbal deficits commonly cooccur, but having aphasia does not automatically imply a nonverbal cognitive deficit (Archibald, Wepman, & Jones, 1967). One study (Fonseca, Raposo, & Martins, 2018) scoured the literature on the relation between aphasia and nonverbal cognition, identifying forty-seven published reports. They found that 61.3 percent of the studies stated that people with aphasia as a group performed more poorly than nonaphasic controls on nonverbal cognitive tests. The complement of this number, 38.7 percent, is in the ballpark of the percentage of people with aphasia who were in the normal range on the Raven's task (42 percent) in Kertesz and McCabe (1975). The authors conclude, "The large variability of results, previously mentioned by other authors, also indicates that the presence of aphasia does not necessarily produce other neuropsychological impairments" (Fonseca et al., 2018, p. 136).

To be fair, some researchers have come to different conclusions. Nina Dronkers, Juliana Baldo, and colleagues have published a series of reports (Baldo et al., 2005, 2010, 2015), in which they argue for "a close relationship" and an "important role of language in reasoning and problem solving." (Baldo et al., 2015, p. 8). But this conclusion overstates the data a bit, in my opinion. In their largest-scale study (Baldo et al., 2015), for example, they report relationships between various intelligence and language tests with correlation values ranging from 0.32 to 0.41 in people with aphasia. This translates to the conclusion that language measures explain only between 10 percent and 17 percent of the variance in nonverbal intelligence tests.

Working with my collaborators at the University of South Carolina—Julius Fridriksson, Lisa Johnson, and Alex Teghipco—I took a look at this issue in our stroke aphasia database. Our sample included 107 people with a left-hemisphere stroke, 78 of whom were diagnosed with aphasia as assessed by the Western Aphasia Battery (Kertesz, 2012). These individuals were also administered the Philadelphia Naming Test, a strong correlate of aphasia severity ($r = 0.92$, in this sample) (Walker et al., 2022), and the Raven's nonverbal intelligence task. Table 8.1 shows the correlation between participants' scores on the Raven's task versus the various language measures, as well as some demographic measures. Figure 8.2 shows a scatterplot of the Raven's score and the Western Aphasia Battery Aphasia Quotient (WAB-AQ), an index of overall aphasia severity.

These simple analyses confirm what others have found. There *is* a statistically reliable relation between language and nonverbal ability, but it is very loose, accounting for about 11 percent of the variance if we take

Table 8.1
Statistical relation between the Raven's Matrices nonverbal intelligence test and various standard aphasia assessment tests

	Raven's Score	
	Pearson's r	BF_{10}
WAB Subtests		
Word fluency	0.408	7.404e +6
Object naming	0.336	14360.263
Responsive speech	0.308	1907.839
Information content	0.302	1287.682
Auditory word recognition	0.296	888.655
Repetition	0.293	711.528
Fluency, grammatical competence, and paraphasias	0.285	439.099
Sentence completion	0.257	84.565
Yes/no questions	0.225	16.18
Sequential commands	0.225	16.283
WAB Composite Scores		
Naming and word finding	0.359	88617.469
WAB-AQ	0.337	15506.482
Spontaneous speech	0.312	2584.273
Auditory verbal comprehension	0.268	154.691
Philadelphia Naming Test	0.348	34098.847

Note: WAB-AQ is an overall index of aphasia severity based on the Western Aphasia Battery subtest scores. Pearson's r correlation coefficients are provided along with the Bayes Factor (BF) value, which estimates the statistical confidence in the correlation value. $BF > 10$ is considered strong evidence for hypothesized statistical relation.
Source: Unpublished data from Hickok, Johnson, Teghipco, and Fridriksson.

WAB-AQ as the gold standard measure of aphasia, and 17 percent if we use the language subtest with the highest correlation value. The scatterplot in figure 8.2 shows what this relation looks like graphically, with the line indicating the best-fit linear relation and the points revealing how substantial the deviation is from the best-fit line. The fractionally bigger story, then, is how much of the variance in intelligence that language does *not* explain.

I suggested earlier that the weak statistical correlation between linguistic and intelligence measures might be due to spurious damage being done to separate systems. But there is another possibility: language might be usable

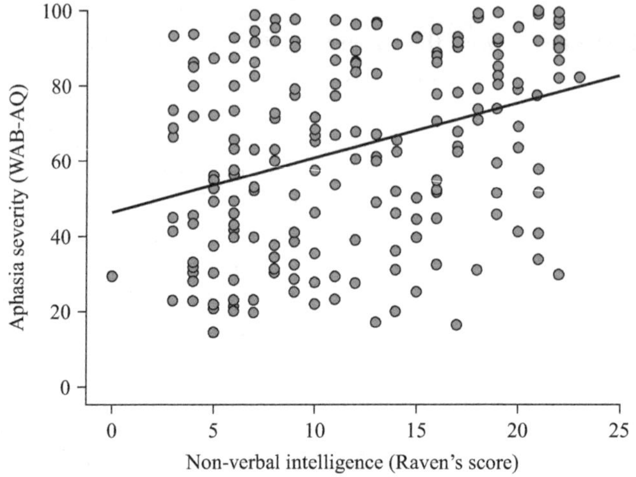

Figure 8.2
Scatterplot of the relation between aphasia severity scores and nonverbal intelligence scores. Unpublished data from Hickok, Johnson, Teghipco, and Fridriksson.

as a working memory crutch to aid performance on nonverbal intelligence tasks. Consider the example problem shown in figure 8.1, which is similar to one of the moderate-difficulty items on the Raven's task. It may be helpful for some people to talk through the relations of elements (it does for me). Left column: "Three gray bars in front, three gray bars in back." Right column: "Two gray bars in back, ___ gray bars in ___." One can then rehearse the answer, "Two gray bars in front" while looking for the correct item in the bottom array. It is a convenient quirk of language that it can be used as a mental scratch pad.

More generally, I think language can aid reason in a similar way—that is, via a kind of mental scratch pad—when ruminating on many logical problems. Darwin expressed a similar view, stating that "a complex train of thought can be no more carried out without the use of words, whether spoken or silent, than a long calculation without the use of figures or algebra" (Darwin, 1871, p. 88). I certainly find the process of writing about the logic of a scientific problem helpful in seeing the flaws in an argument or the paths forward. This use of language extends the intellectual benefit beyond an internal temporary store by offloading information to a text file. This doesn't mean that text files are a core part of intelligence, any more than the numerals and symbols on a whiteboard in an algebraic calculation are themselves the core of algebra. It means that certain aspects of problem solving can be augmented by the use of text files or a whiteboard. If we reframe Baldo et al.'s interpretation of their data—that there is an "important role of language in reasoning and problem solving"—in this more restricted, augmentative sense, I think the conclusion is reasonable. To restate their conclusion in these more restricted terms, I would say that there is an important role for aspects of language in augmenting to a modest degree some kinds of reasoning and problem solving via working memory–like mechanisms.[1]

Caveats and qualifications aside, the evidence from decades of research is clear: intelligence is *not* always lamed in aphasia. More specifically, the severity of language impairment is a rather poor predictor of the severity of intellectual impairment. Nor is language always lamed in intellectual dysfunction. Recall the remarkable language perception and production ability of Geschwind's and Whitaker's case studies (chapter 2) involving people with preserved speech perception, speech production, and syntactic ability despite profound intellectual impairment (Geschwind, Quadfasel, & Segarra, 1968; Whitaker, 1976). Such results tell us rather definitively that linguistic and nonverbal abilities are two distinguishable mental faculties, a conclusion also reached by a recent functional magnetic resonance imaging (fMRI) study by Fedorenko, Piantadosi, and Gibson (2024). Language (in the broad sense) is a conduit for thought, a system for transplanting thoughts from one mind to another. Language is not thought itself.

We reach these conclusions on the basis of decades of studies that focused primarily on measures of *fluid intelligence*, the ability to reason and solve problems. But we started this chapter with an example of a man who lost a different kind of intelligence—the ability to know basic things about the entities that populate our world. What is the neural basis of *this* kind of knowledge, and might *it* be related to language ability?

Loss of Knowledge of People, Places, and Things

Basic memory flubs are commonplace, from forgetting a name to forgetting where we left our phone. Many of us are familiar with the most extraordinary case of chronic forgetting, that of Henry Molaison (H.M.), whose severe, surgically induced amnesia revolutionized our understanding of memory systems (Corkin, 2013). In Mr. Molaison's case, the source of the memory problem was a failure to get new facts and events into long-term storage that resulted from damage to the medial temporal lobe. His previously consolidated long-term memories were relatively unaffected, though, indicating that the neural systems that store memories over the long term do not involve the medial temporal lobe. Where are they?

Carl Lashley (1880–1958) searched for the storage site of the *engram*—Richard Semon's term for the physical change caused by a stimulus such that its memory is "engraved" on the brain (Semon, 1921)—and famously failed in his quest (Lashley, 1950). Lashley found that he could damage a rat's brain so the animal could no longer remember how to perform a previously learned task, but his research suggested that the location of damage didn't matter—only the amount of brain tissue removed. He summarized decades of research on his engram quest as follows: "I believe that the evidence strongly favors the view that amnesia from brain injury rarely, if ever, is due to the destruction of specific memory traces. Rather, the amnesias represent a lowered level of vigilance, a greater difficulty in activating the organized patterns of traces, or a disturbance of some broader system of organized functions" (Lashley, 1950, p. 472).

More than half a century before Lashley wrote these words, Carl Wernicke proposed just such a broader system for the organization of memory traces that together constituted the engrammatic representation of an object, here using the idea of a *bell* to illustrate: "The memory images of a bell . . . are deposited in the cortex and located according to the sensory organs. These would then include the acoustic imagery aroused by the sound of the bell, visual imagery established by means of form and color, tactile imagery acquired by cutaneous sensation, and finally, motor imagery gained by exploratory movements of the fingers and eyes" (1874/1977, p. 179). Assuming that Wernicke was right about such a broadly distributed network underlying conceptual knowledge, it's no wonder Lashley couldn't find engrams in a specific location.

But in the 1970s, neuropsychologist Elizabeth Warrington reported on three patients that would have piqued Lashley's interest (Warrington, 1975). Each came to medical attention due to complaints about poor memory. Clinical evaluation failed to turn up any neurological problems except for an apparent visual object agnosia, an inability to recognize or identify common objects that cannot be explained by sensory deficits or generalized cognitive impairment. Each patient underwent pneumoencephalography, a rather invasive, X-ray-based brain scan technique that involved draining the patient's cerebrospinal fluid via lumbar puncture and replacing it with air to reveal the brain more clearly. The only notable finding from these scans was mild to moderate cerebral atrophy with no clear cause.

Warrington was interested in visual agnosia because it was a relatively rare condition that had not been studied in detail. Previous work had suggested two forms, a lower-level *apperceptive* variant, which involved difficulty perceiving basic shape information, and a higher-level *associative* variant, which was presumed to involve a failure in linking basic visual information to meaning. Warrington was particularly interested in the nature of the associative form, and she enrolled the three patients described here in a detailed study.

A battery of standard psychological tests confirmed the patients' difficulty recognizing objects and ruled out several possibilities as the source of the problem: general intellectual impairment, general memory loss (the patients were oriented to time and place and could recall detailed events in their lives), basic speech and language problems (except for a limited vocabulary, they were not aphasic), and perceptual impairment. Visual shape discrimination was within normal limits, as was the ability to identify the same object from both conventional and unconventional viewing angles, a test sometimes failed by people with right-hemisphere damage. Fascinatingly, the ability to match objects shown from multiple viewpoints was possible even when the object wasn't recognized. Warrington (1975, p. 640) comments, "On being shown the conventional view of a table tennis bat having already attempted to identify the unconventional view of it, [one patient] remarked that he had already been shown a photograph of that object taken from a different angle and that he had already told me that he had no idea what it was."

These were clear cases of difficulty recognizing visually presented objects, but were they specific to the visual modality or something more general? To find out, Warrington probed whether similar recognition problems were evident when the evoking stimulus was an auditory word rather than a picture. She found that the ability to arouse concepts from words was also impaired. In one test, patients were simply asked to provide definitions for spoken words. All three patients performed well on common, high-frequency words, getting nearly 100 percent correct. However, as the following examples show, the patients struggled on the less common words, even though such items were still part of almost every English speaker's vocabulary.

Examiner:	**Patient:**
Hay	I've forgotten
Poster	No idea
Needle	Forgotten
Acorn	Don't know
Cabbage	Eat it (no further information on questioning)
Geese	An animal but I've forgotten precisely

Based on these kinds of responses, we can infer that the word understanding problem is not due to difficulty perceiving speech sounds. Recall that people with word deafness, for example, say things like "I didn't get that," "It's going too fast," or "It sounds like a foreign language." We can further infer from responses like "Forgotten" or correct but vague responses like "Eat it" that the word sound patterns themselves are recognized as familiar, at least in some cases (Rogers & Friedman, 2008). What's lost is the meaning.

More extensive testing confirmed that these patients did not have a *visual* agnosia but rather a moderate-to-severe "selective impairment in semantic memory" (Warrington's title for her report) involving knowledge of common things: animals, objects, or faces. It didn't matter whether that knowledge was probed via pictures or speech—the memories could not be retrieved because only their shell remained. Warrington clarified the nature of those memory shells by demonstrating that superordinate category knowledge tended to be better preserved than subordinate category knowledge: the patients could often correctly identify a picture or word as an animal or plant, say, but fail to home in on the subtype.

Warrington's exquisitely detailed case studies identified selective damage to the engrams for semantic knowledge of entities, as opposed to episodes or events, and showed that they are organized hierarchically. She also showed that such a loss of knowledge could occur in the absence of language deficits. Again, language and intelligence dissociate, this time in crystalized form.

Lashley would have been surprised by these findings, I think, because there is no reason why a "lowered level of vigilance" should affect one kind of memory and not another. I suspect that he would have been eager to know the location and extent of the damage responsible for such memory loss. No doubt he would have predicted it to be extensive and diffuse. It took almost another two decades to get a definitive answer.

Semantic Dementia

Warrington noted that her three cases had a progressive neurological disease that resulted in an overall decline in function over time. In the 1980s, it had become clear that there were several distinguishable types of such progressive diseases distinct from Alzheimer's disease. In 1982, for example, neurologist Marcel Mesulum (1982) identified a progressive nonfluent aphasia without generalized dementia, and several other teams reported progressive fluent aphasias characterized by word-finding difficulty, semantic errors in speech production, and comprehension problems but with preserved phonological and syntactic function (Poeck & Luzzatti, 1988). At least some of these reports likely included additional cases of the selective semantic impairment described by Warrington rather than aphasias. In 1989, the syndrome that she described got a name, *semantic dementia,* and by 1992, enough evidence had accumulated to clearly link it to dysfunction of the temporal lobes (Hodges et al., 1992; Snowden, Goulding, & Neary, 1989).

By the time Patterson and colleagues published their influential "Where Do You Know What You Know?" review in 2007, behavioral studies of people with semantic dementia confirmed Warrington's diagnosis of a modality-independent semantic impairment with a hierarchical gradient. Figure 8.3, for example, shows the performance of people with mild, moderate, and severe semantic dementia on a picture categorization task. When asked to decide whether a given picture represented a *general* category, such as "animal," performance was good, even in severe-stage disease. Categorizing pictures at a *basic* level, such as "dog," became progressively more difficult as the severity of the disease increased. Categorizing at the *specific* level, such as "poodle," showed an even steeper decline. More illustrative is the picture-naming performance of one individual who was followed longitudinally as shown in table 8.2 (reproduced from Patterson et al., 2007). One can almost see this patient's semantic resolution melting away, first for the less common category members (ostrich, peacock), then for the more typical ones (duck, chicken), and finally for the broad category (bird).

Advances in neuroimaging technology in the 2000s also clarified the neuroanatomy of semantic dementia, pinpointing dysfunction of the

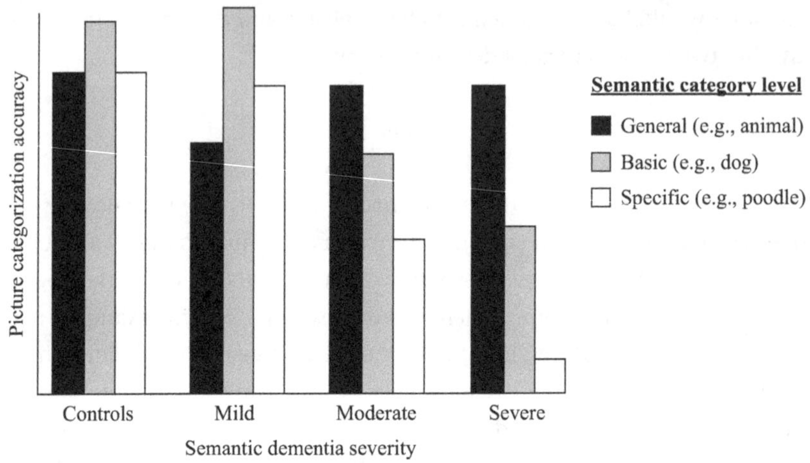

Figure 8.3
Performance on a picture categorization task in control participants and people with different stages of semantic dementia. Redrawn from Patterson et al. (2007).

Table 8.2
Example performance from one individual with semantic dementia on a picture-naming task, showing progression of the disease over 2.5 years

Item	September 1991	March 1992	September 1992	March 1993
Bird	+	+	+	Animal
Chicken	+	+	Bird	Animal
Duck	+	Bird	Bird	Dog
Swan	+	Bird	Bird	Animal
Eagle	Duck	Bird	Bird	Horse
Ostrich	Swan	Bird	Cat	Animal
Peacock	Duck	Bird	Cat	Vehicle
Penguin	Duck	Bird	Cat	Part of animal
Rooster	Chicken	Chicken	Bird	Dog

+ indicates correct responses. Reproduced from Patterson et al. (2007).

mid-to-anterior temporal lobes as its source. This was not predicted by most models of the neural organization of semantic knowledge, which assumed a much more broadly organized system that would be hard to disrupt with focal damage. Yet, to quote Patterson and colleagues (2007, p. 978), semantic dementia "represents a selective impairment to semantic abilities that affects all modalities of reception and expression, for all kinds of concepts, more or less equally . . . , and it is the consequence of relatively focal brain lesions." Lashley would have been stunned.

This doesn't necessarily mean that all this conceptual knowledge is stored in the anterior half of the temporal lobes. Patterson's team proposed a *hub-and-spoke model,* in which, following Wernicke's hypothesis, much of the knowledge of entities is stored in a highly distributed fashion throughout the cortex in regions that code for the various features of the entities like color, shape, sound properties, linguistic labels, and so on, but which are brought together and integrated into unified concepts via a hub network located in the anterior sections of temporal lobes. This idea is quite similar to previous hypotheses regarding the existence of high-level *convergence zones* proposed by Antonio Damasio (Damasio et al., 1986) and by Marcel Mesulam (1990, 1998), which serve to bind information processed and coded by disparate systems in the sensory and motor cortices.

Multiple Hubs: Beyond Knowledge of Things

Knowledge of *things* in the world is a large and important part of our semantic landscape, but it is far from all that we know or need to think about. This broader semantic system includes knowledge of actions, relations, emotion, the self, and more. In 2009, Jeffrey Binder and colleagues reported on a meta-analysis of 120 functional imaging studies of the neural correlates of this broader system. Their analysis focused on language-based studies that modulated semantic processing while controlling for, and I'll quote the authors, "low-level sensory, orthographic, phonological, syntactic, working memory, attentional, response selection, or motor processes" (Binder et al., 2009, p. 2773). So, for example, they included studies that contrasted listening to or reading real words versus nonsense words (controlling for lower-level perceptual and phonological processes) but excluded studies that contrasted, say, real words versus nonlinguistic symbols (like ❖♈⌗➚⭘), which does not control for phonological processes. They identified a semantic network

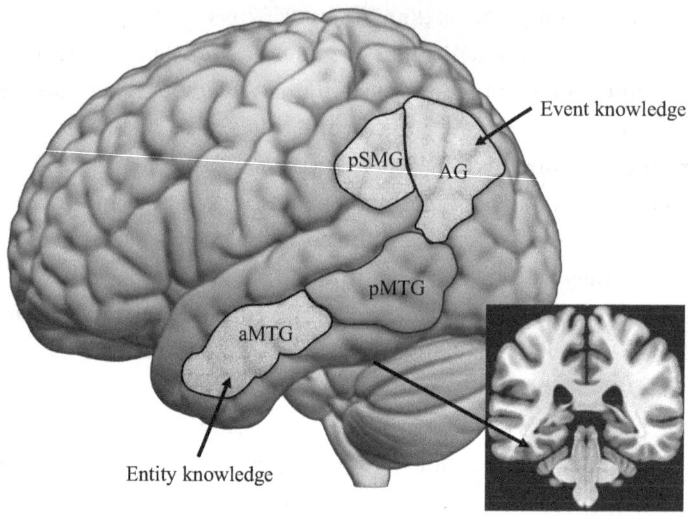

Figure 8.4
Temporal and parietal regions implicated in semantic processing according to a meta-analysis of functional imaging studies. The aMTG region is hypothesized to be important for entity knowledge and the angular gyrus for event knowledge. The role of the ventral temporal region is yet unclear. Although the pMTG region was part of the fMRI identified the semantic network, evidence reviewed in this volume suggests a more linguistic function. SMG, supramarginal gyrus; AG, angular gyrus; aMTG, anterior middle temporal gyrus; pMTG, posterior middle temporal gyrus. Inset arrows identify the ventral temporal cortex/fusiform gyrus. Adapted from Binder et al. (2009).

involving several frontal and posterior brain regions. We are going to focus our attention here on the temporal and temporal-parietal cortex sites, as these are the regions that are at the center of the debate regarding zones critical for the storage of semantic knowledge, as opposed to executive control in semantic tasks (figure 8.4).

Three broad regions were implicated in the temporal lobe and temporal-parietal cortex. One of them is the "lateral temporal cortex," which basically involves all the middle temporal gyrus. We can partition this zone into anterior and posterior subzones (figure 8.4). The anterior subzone overlaps the general regions implicated in entity knowledge, as revealed by analyses of semantic dementia. The ventral temporal cortex zone is also likely to be part of this same network (Forseth et al., 2018; Visser et al., 2012), although it is unclear precisely what the lateral and ventral regions are doing and how they might differ (Shimotake et al., 2015). The posterior middle temporal gyrus is the zone that we previously associated with lemma-level processes, which is not part of the semantic network according to the analysis that we are developing. Rather, it is a linguistic gateway to the semantic network. It's no surprise that Binder and colleagues (2009) identified the posterior middle temporal gyrus as part of their semantic network, for three reasons: (1) they focused on semantic experiments that were mediated by language, (2) semantic networks and lemma networks are easily conflated, and (3) the research team explicitly included contrasts that would have activated lemma networks, such as comparisons between real words and nonsense words.

Another zone that Binder's team isolated is the temporal-parietal cortex, which, in their analysis, included both the angular gyrus and the posterior supramarginal gyrus (figure 8.4). The real hot spot for semantic function in this zone is the angular gyrus, which had the "most dense concentration of activation foci" (Binder et al., 2009, p. 2774) in their entire analysis. Further, as Binder et al. pointed out, the angular gyrus is connected to other high-level cortical regions with little or no direct connectivity with primary sensory areas, and it is situated at the anatomical nexus of visual, auditory, and somatosensory regions, rendering it perhaps "the single best candidate for a high-level, supramodal integration area in the human brain" and "at the top of a processing hierarchy underlying concept retrieval and conceptual integration" (p. 2776). In a later review of the brain's semantic network, Binder and collaborator Rutvik Desai (2011) reviewed the range of functions that implicated the angular gyrus region and summarized its function as a hub for *event knowledge*, contrasting it with more anterior temporal lobe networks involved in entity knowledge.

Why should entity and event knowledge be represented in different neural networks? Likely because they involve different kinds of associations (Binder, 2016; Mirman, Landrigan, & Britt, 2017). Knowledge that codes for *what* something is, a particular kind of animal or plant or object, will need to be built of features that cooccur across exemplars in the entity's category and that are invariant across entity location, pose, or its dynamic/event state. For example, our knowledge of dogs includes information about their shape, size, color, texture, and sound—features that must be extracted independent of the dog's particular situational context. These features will tend to cooccur frequently as we come across dog after dog in different settings and doing different things (sleeping, playing, fighting, eating, chasing, or walking). You might be thinking: Can't different events or actions also be characteristic of dogness? Surely chasing balls, wagging tails, dousing fire hydrants, and going on walks with people are features that tend to cooccur frequently across instances of dogs. If you see a person emerging around a corner holding a leash that is stretched backward, you will surely predict that a dog will appear at the end of it. And doesn't this illustrate that event knowledge is hopelessly intertwined with entity knowledge?

Intertwined? Yes, for sure. But *hopelessly* so? Not at all. Consider the fact that much of our perceptual experience involves identifying what kinds of things are involved *and* what is happening. Note further that knowing about the things involved does not necessarily tell you what's happening, and vice versa. As I look out the window of my house, I see a person, a house, a ladder, a truck, and a tree. This is useful information on its own. If I needed shelter, I could go to the house. If I needed help, I could ask the person, and so on. If instead I saw a different collection of things—a mountain, a deer, an axe, a canoe, and a rock—other opportunities would present themselves. But there is much more information to be gleaned in the world than just the presence of things. The relation between the items and their actions adds another dimension to my knowledge. If the truck were parked by the tree in front of the house, with the person on the roof inspecting the shingles above the ladder, I could infer that some sort of repair was in progress. If, instead, the ladder was in the truck, which was smashed into the tree, which was leaning on the house, which contained a startled person looking out the window, I would infer an accident. Conversely, I might infer a repair or accident event based on observing a particular relation between a completely different set of entities, such as a bicycle, helmet, child, and toolbox.

Knowing only the entities involved doesn't necessarily tell us much about the event, and knowing the event doesn't necessarily tell us much about the entities involved. This point holds even for simple events and entities that are strongly bound together, such as walking a dog. A girl pacing back and forth in a park while her leashed dog walks freely around is not an instance of walking a dog, even though all the prototypical entities and actions are present. Conversely, a droid rolling down Fifth Avenue in New York with a robotic vacuum tethered to it via an electromagnetic beam would no doubt elicit exclamations like, "Look! R2D2 is walking his pet Roomba!," even though none of the prototypical entities or actions is involved.

Research by Franz Heider and Marianne Simmel from the 1940s took this point a step further, clearly illustrating that we are readily able to grasp the semantics of events without the involvement of meaningful entities at all.

The duo made simple films of triangles, circles, and squares moving around in relation to one another. The motions and their relational dynamics were designed to be characteristic of purposeful actions such as pursuing, attacking, fleeing, hiding, pushing, pulling, and trapping one another. (You can view an example yourself by searching "Heider and Simmel animation" on the web.) Observers readily interpreted the movies in terms of social actions carried out by animate beings (Heider & Simmel, 1944; Ratajska, Brown, & Chabris, 2020), even though only inanimate objects were involved. It's as if actions obligatorily impart semantic features onto the entities involved, even when our knowledge of the entities otherwise would contradict such semantic interpretations.

All of this points to the conclusion that, in general, the kinds of features that are necessary to identify a particular entity versus a particular event are (or can be) very different (Binder, 2016; Mirman et al., 2017). In fact, they are often computationally complementary, with entity identification necessarily ignoring features that vary from situation to situation (pose, spatial relationships, dynamics between entities) and event identification making hefty use of exactly those features that must be ignored in entity identification. The situation is similar to what we encountered in our discussion of a ventral sensory-to-conceptual mapping versus a dorsal sensory-to-motor mapping: different computational tasks require different neural networks. We should not be surprised, then, if mapping sensory features onto entity knowledge involves different networks compared to mapping sensory features onto event knowledge.

This line of reasoning qualifies how we think about the relationship between entity and event networks. While Binder and colleagues (2009, p. 2776) suggest that the angular gyrus event semantics network "occupies a position at the top of a processing hierarchy underlying concept retrieval and conceptual integration," we should not take this to imply that event semantics is building on entity semantics, as if in the processing of an event, step 1 is to identify the entities involved and step 2 is to work out what they are doing. Rather, the foregoing discussion suggests more of a parallel organization, with one system keying on features that are important for entity categorization and the other keying on features important for an event representation. A parallel model makes good sense from an adaptive standpoint too, since an organism needs to be able to respond to a situation rapidly even if the entities involved are unrecognized. You don't want to be standing around wondering what that looming, thundering object is before you recognize the threat of the event and take action.

Beyond the functional imaging data described here, the distinction between entity and event knowledge is consistent with evidence from other sources. For example, the ability to visually recognize objects versus actions double-dissociates following brain injury (Ferreira et al., 1998) and processing taxonomic relations (such as "bear"-"tiger" relevant to entity knowledge) dissociates behaviorally and neurally from processing thematic relations (such as "bear"-"forest" relevant to event or situation knowledge), as reviewed in depth by Daniel Mirman and colleagues (2017). These authors discuss evidence that taxonomic and thematic features make separate contributions to judgments of semantic relatedness between items, come online with different timing, change differently in development, and are differentially disrupted by damage to separate brain regions. The latter finding is particularly relevant to anatomical claims: stroke-induced damage to the anterior temporal lobe tends to result in more taxonomic semantic errors on a picture-naming test (e.g., calling a bear a tiger), while temporal-parietal cortex damage tends to result in more thematic errors (e.g., calling a bear a forest) (Schwartz et al., 2011). This predicts that the semantic difficulty in people with semantic dementia (and anterior temporal lobe dysfunction) should lean more toward taxonomic than thematic semantic problems, which appears to be the case (Mirman et al., 2017; Patterson et al., 2007).

Additional evidence for an anterior temporal lobe entity/taxonomic semantic system and a temporal-parietal cortex event/thematic semantic system comes from studies of linguistic manipulations that correlate with this semantic distinction. For example, a recent study of people with primary progressive aphasia (Lukic et al., 2021) indicates that the ability to produce nouns, which are often more entity related, is more affected by anterior temporal lobe dysfunction compared to verbs, which are more event related. The reverse pattern held for dysfunction of the posterior temporal and temporal-parietal cortex, where verb production is more affected than noun production. Functional imaging studies of the perception of nouns versus verbs show similar effects (Elli, Lane, & Bedny, 2019).

The link between verb processing, event semantics, and temporal-parietal cortex runs even deeper. Verbs have long been studied in terms of their *argument structure*, the linguistic reflection of event structures, which, in a sentence context, denote who is doing what to whom (Williams, 2015). For instance, a verb like "attack" typically involves two arguments playing the *thematic roles* of actor and patient. Different verbs have more or less complex argument structures, ranging from simpler verbs like "sleep," which involve only one argument, as in "*They* slept," to more complex verbs like "give," which involve three arguments, as in "*They* gave *the book* to *her*." Researchers have studied the neural correlates of argument structure complexity and found that the temporal-parietal cortex is involved (Meltzer-Asscher et al., 2013, 2015). This makes sense if this region serves as a semantic hub for event knowledge.

A Distributed but Specialized Network for Semantic Knowledge

Our knowledge of the things in the world and the events that they participate in comprises a big part of what we think about, draw inferences from, and communicate to others. The neural basis of this knowledge and whether it is separable from language systems have been questions that exercised the minds of some of the giants in the field. Major roadblocks to progress include the highly distributed nature of semantic knowledge, its multidimensionality, and the covariation of linguistic communication with semantic knowledge use and access. But thanks to what we've learned from people with selective impairments to semantic memory coupled with a variety of other tools and approaches, clear answers are starting to emerge.

Knowledge of the world is indeed widely distributed, as Wernicke argued almost 150 years ago and Lashley seemed to confirm. But newer research has shown that this far-flung information base coalesces around rather well-localized hubs that integrate the more dispersed auditory, visual, somatosensory, motor, emotional, and other features out of which we compose our conceptual-semantic world. Research to date has identified at least two such hubs: one in the mid-to-anterior part of the temporal lobes that integrates features relevant to entity knowledge, and another in the temporal-parietal cortex that integrates features relevant to event knowledge. These networks are distinguishable from neural systems supporting phonological, lexical, and (as we'll see) syntactic processes, which augments behavioral dissociations, demonstrating that intelligence, whether in fluid or crystalized forms, is separable from linguistic ability.

The picture that I've described here, with an anterior temporal lobe entity semantic hub and a temporal-parietal cortex event semantic hub, is not without its doubters (Humphreys & Lambon Ralph, 2015; Martin, 2016; Persichetti et al., 2021) and is, without question, just a very rough sketch of how semantic knowledge is organized in the brain (Fernandino et al., 2016). Recall that Carolyn Patterson and colleagues (2007) proposed not only the existence of an anterior temporal lobe amodal semantic hub, but also modal spokes that the hub integrates. Additional research, however, suggests two major hubs, each presumably with its own spokes that may themselves interact, creating mini-hubs with varying degrees of multimodal properties and levels of abstraction. A 2016 fMRI study led by Leonardo Fernandino, Jeff Binder, and colleagues suggests just this: a patchwork of modality specific and multimodal mini-hubs representing various combinations of conceptual feature information (sound+motion, color+sound, manipulable+sound, sound+motion+manipulable) (Fernandino et al., 2016). Subsequent work by this group has suggested even in the highest levels of the conceptual system, information about low-level, experience-based features is coded (Fernandino et al., 2022). The hubs and the spokes may not be as simple as the hub-and-spoke model would imply. Instead, the anterior temporal lobe and temporal-parietal cortex hubs are more likely the tops of a bushy, bifurcated hierarchy of modal streams and multimodal subhubs coding perceptual and conceptual knowledge features at various levels of abstraction.

Acknowledging that the neural organization of conceptual knowledge is rather messy might diffuse the debate over whether semantic representations are amodal or have modal tendencies. Maybe it's both, depending on the nature of the semantic category and/or how that information is being used. For example, there is debate over whether the ventral semantic region (shown in the inset in figure 8.4) is part of the amodal anterior temporal lobe hub or a high-level node in the ventral visual stream (Persichetti et al., 2021). Suppose that it can be shown that this region is indeed part of the ventral visual "what" stream, and traces of visual features are present in the neural code for some categories of things, particularly those that are experienced primarily visually (buildings, trees). Suppose too, however, that these representations enable semantic inferences, such as seeing a never-before-encountered tree and inferring that is useful for shade, firewood, lumber, and perhaps fruit (Martin, 2016). That region, then, despite having a modal format of representation, is functioning de facto as a semantic system. Other concepts may be inherently more multimodal—such as *dog*, which many of us experience with all our senses—and involve codes in a more amodal format. Similarly, if the size and shape of an entity is the relevant feature for making an inference (Will the dog fit through my fence?), then perhaps the modal code is used. If more abstract features are needed (What kind of pet should I get?), then more abstract amodal codes are used. The two theoretical camps can eat their cake and have it too.

This way of thinking leans toward an adaptive conception of semantic networks: they are systems that allow an organism to make category- or scenario-based adaptive predictions and inferences about the entities and events in our world. The networks are what allow us to encounter a new instance of some entity and be able to use our knowledge structure to predict the properties, uses, or behavior of the entity and what might happen in the situation, even though we've never seen these particular things or events before. Whether the brain uses modal or amodal abstract codes to do so, while an interesting question, is rather beside the point.

Summary and Implications

Semantic knowledge is a complicated topic that we won't be able to fully work out here. Luckily, we don't have to. A rough sketch is sufficient for our purposes. That sketch includes an entity-related semantic network in the mid-to-anterior temporal lobe and an event-related semantic network in the temporal-parietal cortex. The entity network is organized to some extent around taxonomic features, while the event network is organized around thematic features. Each of these networks likely has some sort of hierarchical structure, although the anatomical details of this organization are not yet clear.

From a linguistic perspective, this network of knowledge—the things and events in our world—comprise much of what we end up talking about. (Emotion is glaringly absent from this discussion and a topic that we'll touch on briefly in chapter 12.) As such, the brain's language systems are going to have to interface with the semantic network, and now that we know roughly where the major hubs live, we can use this information to help assemble the puzzle of the neural architecture of language. It's relevant that the semantic network bookends the linguistic network as we've charted it so far. Phonological and lemma systems occupy the mid-to-posterior part of the superior and middle temporal gyri, respectively. The entity semantic network sits anterior and ventral to these language systems, while the event semantic network is situated posterior to them. This puts the language system in an ideal location for connecting with both networks. It also provides some suggestive clues regarding the regions that might support syntax, the most interesting and contentious linguistic ability and the topic of chapter 9.

9 Telegrams and Sentence Monsters

> [Agrammatic] patients only speak in small, primitive mini-sentences, if they continue to create sentences at all. . . . [In paragrammatic patients,] the spoken expression is not simplified overall, instead, . . . it swells to confused sentence monsters.
> —Karl Kleist, 1914

To many pre-1950s scientists, language was something to be understood in terms of psychological constructs such as motor function, perception, memory, and intelligence rather than hierarchies of structure involving phonemes, syllables, morphemes, phrases, and sentences. Words were the linguistic core of what was articulated, perceived, and remembered. Carl Wernicke's model, for example, comprised two language centers: one for motor word memories and one for auditory word memories. To that, he added only the connection between the two and to the conceptual network. That was it.

But other clinician-scientists noticed early on that there was more to language than words. John Hughlings Jackson was among the first to point this out in the context of aphasia when he wrote the following in 1868:

> It is not enough to say that speech consists of words. It consists of words *referring to [one] another in a particular manner*; and without a proper interrelation of its parts a verbal utterance would be a mere succession of names embodying no proposition. A proposition—e.g., gold is yellow—consists of two names, each of which, by conventional contrivances of position, &c. (called grammatical structure in well-developed languages[1]), modifies the meaning of the other. . . . Loss of speech . . . is the loss of power to propositionize. (Hughlings Jackson, 1868/1915b, pp. 66, 114; emphasis in original)

Three years before Wernicke's influential 1874 monograph was published, Heymann Steinthal (1823–1899), a German philologist, also noticed a feature in the utterances of some people with aphasia that didn't line up with the word-centric approach (Steinthal, 1871). He quotes the halting speech of a patient who was undergoing an eye examination (translated in (Tesak & Code, 2008, p. 73), "'one eye—eye is always—tears—been teary—I can't—I could earlier—especially—of course—with the years little . . . writing—glasses'." Steinthal comments that "his wording betrays, it seems to me, more the inability to make sentences . . . than a bad memory" (p. 73). He called the deficit *acataphasia*, referring to a sentence-level problem that could be distinguished from aphasia, which, for Steinthal, was a word memory problem.

In 1877, German physician Adolf Kussmaul (1822–1902) lauded Steinthal for making the important distinction between aphasia and acataphasia and carried the torch a bit further. He outlined three potential sources of sentence-level production deficits. "The correct diction of a sentence," he wrote, "presupposes three things, and when any one of them is wanting, acataphasia results." (Kussmaul, 1877, p. 795). These were the "unbroken flow of words"; "perfect grammatical diction" (morphology, in modern terms); and (3) the "correct arrangement of words." He provided clinical examples of the breakdown of each and coined new terms to describe the deficits.

The flow of words, he wrote, could be broken by severe aphasia, a word memory problem, in his view. He called this *aphasic acataphasia* and used Steinthal's example described previously to illustrate the characteristic halting, broken speech of a such a case. *Grammatical acataphasia* was a deficit in the use of grammatical morphology. He quoted a single example utterance, presumably from one of his own cases: "Toni flowers got, guardian come, Toni hit" (translated in Tesak & Code, 2008, p. 85). The critical feature was that the "patient no longer made use of articles, pronouns, or auxiliary verbs" (Kussmaul, 1877, p. 796). Interestingly, this example did not come from a case of acquired aphasia, but rather from "a feeble-minded girl," suggesting that instances of what we today call *agrammatic speech* were not well represented in the literature at the time. The halting, nonfluent output of aphasic acataphasia and the halting, agrammatic speech of grammatical acataphasia are familiar to us today, of course, as symptoms of Broca's aphasia.

Finally, Kussmaul dubbed word arrangement deficits *syntactic acataphasia*. He provided two examples. The first came from the dissertation of Ludwig Gogol, who described the case of a young man who suffered a severe head injury after being thrown from a wagon. The patient suffered from more than language problems—Wernicke commented on the case and concluded that it involved both visual agnosia and sensory aphasia (Wernicke, 1874/1969)—and the autopsy revealed a complicated pattern of damage involving posterior temporal-parietal regions, parts of the inferior frontal gyrus (IFG), and part of the right occipital lobe. But none of this concerned Kussmaul, who was laser-focused on the case as "an exquisite example of akataphasia" (1877, p. 797). The patient's speech was apparently fluent (he "lacked neither words nor ideas") but was significantly disordered nonetheless, as the following example shows: "I have as 1869 been a coachman and have a coach which me as I have serve the horses me as coachman with the servant been" (translated from the original German by Tesak & Code, 2008, p. 85). Kussmaul's diagnosis was a combination of grammatical and syntactic akataphasia. Today we would call it *paragrammatism.*

Kussmaul also cited a case that he considered to be a pure example of syntactic acataphasia. It was reported by London psychiatrist Forbes Winslow (1844–1913) in his volume *Obscure Diseases of the Brain and Mind.* Aphasia was not the primary interest of Winslow, who was more famous (arguably infamous, in his later years) for his role as an expert witness in several insanity murder cases, including a bit of dabbling in the case of Jack the Ripper (Winslow, 1911). Nonetheless, Kussmaul considered Winslow's description of one of his cases as a likely example of syntactic acataphasia:

> A gentleman was seized with an attack which presented as much resemblance to epilepsy as it did to apoplexy, and for two days his life was in danger. When he recovered consciousness, he was unable to express his wishes in an intelligible manner. He articulated the words correctly, but mixed them up so that it was impossible to understand him, unless his words were first written down and then co-ordinated. (Kussmaul, 1877, p. 798).

Although Kussmaul identified a range of sentence-level speech production deficits in both fluent and nonfluent cases as subtypes of Steinthal's acataphasia, we remember him today primarily for coining the term *agrammatism.*[2]

Research on grammatical deficits in aphasia gained steam in the early 1900s, at least among German authors such as Arnold Pick, Erich Salomon (who devised a receptive test grammatical ability), and Max Isserlin (Tesak & Code, 2008). But it was the contribution of Karl Kleist (1879–1960) that found its way into current concepts (Kleist, 1914). He boiled sentence-level deficits down to two types: *agrammatism* and (introducing a new term) *paragrammatism.* Here is how he characterized them (translation from Papathanasiou & Coppens, 2017, p. 26): "The basic trait of agrammatism is the simplification and coarsening of word sequences. Complicated compound sentences (subordination of clauses) are not built. The patients only speak in small, primitive mini-sentences, if they continue to create sentences at all. All less necessary words, especially pronouns and particles, are reduced or eliminated."

Aphasiologist Harold Goodglass provides an example from his 1993 book, *Understanding Aphasia* (p. 105):

EXAMINER: Can you tell me about why you came back to the hospital?

PATIENT: Yes . . . eh . . . Monday . . . eh . . . dad . . . Peter Hogan and dad . . . hospital.

In contrast to the "small, primitive mini-sentences" of agrammatic speech, Kleist characterizes paragrammatic speech as "confused sentence monsters": "[In paragrammatism,] the ability to create word orders is not abolished, but phrases and sentences are often wrongly chosen and thereby amalgamate and contaminate each other . . . phrases and sentence constructions are not completed. . . . The spoken expression is not simplified overall, instead, also conditioned by a strong overproduction of word sequences, it swells to confused sentence monsters" (Kleist, 1914, pp. 10–11).

Again, Goodglass (1993, p. 86) provides us with some examples of such "sentence monsters":

> Well, all I know is, somebody is clipping the kreples and some wha, someone here on the kureping arm, . . . why I don't know.
>
> I feel very well. My hearing, writing been doing well. Things that I couldn't hear from. In other words, I used to be able to work cigarettes I didn't know how. . . . Chesterfeela, for 20 years I can write it.

The "mini-sentences" and "sentence monsters" that the early aphasiologists documented in the speech of brain-injured people are the two main types of expressive syntactic symptoms that are generally recognized today: *(expressive) agrammatism*, often associated with nonfluent aphasia, and *paragrammatism*, often associated with fluent aphasia.

Noticing the existence of grammatical deficits in aphasia and the fact that there were at least two types of impairment was a big deal. It raised the possibility that grammatical ability—as opposed to word memory or speech articulation coordination—was supported by a separate system in the brain, and this grammatical system had at least two components corresponding to two symptom patterns and different locations in the brain, those associated with nonfluent versus fluent aphasia. But despite a flurry of work in the first two decades of the 1900s, the torch of discovery barely smoldered until the 1950s, when linguists, psychologists, and aphasiologists picked it up again and started thinking about syntactic impairments in light of linguistic theory.

Aphasia as a Linguistics Problem

The brilliant Russian linguist Roman Jakobson (1896–1982) was among the first to reignite the torch, and in a 1956 publication, he shined its light on a new way to think about aphasia: "If aphasia is a language disturbance . . . then any description and classification of aphasic syndromes must begin with the question of what aspects of language are impaired in the various species of such a disorder" (Jakobson, 1956, p. 55).

He was defining "aphasia as a linguistics problem," to quote his chapter title, and went on to describe two basic operations that applied across linguistic levels: *selection* from among a set of possible units (phonemes, morphemes), and *combination* of those units into groups that form larger units (syllables, sentences). He then classified aphasia into two main types corresponding to deficits involving selection versus combination. (Jakobson called these *similarity disorder* and *contiguity disorder*, respectively, but I will stick with the operation-based terms.) A disorder of selection resulted in word-finding errors and circumlocutions, while a combination disorder resulted in agrammatism.

Jakobson made another move that shaped the future direction of the field: he dropped a strong hint that the classical expressive versus receptive dichotomy had seen its day. He reported that "the question of which of these two operations [selection or combination] is chiefly impaired proves to be of far-reaching significance in describing, analyzing and classifying the diverse forms of aphasia. This dichotomy is perhaps even more suggestive than the classical distinction . . . between emissive and receptive aphasia" (Jakobson, 1956, p. 63)

Accordingly, Jakobson discussed how his two proposed deficit types were realized simultaneously in both expressive and receptive modes, switching the focus of language and brain research (at least for some researchers) away from the sensorimotor systems and toward abstract, amodal linguistic levels. Although he doesn't directly say it, Jakobson links selection deficits with fluent aphasias and combination deficits with nonfluent aphasia. In short, he completely overturned the classical view of aphasia. Gone was any distinction between producing and understanding speech. Broca's and Wernicke's aphasia were not distinguished by the different ways that they affect expressive and receptive systems. Rather, the two types of aphasia were distinguished by the types of amodal linguistic operations that they affected, which would be expected to manifest in both production and comprehension.

In the classical model:

- Broca's aphasia is an expressive problem caused by left inferior frontal damage.
- Wernicke's aphasia is both an expressive and receptive problem caused by left posterior temporal damage.

In the linguistic model:

- Broca's aphasia is a disorder of syntax, caused by left inferior frontal damage.
- Wernicke's aphasia is a disorder of access to lexical-semantic information, caused by left posterior temporal damage.

My own hindsight view of Jakobson's move was that it went too far in dismissing the classical *expressive-receptive asymmetry*, the observation that frontal lobe damage causes mostly expressive dysfunction, whereas temporal lobe damage causes both receptive and expressive dysfunction. We will return to this issue shortly.

Jakobson's approach resonated with the intellectual mood of the day, though. In 1957, just one year after the publication of Jakobson's fresh take on aphasia, a young American linguist, Noam Chomsky, burst on the scene with his book *Syntactic Structures* (Chomsky, 1957), which framed syntax as a generative system of abstract rules, separate from semantics, and which laid the foundation for the audacious idea that some core properties of language were unlearnable and therefore innate. This was followed by Chomsky's highly influential critique (a take-down, really) of B. F. Skinner's (1904–1990) magnum opus on the behaviorist approach to understanding behavior. Skinner's book, titled *Verbal Behavior*, was meant to show how the simple learning rules of behaviorism could explain even the most complex of behaviors: human language (Skinner, 1957). Chomsky used properties of syntax to show why Skinner's proposal didn't hold water (Chomsky, 1959). Sandwiched between Chomsky's 1957 and 1959 essays was a 1958 bombshell study by American psycholinguist Jean Berko, who used her iconic

Wug test to show convincingly that young children generated bits of grammatical forms (such as the plural) on the fly, and without the need for rote learning of each application of the rule (Berko, 1958). These publications propelled the study of syntax into the center of what we now call the *cognitive revolution* (Gardner, 1987; Hickok, 2014b) and elevated work on its biological basis to lofty importance.

In this context, Jakobson's claim that in agrammatism, the "syntactical rules organizing words into higher units are lost" (1956, p. 71) became a critical hypothesis to test. Can we localize syntax, the core faculty of language, in the human brain? Is there a special brain circuit unique to syntax and the basis of our supposed innate, uniquely human ability? Can the study of language in the brain reveal fundamental principles about how the mind works? A lot was at stake.

Harold Goodglass (1920–2002) was the next major torchbearer in this story. He was a young clinician-scientist working at the Boston Veteran's Administration Hospital[3] in the mid-1950s when Jakobson, Chomsky, and Berko came on the scene, and he quickly began designing experiments to test these new ideas, some of it in collaboration with Berko. For example, Goodglass and his colleagues confirmed that people who are diagnosed with agrammatism by clinical impression show a quantitatively reduced syntactic range of utterances, made more omissions and substitutions of grammatical morphemes, and have less word-finding difficulty compared to those without agrammatism (Goodglass & Mayer, 1958). They showed further that reduction of grammatical structure was not merely a reflection of "economy of effort" (Goodglass et al., 1972), which was a prominent hypothesis at the time: speech is hard to articulate, so utterances might be reduced to the bare minimum. Overall, Goodglass's data on grammatical deficits in aphasia "serve to justify Jakobson's choice of 'contiguity disorder' and 'similarity disorder' as a significant basis for a dichotomy of types of aphasia" (Goodglass & Mayer, 1958, p. 111). He also established a relationship between the presence of agrammatism and Broca's aphasia and between the absence of agrammatism and Wernicke's aphasia (Goodglass & Mayer, 1958).

Aphasia was no longer about expressive and receptive functions, but rather about linguistic processes that cut across input/output modalities. Given that syntax was (and probably remains) the most interesting and vexing aspect of our linguistic capacity, agrammatism—the apparent loss of syntactic ability—became the focus of attention, particularly in comparison to paragrammatism, in which the ability to produce rich grammatical structure appeared to be spared, if disorganized.

Broca's Aphasia, Broca's Area, and the Seat of Syntax

One nagging problem with this new linguistic perspective remained, however: agrammatism was thought to be an expressive rather than a receptive disorder. My former PhD advisor, Edgar Zurif, and his colleagues summarized the situation as it was commonly understood in the mid-1970s:

> Routine clinical examinations generally suggest that damage to the frontal area in the left hemisphere—Broca's area—disrupts the ability of patients to produce speech much more than their ability to comprehend. Such patients—Broca's aphasics—often seem to fully understand commands and questions and appear to know what they want to say, yet can only produce short, labored, agrammatic utterances. (Zurif et al., 1976, p. 183)

This contradicted the emerging view that aphasia—and grammatical impairment in particular—was to be understood in linguistic terms, not antiquated input/output dichotomies. Yet here was a seemingly clear example of a grammatical deficit that was conditioned by the assessment task, speaking versus comprehension. Some theorists interpreted this situation as an indication that grammatical knowledge (or *competence,* in Chomskian terms) was not actually impaired in agrammatism, only the ability to put it to use (*performance*) in the service of speaking (Weigl & Bierwisch, 1970). But Zurif and colleagues were conducting experiments that would overturn these long-held beliefs about the comprehension abilities of people with agrammatism.

It was already known in the early 1970s that, despite clinical impressions, comprehension was not completely spared in Broca's aphasia. Italian aphasiologists Domenico Parisi and Luigi Pizzamiglio (1970) used a sentence-to-picture matching task to test the syntactic comprehension abilities of people with aphasia. They tested a range of structures in a large aphasia sample (sixty cases) including twenty-eight Broca's and ten Wernicke's types. They found that comprehension impairment was evident in both groups, not just those with Wernicke's aphasia.

Zurif and colleagues probed deeper into the receptive syntactic abilities of people with Broca's aphasia, specifically targeting the question of whether the deficit in agrammatism is "one of expressive performance systems only; leaving the underlying competence intact" (Zurif, Caramazza, & Myerson,

1972, p. 406). In this study, people with agrammatic Broca's aphasia were presented with written sentences of various structural types and then sets of three printed words drawn from the full sentence (which was still in view); the task was to pick the two words that "go best together." These judgments were then fed into a hierarchical clustering algorithm to generate a kind of subjective phrase structure that was compared against that for control participants. People with Broca's aphasia clustered words differently from the controls, largely ignoring grammatical-function words and instead grouping words based on meaning. For example, given a sentence like "The girl was tall" and the triad "girl—the—tall," patients tended to group "girl" and "tall," reflecting the semantic interpretation, whereas controls grouped "the" and "girl" reflecting the syntactic unit.

A seminal follow-up study (Caramazza & Zurif, 1976)—along with a very similar study by Heilman and Scholes (1976) that appeared at the same time—proved to be the clincher, though, and sparked a line of research that continues today. Caramazza and Zurif specifically looked at the supposedly preserved sentence comprehension ability of people with agrammatic Broca's aphasia, asking participants to listen to a spoken sentence and then point to one of two pictures that matched its meaning. There were two key manipulations. One was whether the correct interpretation could be gleaned simply from the conceptual combination of the meaning of the words in the sentence. For example, a sentence like "The wagon that the horse is pulling is green" is semantically constrained: if you know that the situation involves a wagon, a horse, a pulling action, and something green, it is not hard to infer that the horse is doing the pulling, the wagon is being pulled, and it is the wagon rather than the horse that is green. You don't need to fully parse the sentence to figure out what's happening. However, a sentence like "The man that the woman is following is happy" is not semantically constrained. It's not enough to know that there is a man, a woman, following, and happiness; you need the syntactic structure to know who is following whom and who is happy. The results showed that people with Broca's aphasia did almost as well as control participants on the semantically constrained sentences (about 90 percent correct) but were no better than chance on the unconstrained sentences, which required syntactic parsing to interpret.

The other key manipulation was whether the test sentences followed the typical subject-verb-object word order in English, such as *The horse*$_{subject}$ *is pulling*$_{verb}$ *the boy*$_{object}$ or *The horse*$_{subject}$ *that pulled*$_{verb}$ *the boy*$_{object}$ *is brown*, or whether they deviated from this form, as in *The horse*$_{object}$ *that the boy*$_{subject}$ *pulled*$_{verb}$ *is brown*. When the sentences used canonical subject-verb-object order, people with Broca's aphasia performed well (between 90 and 100 percent), even if the sentences were semantically unconstrained, presumably because they were using a strategy of interpreting the first noun of the sentence as the agent of the action (consider how you would interpret the unstructured sequence, *man–follow–woman*). This strategy fails for noncanonical word-order sentences because in those cases, you have two nouns that precede the verb, making it ambiguous with respect to who is the initiator of the action (*man–woman–follow*).

Caramazza and Zurif (1976, pp. 580–581) summarized the conclusions of their study like this:

> It appears that the presumed dissociation between language production and comprehension does not hold for Broca's . . . aphasics: The present analysis of their comprehension skills suggests that such patients are as impaired in comprehension as they are in production.
>
> . . . For the Broca's aphasics, brain damage affects a general language processing mechanism that subserves the syntactic component of both comprehension and production. The implication that follows is that the anterior language area of the brain is necessary for syntactic-like cognitive operations.

The receptive syntactic deficit uncovered by Caramazza and Zurif was later termed *agrammatic comprehension,* and their hypothesis that expressive and receptive agrammatism had a common source came to be known as the *overarching agrammatism hypothesis*. The latter had profound and long-lasting implications. Here are the key points:

- Syntax and semantics are functionally separable, as Chomsky claimed, and localized to different brain regions.
- Broca's aphasia, or agrammatism more specifically, is a modality-*independent* syntactic deficit that equally affects production and comprehension but spares semantic abilities, consistent with Jakobson's view.
- "Syntactic-like" operations were localized to Broca's area, the lesion location responsible for Broca's aphasia, which is the syndrome associated with agrammatism.
- By contrast, Wernicke's aphasia might be viewed as an overarching semantic deficit evident in the more severe comprehension impairment

and the tendency to produce semantically empty speech or inaccurate word selection.

Finally, it seemed, the new field of *neurolinguistics* could rid itself of naive conceptions of language as a simple word-input and word-output system and view aphasia fully in the light of the abstract knowledge systems and processes of modern linguistics. Aphasia is a central linguistic disorder. To the extent that sensory and motor systems are involved in language processing, they are peripheral, and therefore not a part of the language system proper. After a hundred or so years since Wernicke proposed the first functional anatomical model of language, new and significant progress was made toward understanding how the brain creates language. The picture was coming into focus—or so it seemed.

Complications

Four major complications in the late 1970s and early 1980s confounded this new picture of syntax in the brain.

Complication 1: Agrammatic comprehension is not exclusive to Broca's aphasia.

Caramazza and Zurif (1976) tested not just people with Broca's aphasia, but also people with Wernicke's aphasia and conduction aphasia. Findings from the Wernicke's cases were not terribly surprising: they had moderate difficulty (about 70 percent correct) comprehending even the semantically constrained simpler subject-verb-object sentences, like *The girl is pulling a red wagon*, and significant trouble with the more difficult sentences. What was surprising, though, is the performance of those with conduction aphasia. They showed a pattern that was nearly identical to that found in Broca's aphasia. This is important because the lesion distribution in conduction aphasia involves, not Broca's area, but more posterior areas around the temporal-parietal junction. Could it be that "syntactic operations also rely on cortical regions posterior to Broca's area" (Caramazza and Zurif, 1976, p. 581)? Caramazza and Zurif noted and then rejected this possibility by arguing that conduction aphasia represents a disconnection of Broca's area, thus isolating posterior regions from "an area that subserves syntactic processes." (1976, p. 581). But this interpretation raises a second problem that Caramazza and Zurif did not address. Conduction aphasia is not associated with agrammatic production, so while the study successfully demonstrated parallelism in expressive and receptive syntactic deficits in Broca's aphasia, it revealed a new dissociation between expressive (spared) and receptive (impaired) agrammatism in conduction aphasia. Why is the disconnection of the posterior language system from its anterior syntactic processing area sufficient to cause syntactic deficits during comprehension, but not during production? Despite claims to the contrary, the input-output asymmetry persisted, albeit in a different syndrome.

Complication 2: Broca's aphasia is not caused by damage restricted to Broca's area.

The association between Broca's aphasia and damage to Broca's area had not gone without challenge, particularly in the late nineteenth and early twentieth centuries (Brais, 1992; Foulis, 1879), but by the 1970s, many considered the link solid. A report published in 1978 by neurologist Jay P. Mohr and colleagues, however, shattered this view. It is worth our time to digest every word of the article's abstract:

> The speech disturbance resulting from infarction limited to the Broca area has been delineated; it differs from the speech disorder called Broca aphasia, which results from damage extending far outside the Broca area. Nor does Broca area infarction cause Broca aphasia. The lesions in 20 cases observed since 1972 were documented by autopsy, computerized tomography, or arteriogram; the autopsy records from the Massachusetts General hospital for the past 20 years and the published cases since 1820 were also reviewed. The findings suggest that infarction affecting the Broca area and its immediate environs, even deep into the brain, causes a mutism that is replaced by a rapidly improving dyspraxic and effortful articulation, but that no significant disturbance in language function persists. The more complex syndrome traditionally referred to as Broca aphasia, including Broca's original case, is characterized by protracted mutism, verbal stereotypes, and agrammatism. It is associated with a considerably larger infarct which encompasses the operculum, including the Broca area, insula, and adjacent cerebrum, in the territory supplied by the upper division of the left middle cerebral artery. (Mohr et al., 1978, p. 311)

Let me bullet the key points here:

- Broca's aphasia is not caused by damage restricted to Broca's area.
- Damage to Broca's area does not cause persistent language dysfunction.

Modern research using high-resolution magnetic resonance imaging (MRI) lesion mapping has supported and extended Mohr and colleagues' findings (Gajardo-Vidal et al., 2021). A particularly important finding in the context of syntactic function comes from an assessment of the lesions associated with Broca's *aphasia* by Julius Fridriksson and his team at the University of South

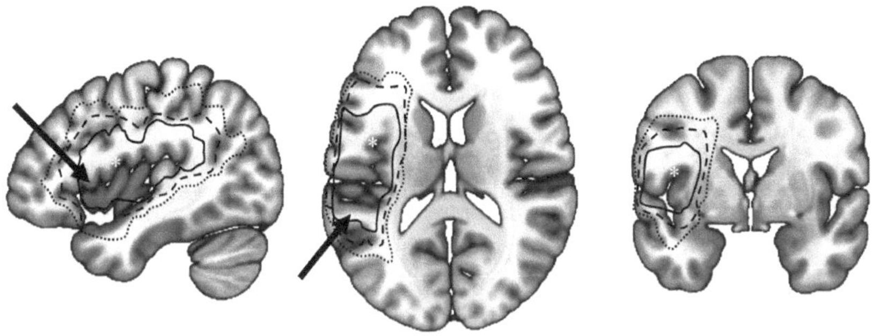

Figure 9.1
Lesion distribution of twenty cases of Broca's aphasia. Outlines show the lesioned area found in nearly all cases (solid lines), a smaller majority of cases (broken lines), and roughly half of cases (dotted lines). The asterisk indicates the point of maximal lesion overlap, which is near the boundary of the precentral gyrus and anterior insula. The arrow in the left image indicates the pars opercularis portion of Broca's area, and the arrow in the middle image indicates the posterior STG, classical Wernicke's area. Redrawn from Fridriksson et al. (2015).

Carolina, who studied seventy people with aphasia, with twenty diagnosed with Broca's aphasia (Fridriksson et al., 2015). Consistent with Mohr's findings, none of the individuals with Broca's aphasia had lesions restricted to Broca's area. Instead, the lesion distribution was much larger, involving frontal, parietal, and temporal lobe areas in nearly all the cases (figure 9.1, solid line). The posterior part of Broca's area, the pars opercularis (figure 9.1, arrow), was included in this lesioned zone, and damage there was statistically associated with the presence of Broca's aphasia, in the sense that patients with Broca's aphasia were significantly more likely to have damage to the pars opercularis compared to people with other types of aphasia. However, there were also several cases in their sample who had damage to the pars opercularis but did not have Broca's aphasia: four individuals had more than 75 percent damage, six had more than 50 percent damage, and twelve had at least 25 percent damage. The team wondered what differentiated those who had damage to the pars opercularis and ended up with Broca's aphasia from those who did not develop Broca's aphasia. The answer was surprising: people with Broca's aphasia also had damage to classical Wernicke's area, the posterior superior temporal gyrus (STG; figure 9.1, black arrow). Their conclusion from this study served as the paper's title: *Chronic Broca's Aphasia Is Caused by Damage to Broca's and Wernicke's Areas*.

Damage to parts of Wernicke's area in the posterior STG is also typical of the lesions associated with conduction aphasia (Buchsbaum et al., 2011; Damasio & Damasio, 1980), which suggests an alternative and very straightforward interpretation of the results of Caramazza and Zurif's seminal 1976 study. People with Broca's aphasia and people with conduction aphasia exhibit similar *receptive* agrammatism deficits because they share a common lesion location involving Wernicke's area. We will consider the nature of this deficit and yet another possible cause later in this chapter.

Complication 3: Grammatical ability is present in people with agrammatism.

As the 1970s gave way to the 1980s, the idea that the syntactic component of language could be selectively damaged and specifically localized to a focal brain region, Broca's area, dominated the thinking and experimental ambitions of many researchers. Among them were Marcia Linebarger, Eleanor Saffran, and Myrna Schwartz, a team in Philadelphia who devised a critical test of the claim that agrammatism represented a selective, amodal disorder of syntax. They reasoned that deficits on a sentence-picture matching task might not reflect a syntactic deficit because a correct response requires processes beyond syntax, such as mapping the structure onto a semantic representation and comparing that to the interpretations of the pictured scenes (Linebarger, Schwartz, & Saffran, 1983). A more direct assay of syntactic ability, they argued, could be achieved with a grammaticality judgment task—listen to a sentence, indicate whether or not it is syntactically well-formed—the bread-and-butter tool of syntacticians. Four people with agrammatism were studied, each with the hallmark, agrammatic comprehension type of difficulty on sentence-picture matching tasks. A range of sentence forms were presented to the participants so they could judge their grammatical acceptability. These included both grammatically correct items (*Is the boy having a good time? How many birds did you see in the park?*) and grammatically ill-formed items (**Is the boy is having a good time? *How many did you see birds in the park?*). Each of the participants performed surprisingly well in discriminating grammatical from ungrammatical sentences, averaging about 90 percent correct overall. The conclusion was clear and stated succinctly in the report: "The comprehension deficit in agrammatism does not reflect loss of the capacity to analyse syntactic structure" (Linebarger et al., 1983, p. 390).

Complication 4: Expressive agrammatism can occur without receptive agrammatism.

The same year that Linebarger and company published their blockbuster findings, another report appeared, which described two interesting cases of expressive agrammatism in Italian-speaking individuals (Miceli et al., 1983). Both patients exhibited classic symptoms of agrammatism in their utterances, yet neither showed difficulty in syntactic comprehension, providing the complementary dissociation to Caramazza and Zurif's conduction aphasia cases. One of the cases proved to have a transient aphasia that resolved fully within a few days. The other had chronic deficits and a computed tomography (CT) scan–documented lesion that largely spared Broca's area but did involve the premotor and motor cortex in the precentral gyrus that extended dorsally to include the posterior middle frontal gyrus. We will return to the involvement of the posterior middle frontal gyrus in syntax-related fundion in chapter 11. For now, the important point is that receptive and expressive agrammatism double-dissociate, which indicates distinguishable neurological sources for the two deficits.

Taken together, these four complications had a disruptive impact on researchers. The tidy picture of agrammatism as a pure, all-encompassing, amodal disorder linked to Broca's area had been fully shattered, and it is fair to say that the field was in a splintered mess by the end of the 1980s. Theories of what is disrupted in agrammatism spanned a wide range. To some, it was a deficit involving just a restricted component of syntax (Grodzinsky, 1986); to others, it was a lexical access problem (Prather et al., 1991; Swinney, Zurif, & Nicol, 1989); and to still others, it was a semantic mapping failure (Linebarger et al., 1983). Some researchers even started seriously questioning the claim that agrammatism was linked to Broca's aphasia (and therefore frontal lobe damage) at all. Summarizing a range of cross-linguistic studies of aphasia, Elizabeth Bates and colleagues wrote that "we have found no evidence to support the idea that Broca's aphasics are unique in the inability to process closed-class [grammatical] morphemes" and argued that the distinction between agrammatism and paragrammatism "has been greatly exaggerated" (Bates, Wulfeck, & MacWhinney, 1991, p. 137). This view was echoed in the handful of papers that even mentioned paragrammatism in that era (e.g., Bates, Friederici, & Wulfeck, 1987b; Butterworth & Howard, 1987; Heeschen & Kolk, 1988). Bates and coauthors went even further, though, noting experiments in people *without* focal brain injuries (including some nonneurological hospital patients on the

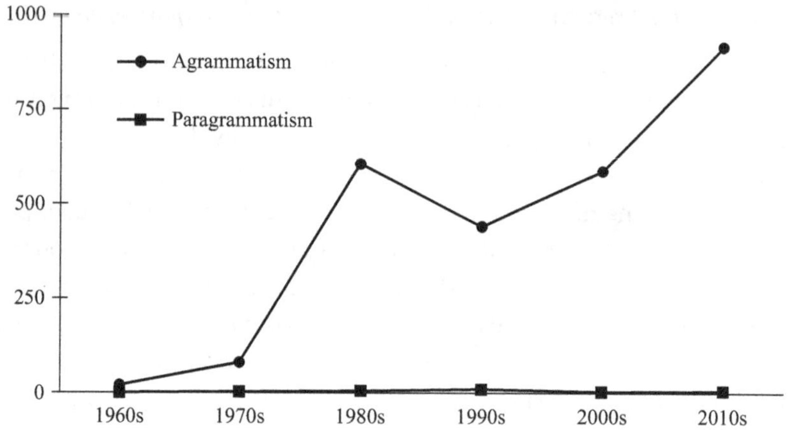

Figure 9.2
Number of scientific papers using the term "agrammatism" versus "paragrammatism" by decade, based on a PubMed search of the two terms. The paragrammatism citations ranged between 0 and 7 per decade.

orthopedic ward!), who were found to show an agrammatic-like pattern of sentence comprehension (Bates, Friederici, & Wulfeck, 1987a).

This was about when I entered the fray, during my doctoral training between 1988 and 1991. Of course, being a student of Edgar Zurif's and hanging out at the Boston V.A. Aphasia Research Center, I was biased toward the view that people with agrammatic Broca's aphasia have a deficit involving some kind of syntax-related process linked to Broca's area. I even published some of my own early experiments and theoretical musings on the topic, arguing for a variant of a more restricted syntactic deficit (Avrutin & Hickok, 1992; Hickok & Avrutin, 1995, 1996; Hickok, Zurif, & Canseco-Gonzalez, 1993). But there were about as many theories of syntactic processing in the brain as there were research groups, and despite my intense interest in syntax, I was pessimistic enough about making progress on its neurology that I decided to give up research in that area and pursue a new focus on sign language aphasia with Ursula Bellugi at the Salk Institute in San Diego. The field that I left behind did seem to agree on one thing, though: agrammatism, not paragrammatism, was the aphasic condition that deserved the most attention in the quest to understand syntax in the brain, as figure 9.2 shows. This move left the field in a theoretically lopsided position, as we'll see soon.

Invasion of the Functional Brain Scanners

And then everything changed. Starting in the late 1980s and early 1990s, functional imaging techniques emerged as a tool to study brain activation patterns noninvasively in healthy people. It wasn't long before the first studies of sentence processing and syntax appeared. Perhaps unsurprisingly, given the history of the study of syntax and the brain, these early functional imaging experiments yielded contradictory results.

One of the very first positron emission tomography (PET) studies reported that a portion of the anterior temporal lobe bilaterally, not Broca's area, was selectively active while participants listened to sentences for comprehension compared to listening to unstructured lists of words (Mazoyer et al., 1993). This should have highlighted syntactic processes, and yet Broca's area was not among the regions implicated. However, another study conducted by a team at Massachusetts General Hospital aimed to specifically target syntactic comprehension as a direct assault on the question of agrammatism and Broca's area (Stromswold et al., 1996). Inspired by the literature on agrammatic comprehension, the manipulation that they used to isolate a critical component of syntactic processing was a comparison of canonical subject-verb-object sentences versus noncanonical object-subject-verb sentences. A contrast between the activation induced by these two types of sentences yielded activation in one brain area: the pars opercularis of Broca's area. The authors interpreted the evidence as "supporting the role of a portion of Broca's area in the assignment of syntactic structure in sentence comprehension, or in operations associated with this process" (Stromswold et al., 1996, p. 471).

The early 1990s also saw some exciting new work on the brain's electrophysiological response to syntactic features of sentences. Using scalp electroencephalography (EEG) recordings while healthy participants listened to sentences containing a range of syntactic and semantic violations, Helen Neville of the Salk Institute and her colleagues reported that syntactic violations elicited a relatively early response, with a more frontal distribution in the left hemisphere compared to semantic violations, which elicited a later, more posterior response (Neville et al., 1991), the previously established N400 (Kutas & Hillyard, 1980). This inspired three conclusions: first, that syntactic and semantic functions during sentence processing where separable, as

Chomsky argued; second, that syntactic operations preceded semantic operations consistent with modular theories of language processing (Fodor, 1983); and third, that syntactic and semantic operations were localized to anterior (think Broca's) and posterior (think Wernicke's) brain regions respectively.

Results like these revitalized the 1970s perspective of a neural organization of language where the major distinction was between syntactic and semantic processes, with the familiar neural epicenter for each. Broca's area was back in the syntax game.

Subsequent functional imaging studies of sentence processing, however, continued to yield a frustratingly mixed bag of support for the relation between syntax and Broca's area. Given Neville's electrophysiological findings on syntactic violations, a reasonable expectation was that Broca's area should be the source of the signal. But, as Angela Friederici of the Max Planck Institute for Human Cognitive and Brain Sciences wrote in a review of early functional magnetic resonance imaging (fMRI) and PET studies on the question, "syntactic violations during on-line processing do not automatically give rise to a specific activation pattern in Broca's area" (Friederici, 2006, p. 208).[4] Other experimental paradigms, however, *did* yield activation in Broca's area, particularly those that manipulated sentence complexity during comprehension; let's call this the *syntactic complexity effect.* Friederici concluded her review of this body of work by indicating that the left IFG did have important roles to play in syntactic processing, specifically that "the frontal operculum [the underbelly of the IFG] appears to come into play when local syntactic structure building is in focus" as indexed by the syntactic violation effects, "and the superior-anterior portion of [Broca's area] seems to support more demanding syntactic processes" (Friederici, 2006, p. 209), indexed by the syntactic complexity effect. In contrast, studies of semantics implicated the left middle temporal gyrus, according to Friederici's evaluation of the data.

Despite the complications that were evident in the stroke aphasia literature, the functional imaging and electrophysiological evidence up through about the mid-2000s was powerful enough (in the minds of at least some researchers) to maintain a view of syntax in the brain that looked a lot like what was proposed in the 1970s. Broca's area and vicinity constituted a key region involved in receptive syntactic processes (and presumably expressive processes as well, although this issue got little attention), while semantic processes were localized primarily to the temporal lobe.

The Syntactic Labyrinth of the Early Twenty-First Century

Despite the neuropsychological complications outlined earlier, Broca's area remained the leading candidate for a syntactic hub in the brain in the first two decades of the twenty-first century. Several new complications and alternative models were starting to emerge, however.

One was the possible role of the anterior temporal lobe in receptive syntactic processing. Mazoyer and colleagues' functional imaging study of receptive language processing had identified this region as the lone area in the brain that was sentence-selective (Mazoyer et al., 1993). Several studies in the following years reported similar results (Friederici, Meyer, & von Cramon, 2000; Schlosser et al., 1998; Stowe et al., 1999), including a couple from my own lab, spearheaded by Colin Humphries (Humphries et al., 2001, 2005). Corroborating lesion data was reported by Nina Dronkers and her team (Dronkers et al., 2004). The anterior temporal lobe seemed like a strong contender to replace Broca's area as a core syntax zone. However, subsequent fMRI experiments (Pallier, Devauchelle, & Dehaene, 2011; Rogalsky & Hickok, 2009), a systematic set of magnetoencephalographic studies by Liina Pylkkänen and colleagues at New York University (Bemis & Pylkkanen, 2011, 2013; Westerlund & Pylkkanen, 2014), and evidence of preserved syntactic ability in people with semantic dementia (see chapter 8) (Mesulam et al., 2015) showed that the anterior temporal region had more to do with combinatorial semantics than basic syntactic processing.

An influential paper by Jared Novick, John Trueswell, and Sharon Thompson-Schill at the University of Pennsylvania argued for a different, nonlinguistic perspective on the role of Broca's area in syntax (Novick, Trueswell, & Thompson-Schill, 2005). They tied together two observations. The first was that the left IFG appeared to play a general role in *cognitive control*, the ability to detect and resolve conflicting bits of information. The second observation was that language processing involves such conflicts, for example, in cases of structural and lexical ambiguities in sentence parsing or in selection from multiple alternative words and phrasings in speaking. The idea, then, is that Broca's area is involved in syntax as part of a more general cognitive operation, cognitive control, on which processing syntax relies. A subsequent fMRI study by this group provided direct evidence for this hypothesis: performance on a nonsyntactic cognitive control task and a syntactic ambiguity resolution task colocalized to Broca's area (January, Trueswell, & Thompson-Schill, 2009).

Yet another perspective invoked phonological short-term memory, which was known to implicate Broca's area (Awh et al., 1996; Smith & Jonides, 1997; Smith, Jonides, & Koeppe, 1996), as a possible explanation of the involvement of the region during complex compared to simple receptive syntactic processing (Caplan & Waters, 1999; Just et al., 1996; Martin, 2003). The idea here is that complex syntactic structures, such as object-subject-verb constructions, are more demanding on working memory than simpler structures (Gibson, 1991, 1998). One means to compensate for this extra mental work while trying to comprehend a long set of boring, out-of-context sentences is to say them back to yourself subvocally while you try to select the appropriate response, or to retain them for a later memory probe. The fact that subvocal rehearsal is a core component of phonological working memory (Baddeley, 1992), which implicates a portion of Broca's area, raised the possibility that the sentence complexity effect in Broca's area had more to do with simple subvocal rehearsal than syntactic computation.

Corianne Rogalsky, William Matchin, and I tested this hypothesis in a combined behavioral and fMRI study that aimed to measure the influence of subvocal rehearsal on performance and activation patterns during sentence comprehension (Rogalsky, Matchin, & Hickok, 2008). We did so by using the *articulatory suppression* paradigm (Baddeley, 1981; Baddeley, Thomson, & Buchanan, 1975), in which participants silently and repeatedly articulated an irrelevant set of syllables (in our case, *ba-da-ga-da*) while performing the task of interest (for us, sentence comprehension). By occupying the articulatory rehearsal mechanism with an irrelevant task, we can take it out of play for the primary task, and thus measure its contribution by comparing it to performing a control irrelevant task (we used finger tapping). In our behavioral experiment, we found that articulatory suppression interfered with comprehension of the more complex object-subject-verb sentences more than the control task, but it had no additional interfering effect on the simpler subject-verb-object sentences. This confirmed that articulatory rehearsal at the phonological level was contributing to comprehension performance on the most complex structures.

We then ran a variant of the task using fMRI to record brain activation. Our participants comprehended more complex object-subject-verb sentences and simpler subject-verb-object sentences either with no secondary task, or while silently articulating *ba-da-ga-da* repeatedly, or while performing the finger tapping task. When there was no secondary task, we found activation in both main subdivisions of Broca's area (pars opercularis and pars triangularis) for the complex sentences compared to the simpler sentences, replicating the standard result. But when articulatory rehearsal was taken out of the equation in the articulatory suppression condition, we found no activation difference between complex and simpler sentences in the posterior part of Broca's area (pars opercularis); only the more anterior activation (pars triangularis) remained. The reverse pattern was found during the finger tapping condition, with the posterior part of Broca's area showing a sentence complexity effect, presumably because articulatory rehearsal was back in play, but the more anterior region showed no difference. It's clear that the receptive sentence complexity effect in the posterior portion of the Broca's area is due to participants using articulatory rehearsal to assist in comprehension.

The status of Broca's area as a syntax region, indeed as a language region more generally, was also challenged by a slew of studies showing involvement of the left IFG in music processing (Hickok et al., 2003; Maess et al., 2001; Patel, 2009), nonlinguistic temporal sequence processing (Clerget et al., 2009; Schubotz & von Cramon, 2004), hand and mouth action imitation (Iacoboni et al., 1999; Leslie, Johnson-Frey, & Grafton, 2004), and more.

All the functional imaging results described so far, though, were based on group-averaged data where each participant's brain scans were transformed into a common template and pooled for statistical analysis. This is a potential problem because brain anatomy and functional organization within that anatomy can vary substantially on the scale of individual sulci and gyri, which means that we lose fine-grained spatial resolution when we average over groups (Petersen et al., 2024). Brad Buchsbaum and I ran into this problem in the early 2000s when we were searching for an auditory-motor interface circuit for speech (which will be discussed at length in chapter 10). The auditory-motor area *Spt* showed up consistently in individual subject fMRI maps but weakened or disappeared in group averages. Could it be that group averaging in fMRI studies of syntactic processing was obscuring more selective networks involved in syntax?

Nancy Kanwisher at the Massachusetts Institute of Technology (MIT) had used individual subject analysis methods for her work in the human visual system, where she had identified a face-selective region in the ventral temporal lobe, the *fusiform face area* (Kanwisher, 2000), although not without controversy (Haxby et al., 2001). In 2011, Evelina Fedorenko and Michael Behr teamed up with Kanwisher to apply similar fMRI methods to the study of "high-level linguistic processing" (Fedorenko, Behr, & Kanwisher, 2011). Their language task involved asking participants to read sentences and then judge whether a subsequent probe word had appeared in the sentence. Activation on this task was then compared to a control condition where participants read a string of nonsense words followed by a nonsense probe word. This contrast defined a broad set of regions of interest (ROIs) in each subject *individually*. Broca's area (or, more accurately, various individually defined subsets thereof) constituted one of the regions. The team then asked: How much activation to nonlinguistic tasks is there in each ROI? They tested the top contenders—music, math, cognitive control, and verbal working memory—and found that the sentence reading task yielded the most activation in each, including Broca's area, suggesting that activation there, when measured in individual participants, is relatively selective for at least some aspect of language.

How do we reconcile the apparent selectivity in Broca's area activation uncovered by the MIT group with the many observations of Broca's area activating in response to nonlanguage stimuli and tasks? A follow-up study addressed this question by showing that in individual participants, part of Broca's area showed a sentence reading–selective response, while other parts showed nonselective responses with strong activation to the nonlinguistic tasks, as well as the language tasks (Fedorenko, Duncan, & Kanwisher, 2012). Fedorenko and colleagues (2012) interpreted this as evidence for adjacent language-selective and domain-general regions lying side by side within Broca's area.

While this is an accurate surface description of the results of the MIT group's experiment, there are other possibilities regarding what might be going on under the hood. For example, it could be that the nonselective areas are actually core language areas that can be coopted for performing nonlanguage tasks, such as when we use subvocal articulation to assist mental arithmetic. Or the reverse could be true, as William Matchin (2018) argued: it could be that some computation (say, that involved in conflict resolution or working memory) is useful for more than one cognitive task (language involves both conflict resolution and working memory), which, for some highly practiced tasks like language, can result in subregions that appear dedicated to that one task via some kind of neural retuning. This is exactly what appears to happen in the case of the *visual word form area,* which is a subzone of the high-level visual cortex that exhibits fairly selective responses to orthographic forms (Dehaene & Cohen, 2011). This is not because we are hardwired for such an area (reading is a relatively new cultural invention), but because some chunk of visual cortex is performing computations that are well suited to visual word form decoding, and with lots of practice, a specialized subnetwork takes shape. In short, interpreting patterns that appear domain-selective versus general is complicated.

Back to the Future: The Reemergence of Lesion Studies of Syntax

Despite early expectations that fMRI could unravel all the conundrums posed by the neuropsychological literature, it turned out that functional imaging had significant interpretation problems of its own. Once this became clear, and once methods were developed to statistically quantify lesion correlates of behavioral deficits (Bates et al., 2003), several groups, including mine, rebooted lesion-based programs to investigate syntax, mostly focusing on sentence comprehension. The results of these experiments yielded a consistent finding: sentence comprehension deficits, involving a range of sentence types, are caused by damage to posterior temporal regions (Dronkers et al., 2004; Magnusdottir et al., 2013; Pillay et al., 2017; Rogalsky et al., 2018; Thothathiri, Kimberg, & Schwartz, 2012). This is not surprising for the standard classical model, given that Wernicke's area is supposed to be a critical node in word comprehension; if you can't comprehend words, you can't comprehend sentences. But recall from part II of this book that word comprehension is minimally impaired following damage to classical Wernicke's area, which means that these sentence comprehension deficits are not merely reflections of a lower-level word comprehension problem. The posterior temporal lobe apparently plays an important role in receptive sentence processing.

Broca's area, in contrast, has not been consistently implicated in sentence comprehension. For example, two studies published in the early 2010s reported conflicting results, with one finding involvement of Broca's area in the comprehension of noncanonical sentences (Magnusdottir et al., 2013) and the other failing to replicate the result with a larger sample (Thothathiri et al., 2012). Neither found evidence for a greater role of Broca's area in comprehending noncanonical versus canonical sentences, as the functional imaging effects would predict.

To help clarify the situation, our team conducted a study, led by Corianne Rogalsky, that approached the question from three angles using two tasks, sentence-picture matching and plausibility judgments, both of which included canonical and noncanonical sentence types (Rogalsky et al., 2018). In our first approach, we identified stroke patients who had lesions to Broca's area, but not the temporal lobe, and asked whether they had any notable deficits in comprehending noncanonical sentences. They did not, performing well above chance level on both tasks. In our second approach, we behaviorally identified cases who showed the typical receptive agrammatism pattern and asked whether their lesions tended to involve Broca's area. They did not. Instead, the lesions associated with the agrammatic comprehension pattern tended to be in the posterior temporal lobe. Finally, we conducted whole-brain correlations between sentence comprehension performance and lesion patterns (separately for each task) to see if Broca's area is implicated in sentence comprehension generally or in noncanonical sentence comprehension in particular. The answer to both questions was no; it was damage to posterior temporal regions that was correlated with sentence comprehension deficits, including noncanonical forms.

These lesion results, together with the four complications enumerated previously, show that Broca's area is playing at most a support role in receptive syntactic processing, perhaps related to situations in which performance is assisted by some form of working memory. There is no convincing causal evidence that it is involved in something more fundamental, such as basic structure building, during comprehension.

The Posterior Lateral Temporal Lobe: The Other Syntactic Hub

While many of us, including me, were obsessing on the role of Broca's area in syntax, the posterior superior temporal sulcus (STS) and middle temporal gyrus in posterior lateral temporal lobe was consistently showing up in both functional imaging and, as we saw, lesion studies of receptive syntax. To illustrate the functional imaging findings, consider a meta-analysis of syntactic processing published by my own group (Hickok, Venezia, & Teghipco, 2023), the results of which are reproduced in figure 9.3. Broca's area is well represented in the activation pattern, as we would expect, but so is the posterior lateral temporal lobe. (We'll tackle that third, more dorsal area in chapter 11.)

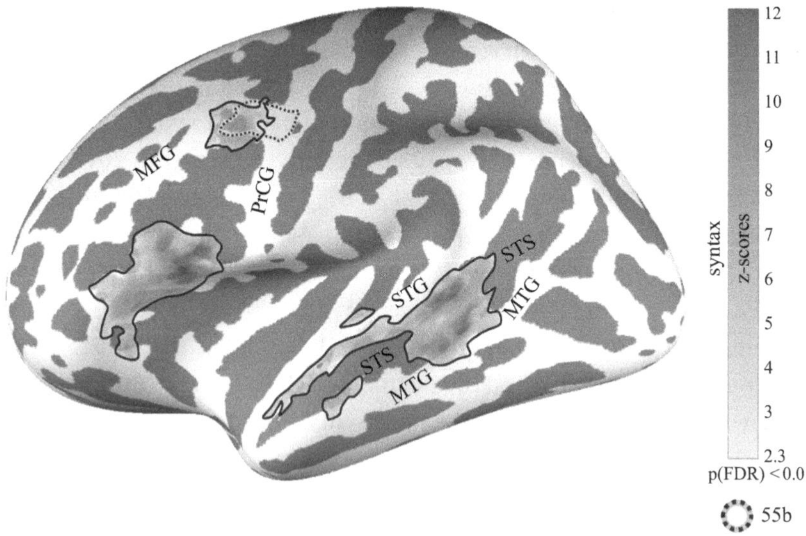

Figure 9.3
Meta-analysis of functional imaging studies using the term "syntax" (neurosynth.org) overlain on an inflated brain. Regions of significant association with syntax are outlined in black. Broca's area and Wernicke's area are both implicated. The posterior part of Broca's area (pars opercularis) and the posterior superior temporal sulcus/middle temporal gyrus have the strongest associations. MFG, middle frontal gyrus; PrCG, precentral gyrus; STG, superior temporal gyrus; STS, superior temporal sulcus; MTG, middle temporal gyrus. Modified from Hickok et al. (2023), with permission. See chapter 11 for discussion of the posterior MFG "syntax" site and area 55b.

A number of leading researchers took notice of the association between syntax and the posterior temporal lobe (Blank et al., 2016; Bornkessel-Schlesewsky & Schlesewsky, 2013; Friederici, Makuuchi, & Bahlmann, 2009; Hagoort, 2003; Matchin, Hammerly, & Lau, 2017; Pallier et al., 2011; Tyler & Marslen-Wilson, 2008; Wilson, Bautista, & McCarron, 2018) and incorporated this zone into their hypotheses about the neural basis of syntactic ability. But my read at the time was that, on balance, the field still put Broca's area on a syntactic pedestal. Even for those who incorporated the posterior temporal lobe in their models, Broca's area remained a core part of their story, including for basic *receptive* syntactic function. For example, Peter Hagoort of the Max Planck Institute for Psycholinguistics proposed that the posterior superior temporal region stored syntactic frames as part of lexical representations, whereas unifying those frames into a full-blown syntactic representation was the purview of the IFG (Hagoort, 2003). Angela Friederici, too, remains committed to a core role for Broca's area in syntactic hierarchy building, with "lexical-semantic information of words stored in the middle temporal gyrus" (Friederici, 2017, p. 225; Zaccarella & Friederici, 2017).

For my part, the lesion studies that had been published in the 2010s had convinced me that Broca's area played a minimal role in receptive syntax, but I had no clear vision of what the right model might be. My eyes were finally opened when my former graduate student, William Matchin, returned to his alma mater and my academic home at the University of California, Irvine to give a lecture on syntax and the brain in which he made a strong case for the role of the posterior temporal lobe in syntax based on the existing literature and some of his own new experiments (Matchin et al., 2017). One of the stumbling blocks for me in accepting a core role for the posterior temporal lobe in syntax was that there was general agreement that the region serves lexical or lemma functions, as reviewed in chapter 7. And according to the 1980s vintage syntactic theories that I grew up with, syntax and the lexicon were two largely different things, the words and rules dichotomy (Pinker, 1991, 1999; Ullman et al., 1997). If a given cortical zone was serving lexical functions, the stumbling block logic goes, it likely isn't a core syntax area except insofar as lexical items encode bits of syntactic information (like grammatical category), which the core syntactic system assembles into a hierarchical configuration (as in Hagoort's proposal).

The linguistic landscape had shifted, however, in the years since I had studied syntax seriously. That shift involved the migration of a good chunk of syntactic machinery to the lexicon. Such lexically heavy grammars had been around for decades (Kaplan & Bresnan, 1982) but started gaining wider acceptance around the turn of the century (Goldberg, 1995; Jackendoff, 2002; Joshi & Schabes, 1997). Ray Jackendoff and Jenny Audring (2016, p. 469) summarized the situation as follows:

> A prominent construct within most linguistic theories has been the distinction between words and rules, or between the lexicon and the grammar, as though they are in different metaphorical "places" in the mind. In parallel, much of *psycho*linguistics has tended to treat storage of words as distinct from rules of grammar. . . . This split is rejected by the Parallel Architecture [(Jackendoff, 2002)], by Construction Grammar [(Goldberg, 1995)], and by Construction Morphology [(Booij, 2010)]. These approaches argue that grammatical rules are themselves lexical items—that is, the grammar is part of the lexicon.

Even Chomsky has shifted more toward an integrated lexical-syntactic system with hierarchical syntactic phrases being built out of lexical features (a "chair" phrase, for example, where "chair" is a particular lexical item with syntactic properties) rather than the traditional abstract syntactic variables (a noun phrase) (Berwick & Chomsky, 2015). Moreover, as I alluded to in chapter 7, some researchers are pushing the idea even further, arguing that what we call the "lemma system" is nothing more than the hierarchical morphosyntactic system itself (Krauska & Lau, 2023). I'm sympathetic to this idea. I find it helpful to think about the middle-word level as a process or computation rather than a set of wordlike representations, where the computation that the middle level performs is the mapping between phonological sound forms and meaning. Perhaps what linguistic research is telling us is that the form-meaning translation requires the computation of hierarchical relations (i.e., morphosyntax).

The neuroarchitectural implication of this shift in linguistic theorizing is significant. It means that our "word search" in chapter 7 was actually an exercise in identifying the syntactic network, at least to a first approximation. And, carrying the logic through, if there is good agreement that the critical region for the middle-word level is the posterior middle temporal gyrus (which there seems to be), then we've got good agreement on the neural basis of syntax. This would be remarkable if true, given all the apparent disagreement.

But what about Broca's area, the historical seat of syntax and a persistent source of squabbling? Even if neural and linguistic theories are converging on the posterior temporal lobe as a syntactic hub, the fact remains: Broca's area is still consistently implicated in some aspects of syntactic processing, as figure 9.3 shows. The breakthrough in my own thinking came with the realization that it is not the classical *versus* the linguistic model, but the classical *and* the linguistic model; both were partially correct. Aphasia *is* a linguistics problem, as Jakobson claimed, but that doesn't mean the classical expressive-receptive asymmetry is wrong. In fact, modern evidence supports the classical expressive-receptive asymmetry in stroke aphasia, as we've seen, as well as in primary progressive aphasia (Lorca-Puls et al., 2024). What's needed, then, is not the abandonment of the classical model, as some have argued (Tremblay & Dick, 2016), but its *integration* with linguistics. This is precisely what William Matchin and I set out to do in the weeks and months after his talk at the University of California, Irvine.

The Hierarchical/Linear Model: A New Neurological Perspective on Syntax

William Matchin and I developed what I'll call here the *hierarchical/linear (HiLine) model*, which partitions the neural organization of syntax into two broad computational functions: *hierarchical structure building* and *linearization*. Both processes are already familiar to linguists (Berwick & Chomsky, 2015; Idsardi & Raimy, 2013). What we added neurologically is to align hierarchical structure building with the posterior middle temporal gyrus, and linearization with the anterior part of Broca's area, the pars triangularis (figure 9.3). We further proposed that the hierarchical structure building component is important for both comprehension and production, while the linearization component is central to production only, playing just a working memory–like support role in comprehension. The HiLine model, then, accounts for expressive-receptive asymmetries, explains why Broca's area shows up in many functional imaging studies of "syntax," and integrates classical aphasiology and modern linguistic insights. I'll unpack some of the details of the model in the next paragraphs.

In chapter 8, I suggested that knowing something about the neural organization of semantics could help inform our search for the neural architecture of syntax. Consider figure 9.4, which shows the anterior middle temporal and angular gyri (key regions in the semantic network); the posterior STG, the auditory phonological area (APA) from chapter 5; and the posterior middle temporal gyrus, the lemma area from chapter 7, which I labeled as the *hierarchical morphosyntactic area* in figure 9.4. Notice how the posterior middle temporal gyrus is perfectly positioned to interface the auditory phonological system to its north with the semantic hub systems to its east and west. This architecture makes sense, given principles of economy in cortical wiring (see the appendix of this book).

Now consider the process of sentence comprehension in terms of basic computational problems that need to be solved. Phonological inputs (sounds, syllables, and phonological word forms) enter the system in a sequential order that is provided "for free" to the listener. By this, I mean that all the system has to do is code the sequence that is already present in the signal;[5] it doesn't have to compute a sequence from a nonsequential signal. What is not given in the signal is the hierarchical structure. This means that a major computational problem for the system that maps sound to meaning is to convert

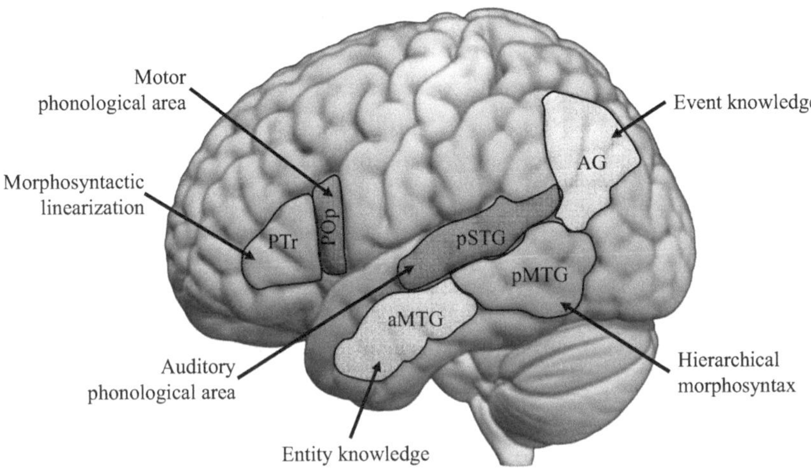

Figure 9.4
Schematic model of the neural architecture of phonological, morphosyntactic, and conceptual knowledge systems. The main components of the HiLine model are the hierarchical morphosyntax area in the pMTG and the morphosyntactic linearization area in the PTr. PTr, pars triangularis of Broca's area; POp, pars opercularis of Broca's area; pSTG, posterior superior temporal gyrus; aMTG, anterior middle temporal gyrus; pMTG, posterior middle temporal gyrus; AG, angular gyrus. Boundaries are approximate. Adapted from Matchin and Hickok (2020).

linear sequences of speech elements into relational hierarchies with no inherent sequence (Berwick & Chomsky, 2015; Idsardi & Raimy, 2013). We saw in the previous section that lesion data implicates the posterior middle temporal gyrus, not Broca's area, in receptive syntactic function, which requires hierarchical representations. The posterior middle temporal gyrus is also well established as the middle-word level, which is increasingly being viewed as a morphosyntactic system. Putting these pieces of the puzzle together, we can conclude that the posterior middle temporal gyrus is critical for deriving the hierarchical relationship between heard sequences of speech elements as the middle stage of the mapping between sound and meaning.

Now consider what is computationally necessary during sentence *production*. Instead of starting with a given sequential phonological input, production starts with a conceptual representation, a message to be communicated (Garrett, 1975), say, the concept of *a child eating some chocolate*. The concept involves a conglomerate representation of entities and events coded in the semantic network. Morphemes (lemmas) that are associated with the entities and events in the message are then selected (Levelt, 1989; Levelt, Roelofs, & Meyer, 1999) and their syntactic features are integrated into a hierarchical representation. It makes sense to hypothesize that the system involved in coding hierarchical relationships in the mapping from sound to meaning is also involved in coding hierarchical relationships in the mapping from meaning to sound, and this is what Matchin and I proposed. Notice that at this hierarchical representational level, there is still no linear, sequential code, just a relational hierarchy. But in order to speak, the hierarchically structured set of morphemes needs to be linearized. This, we argued, is where Broca's area comes in. The pars triangularis, we proposed, is a main hub for coding the linear sequence of morphemes that corresponds to the nonlinear hierarchy. From there, the linear sequence of morphemes can be transformed into a sequence of phonological word forms, syllables, phonemes, and articulatory gestures. That is the basic structure of the model, which has a hierarchical component in the posterior temporal lobe and a linearization component in the inferior frontal lobe.

It's a fairly simple model, but it solves some important problems. For example, the HiLine model can explain the paradox of why Broca's area often activates in fMRI studies during sentence comprehension and yet doesn't seem to contribute much to understanding. Broca's area activates because recoding a heard sentence in linearized form can be helpful as a working memory or attentional crutch, processes that may be recruited in often boring, sometimes stressful functional imaging environments. Indeed, our behavioral study, described previously, indicates that access to verbal working memory improves performance on noncanonical object-subject-verb sentences by about 10 percentage points (Rogalsky, Matchin, & Hickok, 2008). The magnitude of this effect aligns nicely with the observed performance level of around 85 percent on noncanonical sentence-picture matching tasks in people with damage to Broca's area (but sparing the temporal lobe) (Rogalsky et al., 2018) compared to age-matched controls who average 96 percent (Love & Oster, 2002)—an 11-percentage-point difference. These findings support the view that Broca's area is not a critical make-or-break node in the receptive syntactic network; instead, it plays a working memory–related support role that can improve performance on laboratory tasks by about 10 percent. It probably plays less of a role in real-world sentence comprehension, where contextual cues are more readily available.

The HiLine model also proposes a built-in expressive-receptive asymmetry, as classical theorists observed. How strongly does modern evidence support this view? We've discussed evidence that frontal lobe systems do not contribute much to receptive processing compared to temporal lobe regions, which are critical, for both speech perception (chapter 5) and sentence comprehension (this chapter). So the receptive part of the expressive-receptive asymmetry claim stands up to modern evidence. Does the expressive part hold up equally well? Does damage to either the linearization or hierarchical component lead to expressive syntactic deficits? Standard clinical descriptions of aphasia *suggest* that it does. Broca's aphasia typically involves damage to the IFG (in addition to other areas) and is associated with expressive agrammatism, linking frontal lobe systems to expressive syntactic deficits, whereas Wernicke's aphasia typically involves damage to the posterior temporal lobe and is associated with paragrammatism, linking temporal lobe systems to a different kind of expressive syntactic deficit. But we also know that Broca's aphasia involves damage to Wernicke's area too, complicating the picture. Moreover, paragrammatism has received little attention in the literature, and when Matchin and I published our model, its neural basis as a symptom of aphasia had never been directly localized. More direct experimental work is needed to definitively answer the question of whether expressive syntactic deficits are associated with both frontal and temporal lobe regions.

Happily, two groups have studied the neural correlates of expressive agrammatism and confirmed its association with damage to the IFG in both stroke and primary progressive aphasia populations (den Ouden et al., 2019; Wilson et al., 2012). Corroborating these results is a study that used direct cortical stimulation of Broca's area and found that they could induce syntactic production deficits without affecting word production (Chang, Kurteff, & Wilson, 2018). But there were no studies of the neural correlates of paragrammatism, so Matchin and I decided to take a look.

Working with the team at the University of South Carolina, we blindly rated expressive speech samples in fifty-three patients as agrammatic, paragrammatic, or no grammatical deficit. Patients who exhibited features of both agrammatism and paragrammatism were classified as "both" (Matchin et al., 2020). We then checked to see whether people with agrammatism had more damage in Broca's area than those who did not. We also looked at whether people with paragrammatism had more damage to the posterior superior and middle temporal gyri than those who did not. In both

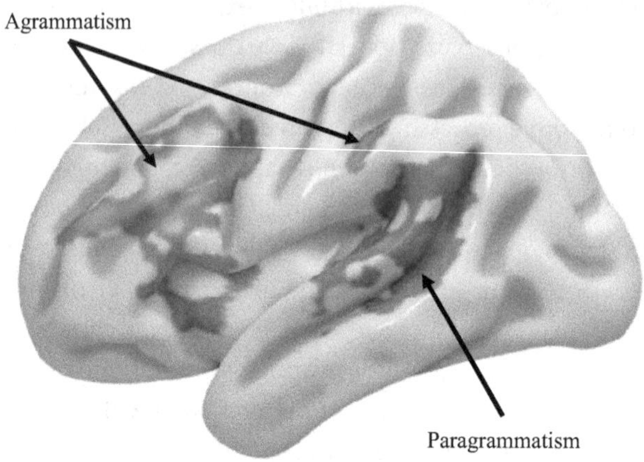

Figure 9.5
Distribution of lesions associated with the symptoms of agrammatism and paragrammatism. The lesions associated with the two conditions are nonoverlapping. Modified from Matchin et al. (2020), with permission.

analyses, we controlled for overall lesion size and the patient's speech rate. Our predictions were confirmed. We then ran a whole-brain analysis to get a wider-angle view of the lesions that were associated with the two types of expressive deficits and found completely nonoverlapping lesion patterns in the frontal (agrammatism) and temporal (paragrammatism) lobes, just as we predicted (figure 9.5). The lesion correlates for agrammatism extended more dorsally than we had expected, however, significantly involving the middle frontal gyrus. This is an important clue, which lines up with the activation in this same region in the functional imaging meta-analysis in figure 9.3. We will take up this issue in chapter 10.

This study did not directly test whether the posterior region that is associated with paragrammatism overlapped the lesions associated with receptive syntactic deficits, although eyeballing the implicated regions suggested that they would. We carried out a follow-up study to confirm the overlap and indeed found a behavioral association between paragrammatism and sentence comprehension deficits, both of which involved damage to the posterior superior and middle temporal gyri (Matchin et al., 2023). We found no behavioral or lesion association between agrammatism and sentence comprehension deficits. In yet another follow-up study led by Danielle Fahey, we mapped the lesions associated with difficulty in making syntactic acceptability judgments, which, you'll remember, is considered by some to be a more direct test of receptive syntactic ability. Again, the lesions were found in the superior temporal lobe (Fahey et al., 2024).

These studies provide strong and direct support for the expressive-receptive asymmetry as proposed in the HiLine model. Frontal lobe systems are implicated in expressive syntactic function in the form of agrammatism but are not causally related to substantial receptive syntactic deficits. In contrast, posterior temporal lobe systems are causally related to both expressive and receptive syntactic deficits. Moreover, this work clearly shows that the old overarching agrammatism hypothesis that linked Broca's area to a central syntactic system is incorrect. If anything is "overarching," it's in the posterior temporal lobe, but it doesn't overarch the entirety of syntactic function, only the hierarchical component (Matchin et al., 2023).

Another benefit of the HiLine model in terms of empirical coverage is that it provides an explanation for the existence and nature of two distinct types of syntactic production deficits, as Kleist observed. The mini-sentences in agrammatism can be understood as reflecting difficulty in translating a hierarchical form into a linear sequence, while the sentence monsters in paragrammatism can be understood as an intact linearization system run amok, generating sequences of morphemes without the constraint provided by a coherent hierarchical representation.

New Questions

New models often raise new questions, and the HiLine model is no exception. Here are a few.

If all sentence comprehension deficits are caused by temporal lobe damage, why do we see differences in severity? Recall that some people, such as those with Broca's aphasia, exhibit an agrammatic comprehension type–receptive deficit, whereas others, such as those with Wernicke's aphasia, have more severe comprehension problems affecting relatively simple sentences (Caramazza & Zurif, 1976), and even single words in some individuals (Baker et al., 1981). Earlier theories had a ready explanation: people with Broca's aphasia and agrammatic comprehension are impaired in syntax, while people with Wernicke's aphasia and severe comprehension problems are impaired in semantics. The modern evidence, however, shows that agrammatic comprehension, more dense sentence comprehension deficits, and even word comprehension problems are all associated with temporal lobe damage. The HiLine model further proposes that the middle-word/morphosyntactic level involves the same posterior middle temporal gyrus zone. What explains the variation in comprehension performance, then?

There are several possibilities that will require further research to fully disentangle. The simplest is that variation in severity is simply a function of how much damage there is to the morphosyntactic network in the posterior middle temporal gyrus region. We know that this will be part of the explanation at least because some of the lesion studies of sentence comprehension ability explicitly calculate the correlation between the amount of damage in an ROI, such as the posterior middle temporal gyrus, and performance on sentence comprehension tasks (Matchin et al., 2023). Thus, the amount of damage to the proposed morphosyntactic system indeed correlates with the severity of impairment. Individual variation in the degree of hemispheric asymmetry in the morphosyntactic system could also play a role. However, the posterior middle temporal gyrus isn't the only region in the posterior temporal lobe associated with sentence comprehension deficits: the lesions also implicate portions of the STG and extend into the temporal-parietal junction (Matchin et al., 2023; Rogalsky et al., 2018). This raises the possibility of other sources contributing to the impairments, which implicates another possible explanation for variation in comprehension ability.

What are these other possible contributors to sentence comprehension performance? We've already mentioned phonological working memory, which, when disrupted behaviorally, can result in worse performance on the types of sentences that cause trouble for those with the agrammatic comprehension pattern. The neural regions associated with phonological working memory and sentence comprehension also overlap in the STG and temporal-parietal junction (see chapter 5), suggesting a likely source of some of the variation. You may also remember from chapter 5 that the phonological network, which supports speech production and phonological working memory, involves at least three subsystems: an auditory-related system in the STG/STS, a motor-related system in the posterior IFG, and an intermediary auditory-motor translation system involving area Spt (further discussed in chapter 10). In this chapter, we are also proposing a syntactic processing architecture involving high-level areas in the temporal and frontal lobes. Might there be some kind of a translation system between these areas at the morphosyntactic level, a higher-level homologue to Spt? And might the existence of such a system be a third source of variation in receptive syntactic function via a syntactic-level production circuit functioning as a working memory crutch for comprehension?

I think the answer to both questions is yes. Regarding the existence of a morphosyntactic interface system, it is relevant that, in 2012, David Gow put forward a *dual lexicon model,* in which he argued for *two* middle-word-level systems: a ventral system in the posterior STS and middle temporal gyrus that maps between phonetic and semantic representations (his terms) and another more dorsal network that maps between phonetic and articulatory representations (Gow, 2012). Now this sounds exactly like the dual stream model that I proposed with David Poeppel. Indeed, the ventral stream component in both models is the same, but Gow's claim about the dorsal stream is different. He argues for a specifically *lexical* (lemma) interface that is localized to the supramarginal gyrus, whereas Poeppel and I proposed a *sublexical* auditory-motor interface, which we localized to area Spt (Hickok & Poeppel, 2007). When Gow's paper came out in 2012, I disagreed that the dorsal stream was specifically lexical (morphosyntactic, in the terms of this chapter), but now I think there is room for both sublexical and morphosyntactic fronto-temporal interface systems in the vicinity of the supramarginal gyrus. Supporting this idea is a 2016 paper that functionally dissected the supramarginal gyrus region using fMRI and reported that the posterior part of the region is more selective for real words than for meaningless words or nonspeech sounds (Oberhuber et al., 2016).

Figure 9.6 shows a schematic of the language system architecture that I'm suggesting. It contains phonological and morphosyntactic levels, each involving posterior and anterior components, and each involving an intermediate interface or transform system. The phonological interface involves area Spt, whereas the morphosyntactic interface involves the posterior supramarginal gyrus. How might this system contribute to variation in sentence comprehension ability in people with aphasia? Just as the network for speech production at the phonological level can be coopted for phonological working memory, the network for speech production at the morphosyntactic level might be coopted for morphosyntactic working memory (see Matchin, 2018, for shades of this suggestion). It is interesting in this respect that the lesions associated with conduction aphasia tend to overlap both Spt and the posterior supramarginal gyrus (Buchsbaum et al., 2011). Perhaps, then, agrammatic comprehension is fundamentally a working memory problem involving a

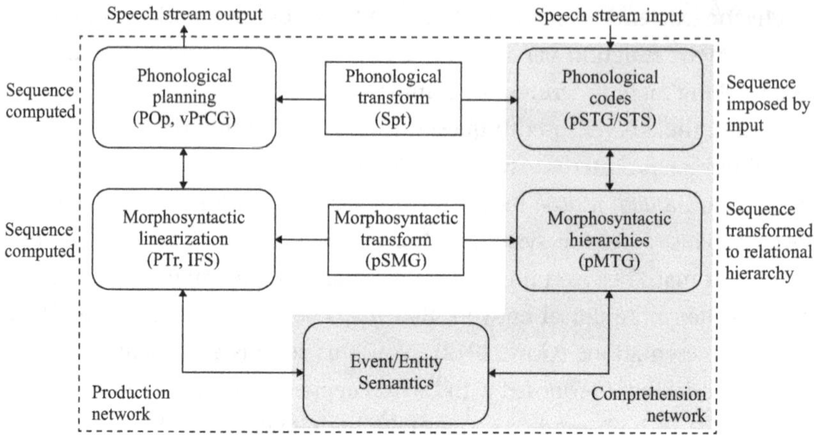

Figure 9.6
Basic architecture of the language system at the phonological and morphosyntactic levels. Both linguistic levels have three components: a posterior sensory-related system that is involved in both comprehension and production, an anterior motor-related system that is mostly involved in production, and an interface system that translates between sensory- and motor-related systems. The shading and dashed lines show the expressive-receptive asymmetry; that is, the comprehension network (shaded) is a subset of the production network (dotted line). Under high processing loads, the three components of each level of processing can be used as a working memory crutch. POp, pars opercularis; vPrCG, ventral precentral gyrus; Spt, Sylvian parietal-temporal area; pSTG/STS, posterior superior temporal gyrus/posterior superior temporal sulcus; PTr, pars triangularis; IFS, inferior frontal sulcus; pSMG, posterior supramarginal gyrus; pMTG, posterior middle temporal gyrus. Expanded upon from Matchin and Hickok (2020).

mix of damage to the phonological and morphosyntactic production systems. The deficit is found in Broca's aphasia because that syndrome tends to have damage to the frontal lobe sequencing (linearization) components of the network, and it is found in conduction aphasia because the damage in that syndrome tends to be in the interface components of the network. Add a dash of damage to the superior temporal lobe, which occurs in both syndromes, and there are now several sources of variation that in some people could mix in just the right way to yield agrammatic comprehension.

Morphosyntactic Conduction Aphasia?

As we considered possible sources of variation in sentence comprehension ability in aphasia, we arrived at the hypothesis that the morphosyntactic level has a three-component architecture, homologous to the phonological level. This architecture makes an interesting prediction for language production: there should be individuals with *morphosyntactic conduction aphasia.*

Recall that conduction aphasia involves spared receptive ability but impaired production at the phonological level—specifically, an increase in the rate of phonological errors in production. This is explained classically and in the dual stream model as a disruption of the interface between auditory- and motor-related phonological codes: receptive function is spared because the entire ventral stream is spared; production is fluent because frontal lobe speech planning systems are intact; but errors occur because the interface system prevents normal auditory-motor interaction (complete details on this are given in the next chapter). If the morphosyntactic level also has a morphosyntactic interface system between the temporal and frontal networks, then people with damage to this interface should have relatively normal sentence comprehension (for sentences or tasks that don't strain working memory too much) and fluent production, but exhibit paragrammatic errors when speaking.

Do such patients exist? This possibility has never been studied, as far as I'm aware, but the available evidence suggests that they do. For example, it has been reported that paragrammatism can occur in people with conduction aphasia (Goodglass, Christiansen, & Gallagher, 1993), a result that our recent study of the lesions underlying paragrammatism has confirmed: we found symptoms of paragrammatism in thirteen out of seventeen cases of conduction aphasia, which amounts to about 76 percent (Matchin et al., 2020). This is telling because, by definition, people with conduction aphasia have relatively spared comprehension, and they tend not to have lesions that extend into the middle temporal gyrus (Buchsbaum et al., 2011). This, in turn, suggests that in at least some of these cases, the cause of paragrammatism is not damage to the hierarchical morphosyntactic system itself but to the morphosyntactic interface system. Moreover, we directly studied the relation between paragrammatism and the comprehension of a variety of sentence types and found a statistical relation between the two, as the HiLine model predicts, but there are also people with paragrammatism who perform very well on sentence comprehension tasks, even for complex sentences (Matchin et al., 2023). This provides preliminary evidence that there are at least two sources of paragrammatism: one involving disruption to the core morphosyntactic system in the posterior middle temporal gyrus, affecting both comprehension and production; and another involving disruption to the morphosyntactic interface system, presumably in the posterior supramarginal gyrus, affecting production more than comprehension. That being said, patients who are paragrammatic in production with normal comprehension might simply have asymmetric lateralization of production (left dominant) and comprehension (more bilateral). This is another area that requires further study to nail down conclusively.

Summary

This chapter summarized research in the late 1800s and early 1900s on syntactic disorders that had identified distinguishable subtypes of grammatical deficits spanning fluent and nonfluent aphasias. We then saw how, in the second half of the twentieth century, researchers became fixated on agrammatism, a presumed overarching deficit of syntax localized to Broca's area. We also saw how several findings seriously challenged this view at every turn. Yet the link between syntax and Broca's area has persisted to this day, partly because of functional imaging studies that show activation in Broca's area during syntactic tasks, and partly because there is a real association between Broca's area and expressive agrammatism. Finally, in the 2000s, researchers started noticing that portions of the temporal lobe were equally if not more important for syntactic function. But no clear picture emerged because, in my opinion at least, there was a failure to recognize the expressive-receptive asymmetry of syntactic function.

I then presented a new model of the neural architecture of syntax that assumes an expressive-receptive asymmetry and attempts to align basic aspects of modern linguistic theory, the distinction between hierarchical processing and linearization, with distinct neural systems. Finally, we discussed some outstanding problems concerning this new model and some possible solutions. One of these solutions suggests a neural architecture for syntax that is homologous to that more firmly established for phonological processing as part of the dorsal stream of the dual stream model. This latter development raises the fascinating hypothesis that the neural organization for language may be homologous to sensorimotor control architectures in general. This is a question that we explore more fully in part IV, where we consider the neural architecture of the talking brain.

IV Talking Brains

10 The Sensory Theory of Speech Production

> Aphasia . . . really, really sucks. Some words . . . I know I know, but . . . I don't know.
> —Carl McIntyre[1]

Speaking, even just a single word, is a deceptively complex task. It's deceptive because it feels so effortless to us; even most eighteen-month-olds can do it pretty well, despite the fact that they won't figure out how to use a toilet for another year or so. Yet speech eludes even the most intelligent and dexterous of our ape cousins and can be desperately impossible for people with aphasia. In this chapter and the next, we will explore the neural architecture behind this remarkable talent.

The Big Picture

Unlike the perception and comprehension of speech, speaking involves all of the brain's language network. Figure 10.1 illustrates this fact. The top row shows the distribution of lesions (outlined) and white-matter connection patterns (stick lines) that are associated with three measures of speech production: *fluency*, a clinical rating of speech production; *articulation rate*, the speech rate after pauses are removed; and *naming*, the percentage correct on a 175-item picture-naming task (Fridriksson et al., 2018). The bottom row shows the distribution of cortical sites that are related to speech arrest and anomia (difficulty producing the name of an object), as determined by direct electrical cortical stimulation (Lu et al., 2021). Modern evidence clearly supports the classical aphasiologists' observation that speech production impairment can arise from both anterior motor-related and posterior sensory-related disfunction (Piai & Eikelboom, 2023).

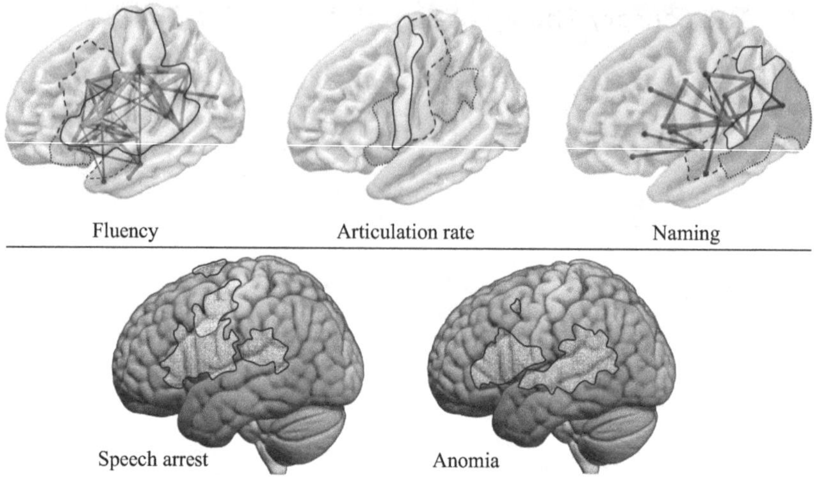

Figure 10.1

Top: Lesion (outlines) and white-matter connection patterns (stick lines) significantly associated with three measures of speech production. The strength of the statistically significant association is indicated by the line type: solid lines indicate the regions with strongest association and dashed lines indicate slightly less strong, followed by the dotted-line zones. Modified from Fridriksson et al. (2018), with permission. Bottom: Distribution of regions where direct cortical stimulation elicits speech arrest or anomia. Subzones indicate distinguishable clusters. Redrawn from Lu et al. (2021).

In addition, the classical theorists noticed that not all regions made the same contribution to speech production. Frontal damage tended to produce more trouble with speech fluency, while posterior damage tended to result in fluent but error-prone speech. Research in the modern era has made tremendous progress in further parceling out the subcomponents of speaking and their neural correlates. This is evident already in the maps shown in figure 10.1, where different speech production tasks or measurements are associated with different lesion distributions. More detailed progress has been made as researchers grounded their neural investigations of speech production in psychological models of how speech is accomplished. These models have emerged out of two rather different traditions: linguistics and motor control. Psycholinguists and motor speech scientists are equally interested in understanding the mechanism of speaking, and both fields have made significant progress. Yet, oddly, the two camps rarely interact (Hickok, 2014a). One goal of this chapter is to show how they can be integrated to provide a better understanding of the speech production system than either has achieved alone. I'll start by describing my own journey into the neuroscience of speech production.

An Auditory Foundation for Speech Production

In the mid-1990s, I had the opportunity to study the neural basis of speech production (picture naming, to be exact) using one of the (then) premier high-field functional magnetic resonance imaging (fMRI) scanners. The 4 Tesla human imaging system at the University of Minnesota's Center for Magnetic Resonance Research was one of only a handful in the world at the time. A few months earlier, I had met the center's director, Kamil Ugurbil, who was giving a talk on some of the advances that he and his team had made in imaging the functioning human brain. We talked briefly about how the brain might be differently organized for spoken versus signed language, my primary research focus at the time, and how interesting it would be to study this with fMRI. One thing led to another, and soon I was in Minneapolis helping to coordinate an fMRI study of picture naming in hearing/speaking participants compared with Deaf native signers of American Sign Language (ASL). We found that naming pictures via speaking versus signing activated similar classical language areas, including Broca's and Wernicke's areas (Kassubek, Hickok, & Erhard, 2004; Petitto et al., 2000). This was highly interesting, of course, given that the two language systems were so different on the surface, but I became obsessed with something else: a tiny region of activation that we observed in the group of hearing/speaking participants. It changed the course of my research focus.

The activation was in the left auditory-related cortex posterior to Heschl's gyrus in the planum temporale (figure 10.2). This was interesting because we had asked participants to name the pictures *silently*; there was no auditory stimulation involved. While the finding was rather surprising to me at the time, it wasn't entirely novel. A positron emission tomography (PET) study on speech production that was published around that time had revealed a similar result, which the authors interpreted as a motor-to-sensory discharge, a kind of "heads up" from the motor system to the auditory system that some self-generated speech input was on its way (Paus et al., 1996). I became interested in an alternative hypothesis, however; namely, that these auditory-related areas were involved in speech *planning* at the phonological level,

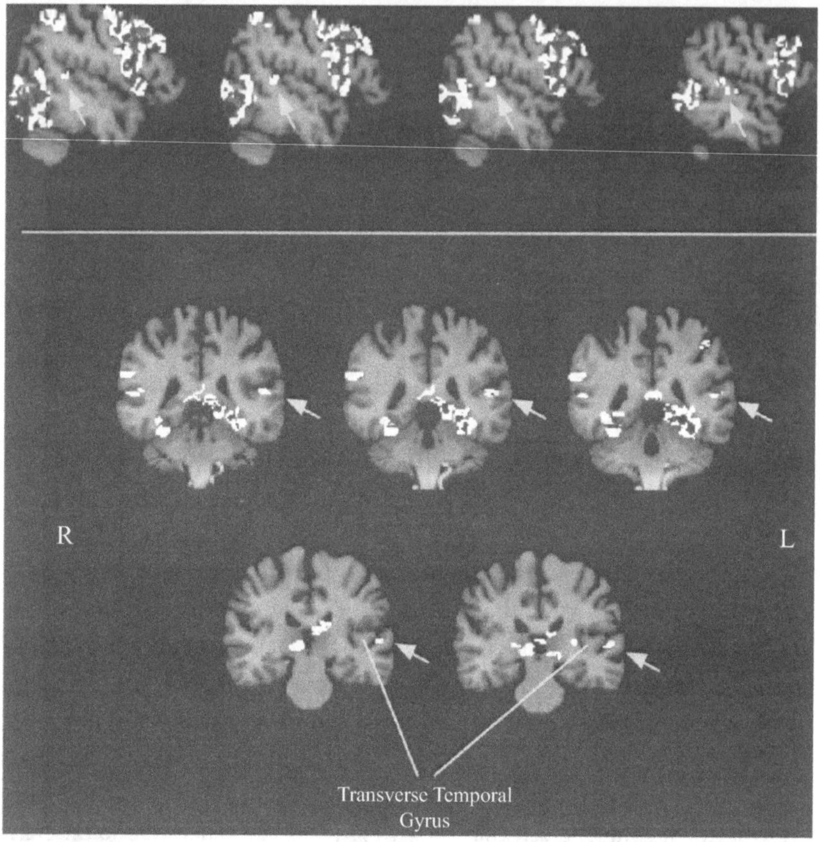

Figure 10.2
An fMRI activation associated with silent picture naming, showing that the left planum temporale is activated (arrows). Modified from Hickok et al. (2000), with permission.

not just low-level feedback from motor speech acts (Guenther, Hampson, & Johnson, 1998). We'll see later in this chapter that motor-to-sensory feedback is one part of the planning process.

Wernicke had hypothesized a role for the auditory system in speech production way back in 1874, albeit not in modern linguistic terms. Writing of the symptoms associated with damage to auditory-related speech areas, he said, "*Apart from his deficient comprehension, the patient has aphasic manifestations in speaking because of the absence of the corrective function exercised unconsciously by the sound images*. These manifestations consist in the tendency readily to confuse words" (Wernicke, 1874/1969, p. 52, italics in original). He was describing the tendency of people with sensory aphasia (later to be called *Wernicke's aphasia*) to produce paraphasic errors in speech. Wernicke noted this symptom also in people with conduction aphasia, where the patient "*can always express himself, but his ability to choose the correct word is disturbed in a manner similar to* [sensory aphasia]." (p. 54, italics in original). The consequence of this failure in conduction aphasia, Wernicke argued, was not due to damage to auditory cortex—after all, he reasoned, their comprehension ability was spared—but to a white-matter disconnection of the auditory speech center from the motor speech center, thereby interrupting the corrective function of the sound images.

Modern evidence on conduction aphasia had refined the picture of what was going on by the 1990s, though. For example, it had become clear that the expressive problems in Wernicke's and conduction aphasia were differentiable, the latter involving a more specific phonological-level production deficit than is found in Wernicke's aphasia (Buckingham, 1992; Goodglass, 1992; Kohn, 1984; Wilshire & McCarthy, 1996). Further evidence had emerged that the damage responsible for conduction aphasia did not primarily involve white-matter tracts and instead implicated auditory-related cortex in the left superior temporal gyrus (STG), at least in some cases (Damasio & Damasio, 1980). Supporting this view was a report that disruptive electrical stimulation of the posterior STG could induce a temporary state of conduction aphasia (Anderson et al., 1999; Quigg & Fountain, 1999). And a magnetoencephalography (MEG) study of picture naming had identified a similar left parietal-temporal junction region in the phonological encoding stage of picture naming (Levelt et al., 1998).

Taken together, these advances suggested a kind of *sensory theory of speech production*—namely, that auditory-related regions in the posterior superior temporal lobe play a role in speech planning at the phonological level—and I became convinced that the activation from our fMRI study had pinpointed the location. The title of our resulting paper reflected this interpretation: "A Functional Magnetic Resonance Imaging Study of the Role of Left Posterior Superior Temporal Gyrus in Speech Production: Implications for the Explanation of Conduction Aphasia" (Hickok et al., 2000). The implications were that auditory-related regions in the STG play a role in speech planning, as Wernicke had proposed, but he was wrong about conduction aphasia: it is not a disconnection syndrome caused by white-matter damage, but rather a cortical disruption to phonological-level systems.

This hypothesis raises a problem, however. If people with conduction aphasia have damage to auditory-related phonological systems, why is their ability to perceive speech not also affected? Wernicke's disconnection model doesn't suffer this complication because it assumes that the phonological system (his "auditory word memories") in the STG are not affected in conduction aphasia. But if the source of the problem is cortical and involves auditory-related systems at the phonological level, then we need to explain the noneffect on receptive speech. There were two potential explanations, one of which we put forward in our report: namely, that speech perception is more bilaterally organized than production, which enables relatively normal speech perception via intact right superior temporal lobe systems. This solution seemed highly probable, given that it was consistent with evidence from word deafness, split brain studies, and investigations into the source of the comprehension problems in Wernicke's aphasia, as discussed in part II of this book. But another possible explanation was that the auditory-related regions involved in speech *planning* and those involved in speech *recognition* at the phonological level were not exactly the same. Perhaps conduction aphasia involved damage to an auditory-related phonological *output* zone but spared the auditory-related phonological *input* zone.[2] Previous studies of speech *perception* had reported activation in the same general vicinity where we found activation during speech *production* (Price et al., 1996; Zatorre et al., 1992), but was it really the same area? Bradley Buchsbaum and Colin Humphries—both graduate students in my lab at the time—performed an fMRI experiment to find out.

We hit on a design that involved participants listening to a set of three multisyllabic nonwords, which they then silently rehearsed for several seconds until a tone cued them to stop. This allowed us to separate the fMRI signal associated with the listening phase of the task from the silent rehearsal phase. We reasoned that if there were an area in the left posterior superior temporal lobe that was involved in phonological aspects of both speech perception and speech production, then we should find that it activated both during listening and during silent rehearsal (our proxy for speech production). This design had the added virtue that it was a kind of speech version of standard visuomotor tasks that were commonly used to define and investigate the primate dorsal stream in vision. A central physiological feature of dorsal stream visuomotor areas was their dual responsiveness to both sensory stimulation (viewing an object, for example) and motor action (reaching for

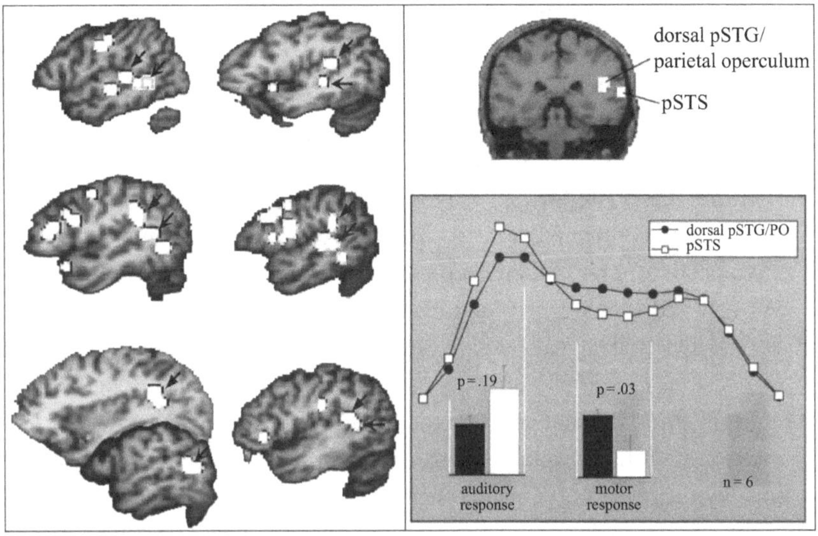

Figure 10.3
Left panel: White blobs indicate locations of fMRI-measured activation to both listening to speech and silently rehearsing it. Filled arrows indicate the more dorsal location in the Sylvian fissure; open arrows indicate the more ventral location in the STS. Each panel shows data from individual participants. Right panel: Brain map shows the location of the two posterior activation sites on a coronal slice in a single participant. The line graph shows the magnitude of activation across the entire 48-second trial; the first 3 seconds is the speech listening phase, followed by 27 seconds of silent rehearsal, and then 18 seconds of rest. The bar graphs represent the mean amplitude with standard error of the response of the two temporal intervals (indicated by the white vertical lines). Modified from Buchsbaum et al. (2001), with permission. pSTS, posterior superior temporal sulcus; pSTG, posterior superior temporal gyrus; PO, parietal operculum.

the object) (Andersen, 1997; Jeannerod et al., 1995). Our task allowed us to look for a homologous dual sensorimotor responsiveness in phonological processing.

We found two posterior superior temporal lobe areas in each of our participants. One was located in the posterior planum temporale region,[3] and the other was more ventral in the posterior superior temporal sulcus (STS; figure 10.3). While both regions responded during both phases of the task, we noticed that the dorsal zone was a bit more strongly weighted toward the rehearsal phase, while the ventral zone leaned more strongly toward the auditory phase (figure 10.3, right-panel graphs). This observation led to a hypothesis: that the posterior STS site "is part of a ventral processing stream involved in processing speech information for the purpose of comprehension, whereas the [more dorsal] . . . site is part of a dorsal processing stream involved in auditory-motor integration in speech" (Buchsbaum et al., 2001, p. 673). A few years later, Bradley Buchsbaum designed a clever experiment that further dissociated the sensory encoding stage from the rehearsal stage of the task and confirmed our early hunch regarding the difference between the two posterior areas (Buchsbaum et al., 2005).

We believed that we had identified two of the components of the dual stream model that David Poeppel and I had proposed: the "sound-based representation of speech" and the "auditory-motor interface" (figure 5.3). We further surmised that conduction aphasia was caused by damage to the interface zone, which explained why perception was unaffected: the cortical *interface* with the motor speech system was damaged (a kind of cortical disconnection), while the perceptual system was left largely unscathed. A brief terminological note: despite our use of terms like *auditory-motor, sensorimotor,* and *sound-based,* which I continue to use here, we assumed that the circuit we uncovered was a phonological-level system because conduction aphasia features a prominent phonological-level deficit in production (Wilshire, 2015).

While our first experiment succeeded in identifying a phonological-level auditory-motor circuit, including a strong candidate for an interface area, several questions remained. One involved the region's speech specificity. Previous experiments had uncovered evidence that perceptual activations in the planum temporale region were not speech specific, which led some to question its role in linguistic, as opposed to lower-level, acoustic processing (Binder et al., 1996; Wise et al., 2001). So we designed a new fMRI experiment that could first serve as a replication of the auditory-motor activation in the posterior planum temporale region, and then assess its selectivity to speech. We decided to use music as our nonspeech comparison because it is a highly familiar auditory signal, and one that humans are quite good at reproducing vocally.[4] Participants listened to simple, novel piano melodies and then silently hummed them back for several seconds, again allowing us to separate the response to acoustic stimulation from that associated with silently humming the tune. For our speech stimuli, we used length-matched nonsense sentences like *There was a glander in my nederop,* which participants heard and then silently rehearsed. We also tweaked the paradigm a little. We first played the sentence or melody; then the participant silently rehearsed or silently

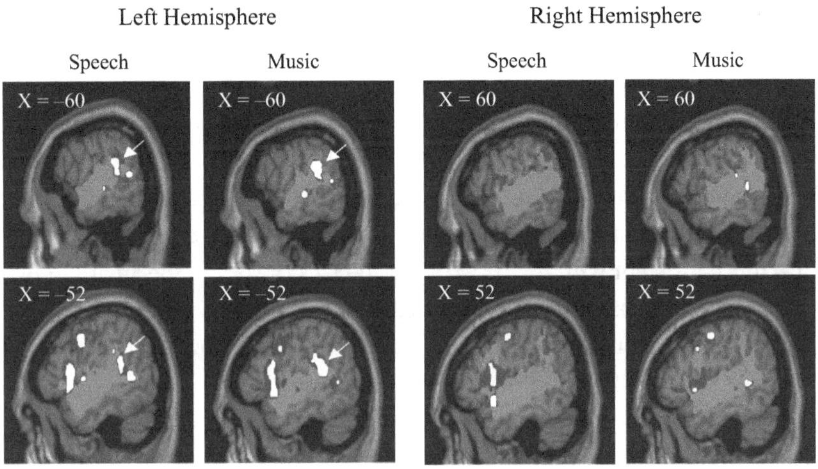

Figure 10.4
Functional magnetic resonance imaging (fMRI) activations in representative slices associated both with listening and silently rehearsing speech or melodies (white blobs). Arrows indicate area Spt. Dark gray blobs, rehearsal-only responses; lighter gray blobs, listen-only responses. Modified from Hickok et al. (2003), with permission.

hummed it for 12 seconds; then they heard the same stimulus again, cuing them to stop rehearsing; and then there was a 12-second rest period. This allowed us to directly compare the activation during the silent rehearsal/hum phase with the equally silent rest phase, both of which followed the same acoustic stimulation; any differences between the rehearse/hum phase and the rest phase, therefore, had to be due to the act of rehearsing or humming.

In the speech condition, we again found auditory-motor activation in both of the posterior superior temporal zones that we observed in our first study, one in the STS and one in the planum temporale region (figure 10.4). Having replicated the involvement of the posterior planum temporale zone in auditory-motor aspects of speech processing, we took the liberty of naming the area. We chose "area Spt," reflecting its location inside the Sylvian fissure at the *p*arietal-*t*emporal junction.[5]

In the music condition, we were rather surprised that we found a remarkably similar network (figure 10.4) that included comparable levels of activation in area Spt and a roughly similar lateralization pattern. There were some differences in the music activations, including slightly different peak activation locations for Spt, defined by the melody condition compared to the speech condition; they were about half a centimeter apart. We also found that when we defined Spt based on the music condition (separately in each participant), speech rehearsal activation in those voxels was not as strong, suggesting some selectivity for humming over speech. The reverse was not true: when we defined Spt using the speech condition, humming-related activation was just as strong as speech rehearsal in those same voxels.

A major conclusion from this study, then, was that Spt is not speech-specific, consistent with earlier work on the lack of speech specificity in planum temporale activation in perceptual studies. Does this mean that Spt and its auditory-motor network are not processing phonological information? Not necessarily. I'll highlight two related reasons.

First, as discussed in chapter 5, although a region's response selectivity for a particular type of stimulus (speech, faces, places) can provide evidence for its involvement in processing that particular type of information, the reverse logic doesn't hold: lack of response specificity in a region for a type of stimulus does not *necessarily* mean that the region isn't involved in processing that type of information. Consider, again, the lungs, larynx, pharynx, velum, tongue, lips, and jaw. All these structures are critically involved in and highly coordinated during speaking, and some of them may have undergone speech-specific adaptation, such as the lowering of the larynx. But they all, together or separately, perform many adaptive nonlinguistic functions.

Second, a system that evolved or developed to perform one function may turn out to be useful for another, different function. Certainly, we didn't evolve fingers to help with arithmetic, but having ten digits on our hands turns out to be helpful for counting. This fact doesn't undermine the theory that fingers evolved for grasping and manipulating objects. In this light, music may be a particularly poor nonlinguistic stimulus to use as a control when looking for speech specificity. This is because phonology includes songlike aspects, specifically prosody and tone. In fact, speech can be so songlike that playing a recording of a spoken phrase on a loop can induce the perception of singing after just one or two repetitions, the *speech-to-song effect* (Deutsch, Henthorn, & Lapidis, 2011), and some researchers have noted important evolutionary connections between features of song and phonology (Fitch, 2019). We will dig deeper into the relation between prosody, song, and Spt in the next chapter. For now, though, the important point is that our music versus speech experiment doesn't rule out an important role for the Spt circuit in phonological processing.

Even though the listen-rehearse paradigms in our early studies identified a *network* of areas showing auditory-motor response properties, including two prominent frontal regions (figures 10.3 and 10.4), I was obsessed with Spt. Just a couple of years earlier, in our first stab at describing the dual stream model, Poeppel and I inferred the existence of an auditory-motor interface for speech based on homology, as you may recall from chapter 5: visuo-motor interface areas had previously been described in the intraparietal sulcus and were argued to perform a coordinate transform between visual, retinotopic space and motor, body-centered space (Andersen, 1997). Speech production, we reasoned, must involve a similar transform between auditory space (the sound pattern of a word) and the motor plans for reproducing that word with the vocal tract (Doupe & Kuhl, 1999). We thought that we had a pretty good idea where the auditory-based phonological-level representations of speech were to be found (posterior superior gyrus, the auditory phonological area) and where the motor speech planning systems were (posterior frontal), but the location of auditory-motor interface system was unclear.

Using functional imaging studies of phonological short-term memory (Jonides et al., 1998)—a kind of auditory-motor task (Wilson, 2001)—we had surmised that the auditory-motor interface was in the inferior parietal lobe. Now, here was Spt, an area with auditory-motor response properties located just below (and, in some people, extending up into the underbelly of) the inferior parietal lobe. I was quite confident that we had found the auditory-motor transformation system of the speech-related dorsal stream. Accordingly, a major focus in my lab for the next several years was to map the response properties of Spt as thoroughly as we could to test this hypothesis.

Mapping Spt

After examining Spt's phonological selectivity, the next major question that we sought to answer was whether the area is motor-effector selective. Up to this point, our Spt studies involved vocal-related actions (namely, speaking and humming). Would Spt be less active if an auditory signal were transformed not into a vocal action, but a manual action? This was theoretically relevant because we had hypothesized that Spt is the vocal homologue to visuomotor interface areas in the intraparietal sulcus, which show some degree of selectivity to motor effectors. If Spt also exhibited a similar selectivity for vocal actions, this would lend support to the idea that it is a homologue to the visuomotor systems.

To test this, Judy Pa, then one of my graduate students, recruited skilled pianists from the University of California, Irvine music department to participate in an fMRI experiment (Pa & Hickok, 2008). The subjects listened to the same melodies that we had used before and then either imagined humming the melodies or imagined playing them on a keyboard. (We had confirmed in the lab prior to fMRI scanning that the pianists could successfully reproduce the melodies on an actual keyboard without error.) We successfully replicated Spt activation in response to listening to and covertly humming melodies, but we found significantly reduced Spt activity during the listen-play condition. A more dorsal parietal area showed the reverse pattern. Spt, like the intraparietal visuomotor interface areas, appeared to be motor-effector selective.

The hum-versus-play-study result was important for another reason. One critique of our interpretation of Spt in the first two studies was the alternative possibility that Spt activity during the silent rehearsal phase of our tasks was "merely auditory imagery," as one reviewer put it. On this view, Spt was fundamentally an auditory area, which could explain why it activated during listening. During rehearsal, according to this theory, Spt's activity simply reflected the "inner voice" echo of the sound associated with the material being rehearsed. But in the hum versus play study, the auditory stimulation was the same in both conditions, and one could argue that skilled pianists should have very strong auditory imagery of the sound echo associated with playing their chosen instrument. Yet we found a notably attenuated response during imagined playing compared to imagined humming.

Other evidence provided further arguments against the imagery explanation. For example, the cytoarchitectonics—the cellular layering pattern in cortex, which varies from region to region—indicated that Spt's location was most likely aligned with cytoarchitectonic area *Tpt* (*T*emporal *p*arietal-*t*emporal), which was not even part of auditory cortex proper and has a cellular architecture like that found in the pars opercularis of Broca's area (Galaburda & Sanides, 1980; Galaburda, 1982; Hickok & Saberi, 2012).

In addition, we conducted an fMRI study aimed specifically at trying to measure the joint auditory and motor responsivity of Spt, as opposed to the auditory-only responses predicted by the imagery account (Hickok, Okada, & Serences, 2009). To do this, we first localized Spt and then, using independent data, evaluated the pattern of activity within it during the listen versus rehearse phase of the task, looking to see if we could decode perception versus rehearsal from the activation pattern after controlling for overall amplitude.

The underlying neurophysiology behind this test is as follows: We know that visuomotor areas in the intraparietal sulcus contain different classes of neurons, some more sensory-weighted (they respond more during perception) and some more motor-weighted (responding more during action). If Spt is a homologue to these areas, it should also have cells that are more auditory- versus motor-weighted. It's true that fMRI can't measure individual neuron activity, but it can pick up on subtle differences at the voxel level (representing 1–2-mm cubes of brain tissue), assuming that the distribution of the various cell types isn't perfectly uniform across voxels. The listen phase of the task should drive more of the auditory-weighted units, while the rehearsal phase of the task should drive the more motor-weighted units. We used a machine learning algorithm applied over the pattern of activity across the pool of Spt voxels during the listen versus rehearse phase of the task to see if different patterns were present, reflecting differential activation of auditory- versus motor-weighted neuron activity. We found strong evidence for different patterns (figure 10.5).

Finally, in a joint project with Bradley Buchsbaum, Nina Dronkers, and Juliana Baldo, we sought to determine whether the lesions associated with conduction aphasia likely involved area Spt. The reason why we were interested in this possibility is that, as noted, our thinking regarding the computational source of conduction aphasia had evolved. My original thought was that it was a consequence of damage to phonological networks in auditory-related cortex, the same networks involved in speech perception. But after the discovery of area Spt, it seemed more likely that damage to the interface between auditory and motor speech systems was the culprit. This preserved the spirit and explanatory benefits of Wernicke's disconnection hypothesis, while acknowledging that it is a cortical disruption disorder. Furthermore, providing evidence that damage to Spt caused a phonological-level deficit in speech planning would add some causal support for our hypothesis regarding Spt function (and against the imagery hypothesis). We assembled data from over 100 fMRI participants in studies of Spt from both my lab and Buchsbaum's and aligned those activations with the lesion distribution of several cases of conduction aphasia from Dronkers's lab. We found that they lined up quite nicely (figure 10.6, top row) (Buchsbaum et al., 2011).

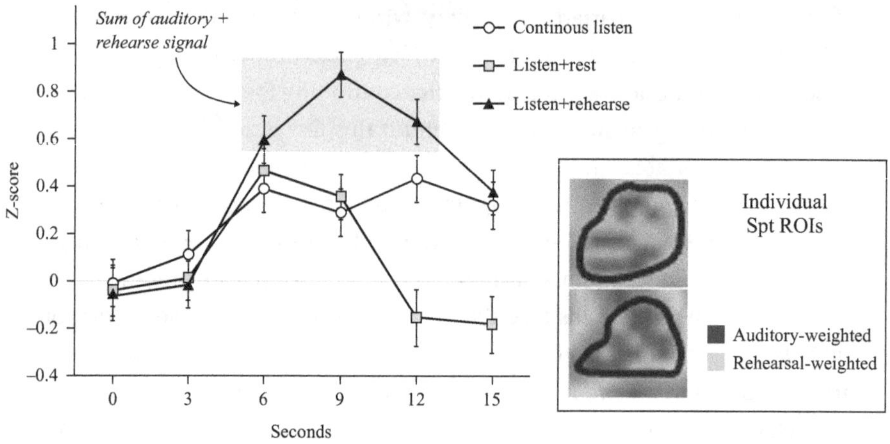

Figure 10.5
The line graph shows the average fMRI activation amplitude in individually defined Spt regions of interest (ROIs) for three task conditions as a function of time. "Continuous listen" is listening to speech for the entire 15-s trial; "Listen+rest" is listening to speech for 3 s, and then resting for the remainder of the trial (scanner noise only); "Listen+rehearse" is listening to speech for 3 s, and then silently rehearsing the speech for the remainder of the trial. Notice that in the two listen-only conditions, the amplitude maximum (Z ~ 0.4) is roughly half of the amplitude maximum for the Listen+rehearse condition (gray-shaded area), which then decays toward the end of the trial (final triangle point). This is explainable if there are distinguishable auditory-weighted versus rehearsal motor-weighted neurons in Spt, which sum in the middle of the trial before the auditory signal decays at the end of the trial. The images show individual Spt ROIs from two participants. The dark lines in the images define a portion of the Spt ROI on a single sagittal slice. Darker and lighter gray-shaded regions within the ROI identify auditory- versus rehearsal-weighted voxels showing interdigitated subzones of Spt. Redrawn from Hickok et al. (2009).

In a subsequent larger-scale lesion-symptom mapping study spearheaded by Corianne Rogalsky, we zeroed in on speech repetition ability, a task that has proven particularly difficult for people with conduction aphasia and that seems to drive Spt activity in healthy people. We found that the damage associated with speech repetition deficits indeed involves area Spt (crosshairs in figure 10.6); see also Baldo, Katseff, and Dronkers (2011) and Dell et al. (2013) for similar results. Further analyses revealed that the Spt region was implicated even when receptive speech ability and white-matter damage was controlled for in the analysis, confirming a cortical locus for repetition deficits that are not attributable to perceptual difficulty (Rogalsky et al., 2015).

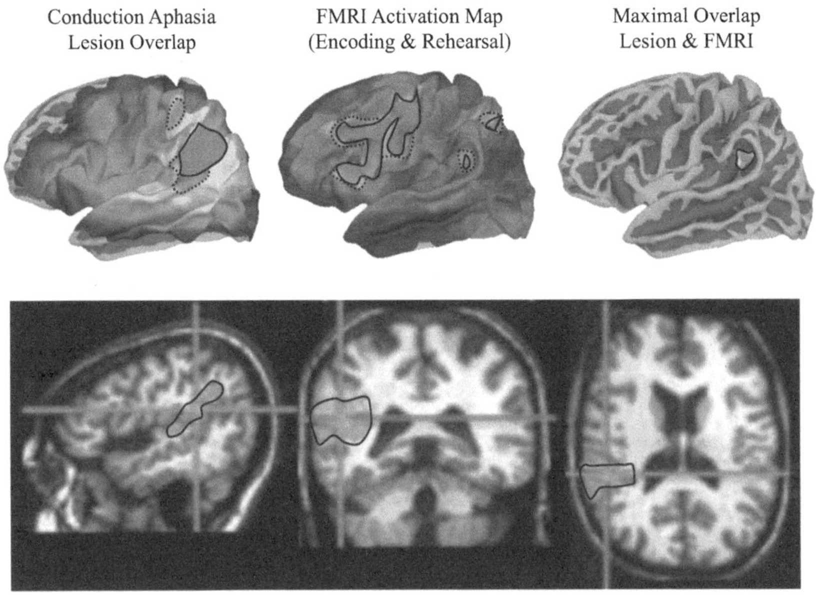

Figure 10.6
Top row: Left, overlap map of the lesion distribution in fourteen cases of conduction aphasia. Solid line shows the region of maximal overlap. Middle, fMRI activation map aggregated from 105 participants in several Spt studies. Solid line indicates maximal overlap. Right, solid line indicates the maximal overlap between the two datasets. Modified from Buchsbaum et al. (2011), with permission. Bottom row: Lesion-symptom mapping study showing lesion distribution associated with deficits in speech repetition (black outlines). Crosshairs indicate the average location of Spt from functional imaging studies. Modified from Rogalsky et al. (2015), with permission.

After almost fifteen years of work, we had reached a point where I felt that Spt function had been reasonably well characterized as an auditory-motor interface zone involved in transforming auditory-based signals into vocal actions at the phonological level. But a core question remained: What is the computational nature of this interface?

Computational Accounts of Speech Production

Work on the Spt circuit provided a neuroscientific characterization of the functional architecture of the auditory dorsal stream that Poeppel and I had hypothesized. It involved a constellation of areas that exhibited auditory and motor response properties in the listen-rehearse task: the posterior

STG, which stores auditory-based representations of speech (and melodic/prosodic information); the inferior frontal gyrus (IFG) and the dorsolateral premotor cortex, which stores motor-based representations; and Spt, which serves as an interface between the auditory and motor systems, translating between the two. But this was a coarse functional characterization in want of some detailed understanding of what it means to "translate between the two."

There were two directions that one could look when thinking about what might be going on computationally in the dorsal stream. One was toward psycholinguistic models of speech production, and the other was toward motor control models. I looked both ways.

The Psycholinguistic Approach

The psycholinguistic approach starts with the question of how we get from the message that we want to communicate to a sequence of phonemes that codes the intended idea. As we've already seen in chapters 7 and 9, going from concept to speech sound is not a one-step process, but rather a series of steps encoding a hierarchy of representational levels involving morpho-syntactic and phonological features (Levelt, 1989; Roelofs & Ferreira, 2019). Applying this kind of model to the neural systems that support it, the goal becomes to find the set of subregions or subnetworks of the speech production system that correspond to these various hierarchical levels and processes. This has been the default assumption of neurolinguistic research since the time of Roman Jakobson (see chapter 9), and we already have seen plenty of evidence to support the idea that different linguistic levels map to distinguishable areas or networks.

In chapter 7, I introduced Indefrey and Levelt's classic meta-analysis of functional imaging studies on speech processing, a discussion that focused on their localization of the lemma level to the posterior middle temporal gyrus (Indefrey & Levelt, 2004; also see Indefrey, 2011). Their investigation included several stages of speech production, which are now relevant to this chapter. Figure 10.7 presents a simplified version of their model, highlighting the psycholinguistic levels that they localized in the brain. The stage labels on the brain image indicate the rough location of these processes, which include a lemma retrieval and selection system in the middle temporal gyrus, a phonological retrieval system in the posterior superior and

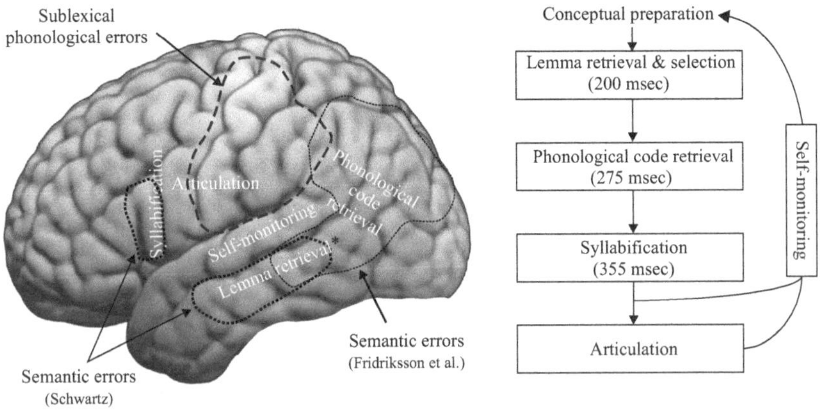

Figure 10.7
A simplified depiction of the Indefrey and Levelt (2004) neurolinguistic model of speech production. The box and arrow diagram shows the major stages of the processing architecture, with estimated timing information associated with each stage. The neural correlates of each stage is indicated roughly by the location of the text labels on the brain image. Also shown on the brain image are the rough location of lesions associated with semantic errors (dotted lines, from (Fridriksson et al., 2018; Schwartz, 2014)) versus nonword phonological errors (dashed line) in aphasia, from Schwartz (2014). Finally, the black asterisk shows the approximate peak location of lesions associated with the semantic interference effect in naming, from Harvey and Schnur (2015).

middle temporal gyri, a syllabification system in the posterior part of Broca's area, and an articulation system in the inferior precentral and postcentral gyri. They also posited both an internal (prearticulation) and external (post-articulation) *self-monitoring loop* that engages the STG and partially overlaps the phonological retrieval system in the posterior STG that is in the vicinity of the area Spt. It is worth pointing out that their localization of the lemma network is a bit more anterior in the middle temporal gyrus than the location that we found in chapter 7 and 9.

Myrna Schwartz, Gary Dell, and colleagues also used the psycholinguistic approach to study word production, but in the context of the lesion method (Dell et al., 2013; Mirman et al., 2015; Schwartz, 2014; Schwartz et al., 2009, 2012; Walker et al., 2011). One set of studies looked at picture-naming ability in aphasia and mapped the brain regions that, when damaged, resulted in different types of naming errors. This approach of using error types to divvy up the computational neural networks underlying speech production relies on the assumption that different error types emerge from disruption of different components of the speech production process, an idea that has a long history of productive research in psycholinguistics (e.g., Fromkin, 1971). It proved useful in neuroscience as well: Schwartz, Dell, and the rest of the team found that different lesion distributions were associated with the tendency to make nonword phonological errors (such as saying "cag" for "cat"), which implicated a dorsal fronto-parietal-temporal network, versus semantic errors (such as saying "dog" for "cat"), which implicated a more ventral middle temporal area, along with a part of Broca's area.

Another study, one that I was involved in, also looked at the lesions associated with semantic errors and found partial overlap in the middle temporal gyrus, but with a posterior extension (Fridriksson et al., 2018). Yet another lesion study used psycholinguistic methods to isolate lexical (lemma) access via the *semantic interference effect*, which is known to be exaggerated in some people with aphasia. The effect was localized to the posterior middle temporal gyrus (Harvey & Schnur, 2015). Figure 10.7 summarizes these findings.

Although the picture is not entirely in focus, the psycholinguistic approach has made progress, identifying at least one major distinction between sublexical processing in more dorsal parts of the production network and lexical-semantic processing in more ventral parts. Notice, though, that the psycholinguistically based neural models tend to emphasize a one-to-one alignment of brain regions with psycholinguistic processing stages. This differs from the stance that I've adopted in thinking about the dorsal stream, Spt, and phonological processing. From the start, my approach to mapping the dorsal stream was inspired by homology to visuomotor dorsal stream circuits, which are argued to perform a kind of coordinate transform between sensory- and motor-related representations of speech, and therefore to involve (minimally) a three-component architecture: a sensory system, a motor system, and a translation system. Yet, at the same time, I've suggested that the auditory-Spt-motor circuit that we've identified *is* the phonological-level system for speech production. Is this suggestion valid? I think so, for three reasons.

The first is that the distribution of the auditory-Spt-motor circuit overlaps regions that the psycholinguistic approach has identified as phonological, including the posterior STS (phonological code retrieval), the planum temporale (implicated in nonword phonological errors following stroke), and the posterior inferior frontal lobe (syllabification) (figure 10.7). Second, people with condition aphasia, which seems to involve a phonological production deficit (Wilshire, 2015; Wilshire & McCarthy, 1996), have lesions that overlap area Spt (Buchsbaum et al., 2011), as described earlier. Third, functional activation and lesion distribution patterns associated with psycholinguistically localized phonological processing implicate sensory-related (posterior temporal-parietal) and motor-related (Broca's area, premotor cortex) areas. You can see this in Indefrey and Levelt's model with a posterior area involved in phonological code retrieval and a subregion of Broca's area involved in syllabification, and you can see this in the lesion data with sublexical phonological errors associated with damage to the precentral gyrus, a motor area, and several sensory-related areas including the postcentral gyrus, the inferior anterior parietal lobe (secondary somatosensory cortex), and the posterior planum temporale (Schwartz et al., 2012). Yet unpublished work by Grant Walker and our team has also demonstrated that a substantial chunk of the STG (not just the planum) is implicated in *lexical* phonological processing (e.g., real-word phonological errors as opposed to nonword phonological errors). In other words, the neural architecture of phonology looks a lot like sensorimotor control architectures in general.

It was this observation—that the neural network involved in phonological processing appeared to be both an abstract linguistic system as uncovered by psycholinguistic research *and* exhibited sensorimotor-like properties in its neural architecture—that turned me on to research on motor control approaches to speech production. I wanted to know whether there were commonalities with the psycholinguistic approach and whether an integrated approach might lead to new insights.

The Motor Control Approach

The motor control approach to understanding speech production takes as its starting point the motor speech organs in the vocal tract and asks how these are controlled to produce the target speech sounds. A fundamental question of this approach is: What is the reference frame or goal for planning a speech act? Is it a particular configuration of the vocal tract (a motor goal), or is it a particular sound pattern (a sensory goal)? Frank Guenther and colleagues (1998, p. 611) stated the problem this way:

> When the articulators are moved to produce a speech sound, is the sole invariant goal of the movements an auditory goal (i.e., a goal defined by parameters of the acoustic signal as transduced by the auditory system . . .) or does the brain

> effectively equate the auditory goal to a motoric goal, such as an intended vocal tract shape or target set of vocal tract constrictions, that serves as the target of movement?

Guenther and colleagues then reviewed a wide range of evidence coupled with computer simulations of speech production, concluding that the articulation system plans speech in auditory space. The target, the ultimate goal of speech production planning is a specific *sound* pattern, not a specific vocal tract shape.

What's the evidence for this claim, and why does it matter? The evidence is wide-ranging, but you can get an intuitive grasp of a key observation by putting a pencil crosswise in your mouth, biting down on it with your front teeth, and then rereading this sentence out loud. Chances are that immobilizing your jaw and obstructing your ability to bring your lips together had little effect on your ability to articulate reasonably clear speech. You were easily and instantly able to adjust your motor articulation to achieve the goal of producing an intelligible sequence of speech sounds. If the goal of speech planning was to achieve a particular vocal tract shape, and if the vocal tract was prevented from achieving that shape in one way or another, then how would the brain know, almost instantly, what alternative shape could be used to achieve the same sound pattern instead? Answer: the system uses acoustic goals to plan speech in the first place.

Other evidence includes the well-established fact that messing with sensory input during speaking affects one's speech output. *Delayed auditory feedback* is a familiar example that you've probably experienced when you have a poor phone connection and can hear your voice echo in your ear with a short delay. You may have noticed that this makes it hard for you to speak fluently, an effect that has been documented experimentally many times (Fairbanks, 1955; Lee, 1950; Stuart et al., 2002; Yates, 1963). This makes sense if speech is planned in auditory space where delayed feedback is perceived and interferes with planning.

Other forms of altered auditory feedback have also been shown to affect speech production automatically, often unconsciously. In one paradigm, participants are asked to phonate a vowel (say, "aaahh") or produce a word with a long vowel (such as "dog") while wearing headphones. Unbeknownst to the talkers, their speech is digitally recorded, altered, and played back to them through headphones so what they actually said and what they heard themselves say is different. For example, researchers can shift the fundamental frequency (F0 shift) of the voice pitch up or down, or the first formant (F1 shift) up or down, changing the vowel. Experiments like these show that talkers compensate for these shifts automatically and rapidly (the term *reflex* is sometimes used) within about 100 msec (Burnett et al., 1998; Burnett, Senner, & Larson, 1997; Houde & Jordan, 1998; Purcell & Munhall, 2006). If such shifts are introduced in small increments over a set of trials, participants may not even notice that they have changed their articulation to make the sounds that they hear themselves speak match what they intended to produce. Studies such as these provide strong evidence that speech production planning is grounded in auditory representations, which serve as the targets for articulation.

It's worth mentioning that the auditory system is not the lone sensory system involved in speech production. You've probably experienced the disruptive effects on speech articulation when parts of your mouth are numb following a dental procedure. Controlled experiments targeting anesthesia to the lingual nerve have confirmed the effect of somatosensory feedback on speech articulation ability (Niemi et al., 2002), as have clever behavioral experiments using altered somatosensory feedback (Tremblay, Shiller, & Ostry, 2003). The observation that damage to primary and secondary somatosensory regions is associated with phonological errors in speech production (figure 10.7), therefore, makes perfect sense from the motor control perspective.

These and other experiments that have demonstrated a role for sensory systems as the goals or targets for motor speech planning have led to the development of computational models of speech production based on *feedback control* principles (Guenther, 2016; Houde & Nagarajan, 2011). The basic idea is that motor planning uses sensory feedback as part of the planning process. Some of these models use only external feedback to update motor plans, while others also use internal feedback to predict the sensory consequences of a motor plan prior to its execution, sometimes called a *forward prediction* using an *internal model* (for more details, see Guenther & Hickok, 2015; Houde & Nagarajan, 2011). Forward predictions are typically assumed to be implemented neurally as an inhibitory signal, suppressing activation of the sensory-related target, resulting in an attenuation of the perceptual signal if the generated action correctly hits the target. Indeed, hearing your own speech while talking produces less auditory cortex activation than listening to a playback of that same speech, the *motor-induced suppression effect* (Aliu, Houde, & Nagarajan, 2009; Okada, Matchin, & Hickok, 2018). This effect is not specific to speech, explaining phenomena like why we don't see the world moving around us when we move our eyes (saccadic suppression of motion signals; Bremmer et al., 2009) and why we can't tickle ourselves (suppression of self-generated somatosensory signals; Blakemore, Wolpert, & Frith, 2000). A breakdown of this suppression mechanism may be the source of auditory hallucinations in schizophrenia, in which inner thoughts are experienced as external sounds, including voices (Heinks-Maldonado et al., 2007).

Such internal prediction or feedback is potentially useful for detecting and correcting errors *during* the planning process and for optimizing correction ability when external feedback errors are detected (Hickok, 2012a; Hickok, Houde, & Rong, 2011; Houde & Nagarajan, 2011). We will return to these ideas shortly. For now, the point is that the motor control approach has advanced sufficiently to work out a very rich set of computational ideas regarding how sensorimotor systems interact to produce speech, including fully implemented models (Guenther, 2016).

This is impressive progress, grounded in an extensive set of empirical results. But how does the motor control approach relate to psycholinguistic models? The typical assumption is that these sensorimotor planning systems for speech production are separate from psycholinguistic planning systems, such that the output of the psycholinguistic models provides the input for the sensorimotor control models. This seems reasonable, at least for some of the sensorimotor effects that have been studied, such as the lingual nerve block, which results in more of an articulatory dysarthria than a problem at the phonological level. But lumping all these sensorimotor effects and computational principles into a postlinguistic level of speech planning goes too far, I will argue.

Similarities between Psycholinguistic and Motor Control Models

One clue indicating that psycholinguistic and motor control models may be studying the same system from different theoretical perspectives is that several curious similarities have emerged in the two traditions. For example, both approaches have discovered and incorporated feedback loops into their models. Feedback control is a core concept in motor control theory, of course. But psycholinguistic research has provided evidence for both internal and external *self-monitoring* loops, as they are called (figure 10.7), even if they are not considered core linguistic processes. The existence of external feedback is self-evident in the fact that we often consciously recognize and correct speech errors that we commit. Psycholinguistic evidence for *internal* feedback required some sleuthing but turned up some solid empirical clues.

One is that some naturally committed speech errors are sometimes caught too quickly to have been based on external feedback. The classic example is the error reported by Willem Levelt in which a talker said, "It is a v– a horizontal line" (Levelt, 1983, p. 64). These sorts of errors cannot be detected from external feedback because there is not enough time to perceive the error and hit the brakes on the utterance before the very next phoneme is uttered (Nooteboom & Quene, 2017). Sometimes such errors are even corrected with almost no delay from the point of interruption, indicating that the correction process had commenced prior to articulation (Blackmer & Mitton, 1991). The other source of evidence for internal error detection comes from experiments in which participants are asked to produce tongue twisters, causing them to make errors more frequently than is found in natural speech samples. These experiments have the advantage that researchers know exactly what the targets were and can manipulate the type of error that might be induced. Let me illustrate.

Suppose that you are presented with a sequence of word pairs and asked to read out loud certain cued target pairs, as in a classic study by Bernard Baars and colleagues (1975). The words preceding some cued word pairs generate a tongue twister situation on the target pair, such as *barred dorm*→*bought dog*→*darn bore* (target). In these tongue twister trials, participants make an error about 30 percent of the time, often swapping the onsets of the target words to follow the pattern of the preceding items, as in the error *barn door.* Notice that in this example, the resulting onset phoneme slip results in a pair of real words. If, however, a different target is used, like *dart board,* where a slip would result in a nonsense phrase, *bart doard,* errors are much less common. This is the *lexical bias effect* in speech errors, and it indicates that some kind of process is operating that reduces nonword errors, presumably unconsciously. Some have proposed a post–word selection "editing" process that is heightened for nonwords because they are more salient to the system, making them easier to detect and correct prior to articulation (Baars et al.,

1975). This idea was tested in one study that ramped up the salience of the slips considerably using word pairs that, if slipped, would result in taboo words (e.g., *hit shed*) (Motley, Camden, & Baars, 1982). Taboo word errors were dramatically reduced compared to "safe" word errors.

Another source of similarity between psycholinguistic and motor control models is simply the levels of representation that they include. For example, phonemes and syllables are core parts of psycholinguistic models, but they are also featured in motor control models, such as Guenther's DIVA model (Guenther, 2016). One might be tempted to argue that these "phonemes" and "syllables" in motor control models are merely motor programs that correspond to higher-level amodal linguistic representations of phonemes and syllables. But an examination of the neural correlates of various levels of a motor control model, such as DIVA, shows that components of motor control models overlap regions typically associated with aphasia and with components of psycholinguistic models such as the posterior STG (classic Wernicke's area) and posterior IFG (classic Broca's area) (Golfinopoulos, Tourville, & Guenther, 2010; Tourville & Guenther, 2011).

I take this evidence of overlap between psycholinguistic and motor control models as good reason to explore a possible integration.

Toward an Integrated Model of Speech Production

The direction that my colleagues and I took in exploring a possible integration of psycholinguistic and motor control models was to assume, on evolutionary grounds, the existence of a basic sensorimotor control architecture for speech. (I'll use the term *sensorimotor* rather than *motor* from here on to reflect the fundamentally sensorimotor nature of motor control.) What this claim means specifically is that every level of representation in the architecture of speech production has three basic components: a frontal lobe motor-related planning component, a temporal or parietal sensory-related target component, and an intermediary computational system that translates between the two (figure 10.8). This is the basic architecture for sensorimotor control generally; what I'm suggesting is that every level of the speech production process follows this basic architecture, from phonetics to morphosyntax (e.g., see figure 9.6).

This assumption is evolutionarily grounded because it builds on the principle of evolutionary tinkering with existing systems rather than de novo

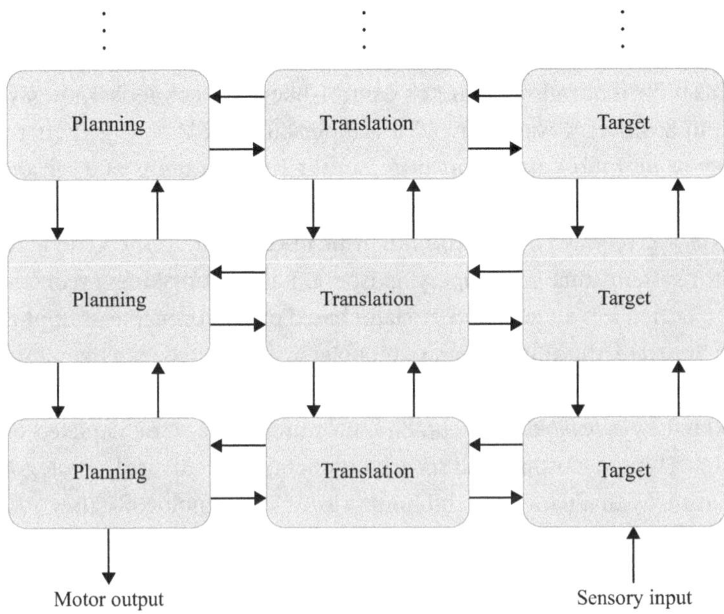

Figure 10.8
A generalized architecture for hierarchical sensorimotor control. Each level contains three major components: a motor-related planning component, a sensory-related target component, and a translation component. The different hierarchical levels represent planning on different scales and, at higher levels, can be quite abstract. In the language domain, for example, the levels might correspond to phonetics, phonology, and morphosyntax. One can think of sensorimotor control generally as a bank of such hierarchies running in parallel, each for its own domain of control.

invention. Prior to our becoming a verbal species, our ancestors had sophisticated sensorimotor control systems that had this basic architecture (it appears to be common to us and our primate relatives). Given that speech production is a kind of motor control (the end point is coordinated action of the lungs, larynx, and vocal tract), the most likely architecture for speech production should involve the duplication and shaping of this basic motor control architecture (Chakraborty & Jarvis, 2015), presumably starting from some basic vocal ability to more and more elaborate planning layered on top. Parallel hierarchical subsystems likely exist even *within* the speech production system, as I'll argue in the next chapter.

Assuming a basic motor control architecture does not strip the validity of traditional levels of linguistic representation as identified by decades of linguistic and psycholinguistic research. Rather, it hypothesizes that each of these levels (phonetics, phonology, morphosyntax) has an *internal* organization that follows a motor control–like architecture. Let me say this again in a different way to avoid misunderstanding: When I say that phonology or morphosyntax is organized in the brain in terms of a sensorimotor control architecture, I do not mean that phonology or morphosyntactic principles as revealed by (psycho)linguistic research reduces to generic sensorimotor systems that control, say, grasping. I am not making a reductionist claim. Rather, it is an *evolutionary* claim based on the principle of homology.

To illustrate, the ability to grasp an object is supported by a neural system that follows a sensorimotor architecture. Speech production, I'm claiming, is supported by a *homologous* neural architecture in the sense depicted in figure 10.8. Human grasping and speech production systems are homologous in the same way that bat wings and human arms are homologous: they evolved from a common ancestor. This does not mean that human arms reduce to bat wings, bat wings to human arms, or arms and wings to their common ancestor. In fact, a lot of specialized and unique evolution has taken place in both species' forelimbs to serve different functions, even while both follow a common anatomical plan. In the same way, sensorimotor control systems for grasping and those for speech production evolved from a common motor control architectural plan, but they have clearly evolved considerably to serve different adaptive functions, which are as abstract, specialized, and unique as they need to be to serve their purpose. Abstract, special, and unique linguistic entities and processes can exist within the context of a sensorimotor control-like architecture.

Assuming a sensorimotor-like architecture of linguistic levels of processing has two major explanatory benefits. It explains the involvement of both motor- and sensory-related brain areas at a given level of linguistic processing, and it explains asymmetries regarding the effects of damage to different portions of the network, including the expressive-receptive asymmetry. I'll risk redundancy and remind you what the evidence is for both of these points.

Motor- and sensory-related involvement at each level of linguistic processing. Take a look back at figure 7.3, the automated meta-analysis of functional activation associated with the terms "phonological" and "syntactic" (see figure 9.3 too). Notice that both frontal and temporal lobe regions are implicated for both domains. Indefrey and Levelt's more targeted meta-analysis implicates the posterior STG and portions of Broca's area in different subprocesses of phonology (retrieval and syllabification). Lesion studies report phonological and syntactic deficits associated with frontal and temporal-parietal areas. In the next chapter, we'll consider lower levels of speech planning and coordination, where we see evidence of both motor cortex and somatosensory cortex involvement (Hickok et al., 2014). Direct cortical stimulation of both frontal regions and posterior temporal regions can cause speech arrest or word retrieval problems (Lu et al., 2021; also see figure 10.1). Just about everywhere you look, there is evidence for frontal and temporal/parietal involvement at every level of linguistic processing.

Asymmetric effects of frontal versus temporoparietal disruption. This multiregional, fronto-parietal-temporal organization at each linguistic level of processing might be interpreted as evidence for a broadly distributed system in general, meaning that it is just one big interactive network rather than involving computationally distinguishable subcomponents subserving planning, targets, and translation. But such a view cannot explain the differences that one often sees following disruption to motor-related versus sensory-related components within a linguistic domain. One clear example is the distinction between frontal lobe–related agrammatism and temporal lobe–related paragrammatism (chapter 9). Another is the clinically distinguishable phonological difficulties of those with apraxia of speech, which implicates frontal areas (Basilakos et al., 2015; Dronkers, 1996; Duffy, 2013; Hickok et al., 2014; Patira et al., 2017), versus conduction aphasia, which implicates temporo-parietal areas (figure 10.6) and the translation system in particular. And then there's the expressive-receptive asymmetry, which has been a reoccurring observation in the history of language neuroscience and a theme of this book. A tidy interpretation of these differences is that damage to the planning, target, and translation components of a sensorimotor architecture have demonstrably different effects on processing within a given linguistic level.

The Integrated State Feedback Control Model of Speech Production

With these considerations in mind, I teamed up with John Houde to build on his ideas regarding state feedback control and integrate it with

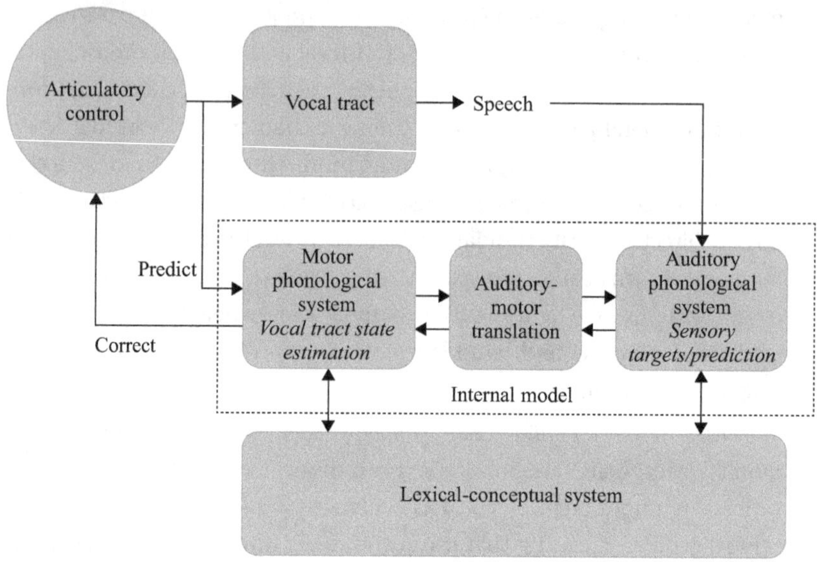

Figure 10.9
The integrated state feedback control model of speech production. Redrawn from Hickok, Houde, and Rong (2011).

psycholinguistic models. Feng Rong, a postdoctoral scholar in my lab at the time, also joined the team. We proposed the architecture for speech production shown in figure 10.9. At first glance, it looks like a motor control model. Indeed, it was modeled after Houde's depictions of his state feedback control model (Houde & Nagarajan, 2011): it sports an external feedback loop; an internal feedback loop, including a forward prediction motor-to-auditory pathway; and an auditory-to-motor correction pathway. But a closer look reveals the mashup. The "auditory" and "motor" components are pitched at the *phonological* level, and there is input to that phonological system from a lexical-conceptual system. In essence, the horizontal axis of figure 10.9, with its "sensory" and "motor" components and feedback loops, follows a sensorimotor control architecture, while the vertical axis is a standard psycholinguistic model with semantic and word levels (collapsed into one box) and a phonological level. From the perspective of a psycholinguistic model, all we really did was partition the phonological level into "auditory," "motor," and translation subcomponents and then endow those subparts with standard motor control computational abilities, like forward prediction, error detection, and correction.

You'll notice that there are two input pathways from the lexical system to the phonological system: one with the auditory-phonological network and one with the motor-phonological network. Although this feature is obviously not part of psycholinguistic models that posit a unified phonological level, it is found in models that posit separate input and output phonological lexicons (Martin & Saffran, 2002) and, of course, it was featured in nineteenth-century models by Wernicke and Lichtheim (Lichtheim, 1885; Wernicke, 1874/1969). I find this connection pattern to be critical for the explanation of why people with damage to the posterior superior temporal lobe can produce fluent, if error-prone, speech. If access to the phonological level was only via the auditory-phonological system, then damage to that system should severely disrupt speech fluency because there would be no way to activate the motor-phonological system at all. Clearly, there must be a way to activate motor-phonological plans directly from the lexical system to enable fluent speech.

We are now in a position to see how this integrated model explains some basic expressive symptoms of aphasia. Damage to the motor-phonological network causes deficits in motor speech planning at the phonological level, disrupting fluency. Damage to the auditory-phonological system and/or the auditory-motor translation system spares fluency because motor-phonological systems can be activated directly from lexical networks; instead, posterior damage results in an increase in speech error rate because the ability to detect and correct motor-phonological planning errors via the internal feedback loop is disrupted. In some people, extensive damage to posterior temporal areas may disrupt the ability to activate phonological targets for speech planning and prevent the detection of errors even in overt speech feedback, often a feature of severe Wernicke's aphasia. More restricted damage affecting mostly the auditory-motor translation component will spare the ability to activate auditory targets for speech planning, even if they can't be used for internal error correction. Overtly produced errors will be detected, however, because the auditory-phonological targets were activated normally and the mismatch between the target and produced speech is readily detectable; this is a common feature of conduction aphasia. Speech errors are exacerbated by low-frequency, phonologically complex words or sequences because such items are less practiced (i.e., less automated) and more susceptible to planning errors, which can't be corrected internally.

How is this internal error correction happening? Sensorimotor control theory provides us with at least the sketch of an answer, and it involves inhibitory motor-to-sensory feedback. Auditory-phonological targets and motor-phonological plans are activated in parallel from lexical-level input. Activation of the auditory-motor target sends excitatory input to its corresponding motor plan, while at the same time, activation of the motor-phonological plan sends an *inhibitory* signal[6] to its auditory-phonological equivalent (Ventura, Nagarajan, & Houde, 2009). This inhibitory signal is the neurocomputational instantiation of a sensory prediction. Now, if the initially activated motor-phonological plan is correct, it will match the target that was activated from lexical input to the auditory-phonological system and essentially cancel the activation of the target. This is the neurocomputational instantiation of an error-free motor-phonological plan: activation of the motor-phonological plan can continue unchecked by a correction signal.

If the motor-phonological plan is inaccurate, however, it will send an inhibitory signal to a nontarget auditory-phonological representation, which will fail to cancel the target auditory-phonological activation, which will then continue to send an excitatory signal to the correct (target-corresponding) motor-phonological plan. This is the correction signal, which will eventually overcome the activation of the incorrect motor plan. I tested the computational feasibility of this error detection and correction system in a set of simple simulations and found that it can actually work (Hickok, 2012a).

Thus, in addition to potentially revealing the possible relation between motor control and psycholinguistic models, there are several advantages of an integrated sensorimotor control/psycholinguistic model of speech production. To summarize, it provides the following:

- Evolutionary continuity between motor control and speech production, while preserving linguistic levels of representation
- A principled explanation of the neural distribution of linguistic processing in sensory- and motor-related systems
- An explanation of expressive-receptive asymmetries
- An explanation of common speech production difficulties in aphasia
- A theory of speech error detection and correction, which is a topic of much recent psycholinguistic research (e.g., Gauvin & Hartsuiker, 2020)
- Integration with neural circuits shown to be involved in speech processing; specifically, the posterior STG (auditory-phonological system), posterior IFG (motor-phonological system), and Spt (auditory-motor translation) (Hickok & Poeppel, 2007)

These are nice conceptual advances, but the new integrated approach lacked a quantitative test. Grant Walker, my then–graduate student and now longtime collaborator, and I came up with a plan. We reasoned that if a sensorimotor control architecture were on the right track, then approximating that architecture in an implemented model of naming should lead to quantitative improvements in the model's performance over a standard psycholinguistic architecture. We chose to pursue this by building on a dominant psycholinguistic model of naming that had been proposed and developed by Gary Dell and colleagues (Dell et al., 1997; Foygel & Dell, 2000). Their semantic-phonological (S-P) model is implemented in a spreading activation connectionist network with three representational layers: semantic, lexical, and phonological (figure 10.10). Input to the model comprises the activation of a set of semantic feature nodes corresponding to a particular concept, such as CAT, which spreads activation to the lexical level according to a pattern of semantic connection weights, and then on to the phonological level via phonological connection weights. Activation is allowed to spread forward and backward throughout the network for several time steps, and then the most active lexical node is selected. A new jolt of activity is introduced to the selected lexical node, and this activity is again allowed to spread for several time steps, followed by phonological selection. Successful naming occurs when the semantic features for CAT results in the selection of the lexical node, "cat," which results in selection of the phonemes /k/ /a/ /t/. The model can then be applied to naming error data in aphasia, where the game is to "lesion the model"; that is, adjust the semantic and phonological weights so the model's correct and error-type responses (semantic, phonological, nonword, etc.) best approximate the error pattern of a given patient's naming ability. By only adjusting two free parameters, the S-P model does a remarkably good job at approximating a wide range of naming error patterns found in people with aphasia.

The question that Walker and I wanted to answer was whether model performance would improve if we morphed the S-P architecture into a sensorimotor architecture; that is, duplicating the phonological layer so there are auditory- and motor-phonological sublayers. We called it the SLAM model, given that it had weight parameters associated with *S*emantic, *L*exical, *A*uditory, and *M*otor levels (figure 10.10). SLAM actually outperformed the S-P model in fitting naming error patterns from aphasia and made the most notable gains in people with conduction aphasia (Walker & Hickok, 2016).[7]

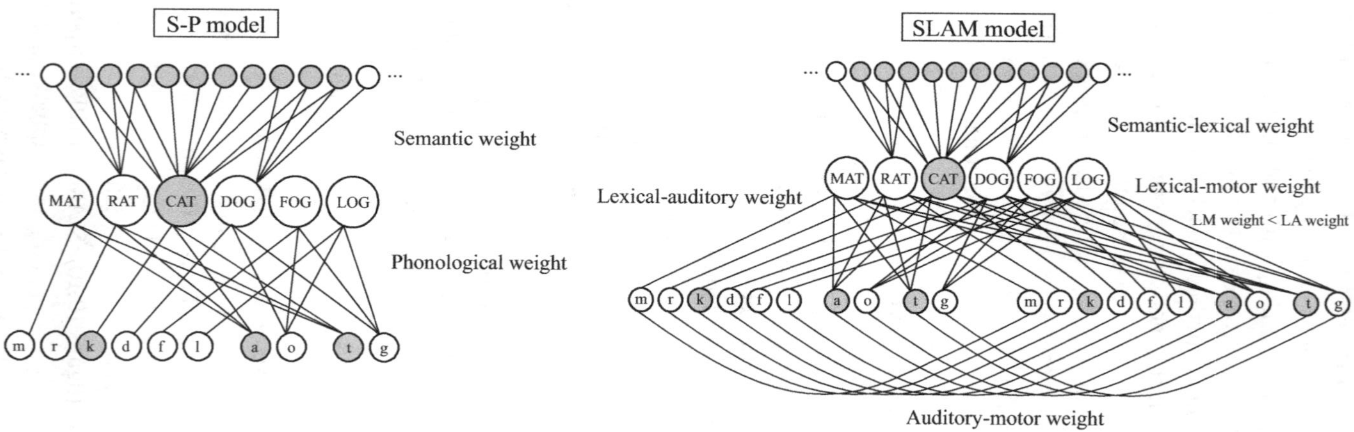

Figure 10.10
The S-P and SLAM computational models of speech production. Reprinted from Walker and Hickok (2016), with permission.

This latter finding is particularly interesting, given that conduction aphasia appears to be a deficit implicating precisely the relation between auditory-phonological and motor-phonological subcomponents, which SLAM explicitly has built into it.

The Hierarchical State Feedback Control Model

The integrated state feedback control model that John Houde, Feng Rong, and I proposed seemed to have reasonable support from both conceptual and quantitative perspectives, but it dealt only with auditory-related phonological circuits. As I mentioned earlier, it has been recognized for decades now that somatosensory systems also participate in speech motor control at some level (Tremblay et al., 2003), and prominent motor control models of speech production, such as DIVA, include somotosensory feedback circuits (Guenther, 2016). The hierarchical state feedback model represents an extension of the integrated state feedback control model to include somatosensory circuits, and also provides some additional tweaks (figure 10.11) (Hickok, 2012a).

The darker-shaded boxes in the diagram represent the integrated state feedback control model with the word and conceptual levels separated out.[8] What's new is the lighter-shaded level above it, which represents a lower level of somatosensory-related control. (I realize that it might be confusing to put the lower levels at the top of the diagram and the higher levels at the bottom. I drew this, however, with the brain's architecture in mind, where word/lemma levels are in the middle temporal gyrus and phonological systems involve the STG.) Although I hypothesize that this somatosensory circuit is lower level, I do not think that it is subphonological. Rather, my hypothesis, following similar proposals by Gracco and colleagues (e.g., Gracco & Lofqvist, 1994), is that phonemic segment targets (the level of individual "phonemes," in common terms) are coded in the somatosensory cortex and serve as targets for segment-level planning in the lower-level motor cortex. Truth be told, I don't find the traditional distinction between linguistic and motor-speech systems particularly helpful; it's all just different levels of the hierarchy.

Following the basic architecture of sensorimotor control systems (figure 10.8), I also proposed a go-between translation system involving the cerebellum. The evidence for this hypothesis comes from three main sources. The first is that there is good evidence that the cerebellum is involved in coordinate transformations, including forward prediction, in the visual domain

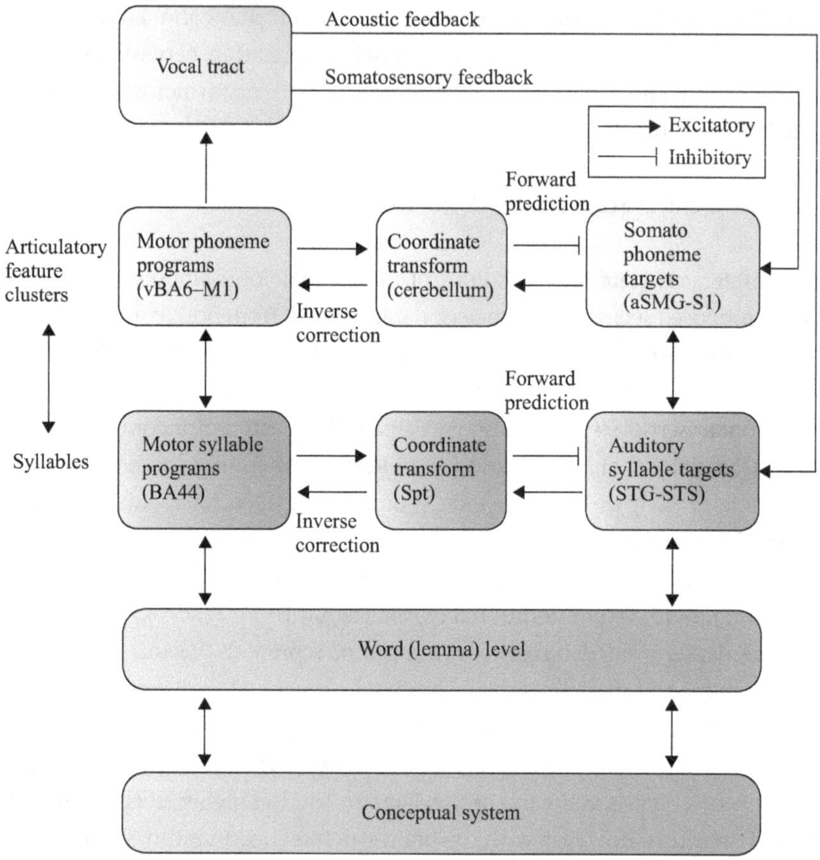

Figure 10.11
The hierarchical state feedback model. From Hickok (2012b).

(Wolpert, Miall, & Kawato, 1998). The second is that damage to the cerebellum produces a characteristic speech disorder affecting rather low-level coordination, called *cerebellar (or ataxic) dysarthria,* in which speech sounds are imprecisely articulated (slurred), speaking rate is reduced, including pauses between syllables and isochronous syllable durations ("scanning speech"), and abnormalities of prosodic stress are present; the disorder appears to affect single vocal tract gestures (consonant and vowel articulation), as well as sequences of gestures, the latter evident in reduced anticipatory coarticulation effects (Ackermann, Mathiak, & Riecker, 2007; Spencer & Slocomb, 2007). The third piece of evidence is that cerebellar dysfunction results in abnormally exaggerated compensation to altered auditory feedback during speech production, as we and others have shown (Houde et al., 2019; Li et al., 2019). This can be explained if the cerebellum is part of the somatosensory-related feedback control system that, when damaged, makes the system over-reliant on auditory feedback, leading to exaggerated compensation responses (Houde et al., 2019).

One way to think about the hierarchy of phonological planning that the hierarchical state feedback control model is intended to capture is that the higher-level, auditory-grounded circuit codes information at the syllable and syllable sequence levels in a context-free format, while the lower-level, somatosensory-grounded circuit codes finer-grained plans at the segment and segment sequence levels in a context-dependent way (i.e., articulation of one phoneme depends on the surrounding phonemes). Again, while I've labeled the function of the boxes at each level of the phonological hierarchy in figure 10.11 as "motor programs" and "auditory targets," I don't mean to imply that these systems support nonlinguistic motor control functions. Rather, I believe that when mainstream generative phonologists and psycholinguists study phonology, they are studying the operation of a hierarchical, sensorimotor-like (homologous) system.

Still, from the perspective of psycholinguistics, this view of the neural architecture supporting phonological planning for speech production may *seem* overly weighted toward a motor control–like view, with its emphasis on sensory-based targets, sensorimotor coordinate transforms, forward prediction, and internal error correction. For example, in a commentary on a follow-up paper to my hierarchical state feedback control model (Hickok, 2014a), neurolinguist Brenda Rapp and colleagues (2014) wrote that "rather than an integration of psycholinguistic and motor theories, the proposal appears largely to be a reduction of the former to the latter" (p. 26), and that "there is a significant 'flattening' of the representational structures, eliminating the various levels of phonological representation that psycholinguistic and cognitive neuropsychological research has posited" (p. 25). But this is not what I was or am claiming. Again, my hypothesis is that the various levels of phonological representation are there, it's just that the neurocomputational architecture in which they live is *homologous to*—not equivalent to—nonlinguistic sensorimotor control architectures.

Interestingly, researchers in the motor control tradition lean toward the reverse opinion of my integrated approach noting, correctly, that my proposed syllable and phoneme categories—traditional in the psycholinguistic approach—don't help much with planning for the actual dynamics of speech motor planning. In Guenther's words, "phonemes are not static vocal tract configurations but instead are inherently time-varying entities" (Guenther, 2014, p. 45). In other words, articulating speech isn't a sequence of stroboscopic, digital transformations from one vocal tract position to the next; rather, it's about controlling the movements through that space dynamically. This is obviously true, but it doesn't imply that static phoneme or syllable representations aren't real entities that exist somewhere in the speech production network. For example, a static representation of a phoneme, a particular vocal tract configuration, could very nicely serve as a target end point for a dynamic action plan, just as the static location and shape of an object can serve as target end points for a dynamic grasping action.

The hierarchical state feedback control model integrates these approaches by finding room for both perspectives in the same framework. There are abstract, static representations of linguistic elements, which serve as the targets for speech production planning, and there are also dynamic, time-varying plans for hitting those targets by transitioning from one element to the next. We might be tempted to say that the existence of both sorts of representations looks a lot like the classic distinction between the psycholinguistic and the motor control approaches—in which case, have we done any real integration? I would argue that the integration is found in at least two differences from the typical thinking.

First, the typical thinking holds that the entirety of the linguistic hierarchy from syntax down to the phonological score is on one side of the divide, the output of which serves as the input to the motor control side. The hierarchical state feedback control model, in contrast, proposes interaction between targets and plans at different hierarchical levels of representations from syllables to phonemes, and we've seen that this concept can be extended all the way to the morphosyntactic level (chapter 9). Second, the interactions between targets and plans at the different levels of representation take the form of feedback control. Recent evidence from my team on functional activations associated with avoiding slips of the tongue, the proposed role of feedback control for speech, applies not only at the phonological level, but also at the middle-word level involving the posterior middle temporal gyrus (Teghipco et al., 2023). This imports a fundamental computation from motor control into the activation and selection of linguistic representations at the phonological and morphosyntactic levels.

Toward a More Nuanced View of Planning Hierarchies

The hierarchical state feedback control model of speech production proposes a hierarchy of linguistic planning. It aligns lower-level phonological planning at roughly the segment level with a somatosensory-motor loop, and a higher-level of phonological planning at roughly the syllable level with an auditory-motor loop. But the picture isn't this clean. To give one example, results of research involving altered auditory feedback, discussed earlier, show that auditory information is used to control low-level features such as voice pitch or within-phoneme formant spacing (Guenther, 2014); this indicates that auditory signals have access to fairly low levels of motor control (Tourville, Reilly, & Guenther, 2008). Conversely, it also appears to be the case that somatosensory signals have access to relatively high levels of motor-related speech planning. For example, an fMRI study showed that altered *somatosensory* feedback activates not only low levels of motor and somatosensory cortex, but also portions of Broca's area during speaking with altered feedback compared to speaking without it (Golfinopoulos et al., 2011).

It appears, then, that my proposed strong alignment of sensory input types with levels of the planning hierarchy in the hierarchical state feedback control model was rather too strong. A closer look at the evidence has pointed to a better way of framing the situation. Namely, there is a hierarchy of speech planning (this point is uncontroversial) and the lowest levels of control, such as in primary motor cortex, are dominated by somatosensory inputs on the post-central gyrus. Moving up the hierarchy into the level of the premotor cortex, we find evidence for intermediate-level somatosensory- and auditory-related inputs, but with a bias toward different subregions receiving the two types of signals; we'll tackle the evidence for this claim in chapter 11. Above this level of planning—for example, at the level of Broca's area—we see segregated inputs both from high-level, auditory-related zones (regions surrounding the posterior STS), and from high-level somatosensory areas.

This hierarchy is nicely illustrated by an fMRI-based functional connectivity map generated by my former student, Alex Teghipco, in his dissertation work (figure 10.12). In this map, areas that are functionally connected are shown in like colors. You can clearly see a rough hierarchy of low-level sensorimotor areas on the precentral and postcentral gyri (light green): an intermediate level on the precentral gyrus extending down into the operculum and insula, with more dorso-lateral regions connecting to auditory-related cortex (yellow, black) and more ventral regions connecting to areas in the

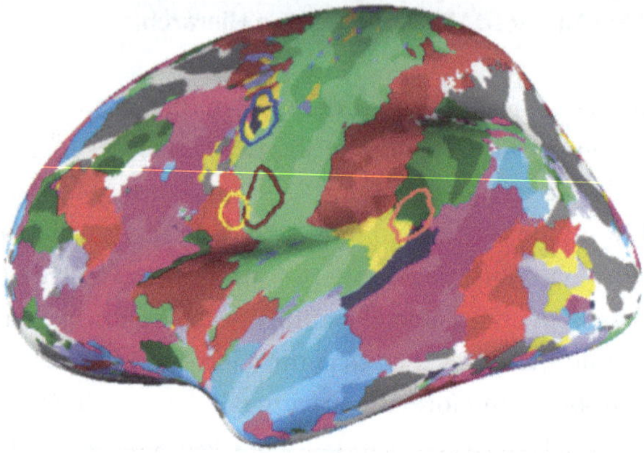

Figure 10.12
The Teghipco functional connectivity map, which was generated in an effort to parcellate language-related cortex in and around the planum temporale by looking at functional connectivity to the rest of the brain. The map here shows this connectivity pattern, where like colors indicate connectivity. Outlines show the locations of area Spt (salmon), the dorsal precentral speech area (blue), the ventral precentral speech area (yellow), and the ventral laryngeal motor cortex (red); the dorsal laryngeal motor cortex is just posterior to the blue outline. These areas are explained in chapter 11.

parietal lobe (red); and a high-level network involving regions anterior to the precentral gyrus functionally connected to the lateral posterior superior and middle temporal gyri (pink). Thus, there seems to be a gradient of input (or target) types as one moves from low to high in the planning hierarchy, with somatosensory dominating the lowest levels, a mix of somatosensory- and auditory-related inputs at intermediate levels of planning, and auditory-related representations dominating the high levels of planning.

Were We Wrong about Spt?

In the late 2000s, Lisette Isenberg started her dissertation experiments in my lab, which aimed to clarify the relation between Spt and spatial hearing functions of the planum temporale. Her work ended up making us question whether we were wrong about Spt all along. Here's the backstory.

As discussed earlier, we had identified Spt as an auditory-motor region in the posterior planum temporale that we believed played a role in transforming auditory-related phonological representations of speech into motor-related phonological plans for speech production. But other researchers had found the planum temporale to be involved in other functions, notably *spatial hearing* (i.e., localizing or tracking acoustic signals in space; Warren & Griffiths, 2003; Warren et al., 2002), and *auditory stream segregation* (i.e., figuring out which bits of sound are coming from which sources in a noisy environment; Bregman, 1990; Zatorre et al., 2002). Some had even proposed that the planum temporale was a "computational hub," a kind of router that took all sorts of auditory signals as input and then shuttled them off to separate, specialized systems (such as speech, space, or melody) for further analysis (Griffiths & Warren, 2002). We wanted to know whether Spt was separable from these other functions.

A series of fMRI studies led by one of my graduate students, Kevin Smith, pointed to the conclusion that the spatial hearing–related activations in the planum temporale were not tracking spatial information per se, but rather using spatial information for auditory stream segregation (Smith et al., 2004, 2010; Smith, Saberi, & Hickok, 2007), as Robert Zatorre and colleagues (2002) had proposed. Isenberg built on this foundation, and with fellow student Kenny Vaden, she designed the critical experiment to test whether Spt was separate from this stream segregation system (Isenberg et al., 2012). She used fMRI to localize Spt using our standard paradigm and then compared those activations in each subject with regions activated while listening to a moving white noise sound source compared to a stationary one. She found that the regions were distinct. Spt localized to a more posterior-lateral zone than the more anterior-medial zone activated by the auditory motion condition. She then used the regions of activation for Spt and auditory motion localizers to map the white-matter connectivity of those ROIs, which revealed distinct patterns of structural connectivity. Spt connected to the posterior STS, the inferior parietal lobe, and the ventral precentral gyrus, while the auditory motion ROI connected to the middle part of the STS and a more ventral zone of the parietal lobe. Spt occupies a distinct portion of the planum temporale region compared to other auditory-related functions.

Having demonstrated the separability Spt activations from spatial hearing and stream segregation systems, we turned our attention to the possible interaction of these systems. And that's when things got weird. Here was the idea. Spt is supposedly the hub in a network that allows auditory signals to guide vocal tract actions and is distinct from nearby areas involved in processing spatial auditory signals for some purpose, such as stream segregation.

But what if a spatial auditory signal were used for guiding vocal tract actions? Would Spt get involved then? Isenberg's second dissertation experiment was designed to find out. She presented listeners with a sound source that moved from left to right, back and forth. We then asked participants to "point" to the source with either their finger, their eyes, or their tongue. We expected to see Spt activation while tracking the location of the source with the tongue, but not the finger or eyes. What we found, however, was no activation for tongue tracking in Spt. This surprised us. Maybe, we thought, it was because pointing with your tongue is a weird task or too simple compared to control of speech, the prototypical driver of Spt activation. But we knew that Spt also activated during imagined humming, which is also a relatively simple motoric task that doesn't even involve the supralaryngeal articulators.

And then it hit us. What if the (or *a*) major driver of Spt activation was more related to the prosody of speech than all the phonetic detail? This would explain why we saw no activation during tongue-tracking, and it also would explain a (retrospectively appreciated) puzzling fact about our humming experiments: Why does listening to and silently humming a melody activate Spt just as much as full-blown speech, with all its added phonological complexity? If it was the melodic bits of speech that mostly drove Spt, this would make sense. So Isenberg's third dissertation experiment involved an even more bizarre task: tracking a moving sound source with the *larynx*.

We trained participants to hum a lower or higher pitch according to how far left (low pitch) or right (high pitch) a moving sound ranged. Once they were trained, we asked them to silently perform the task while in the fMRI scanner and compared that to the tongue-tracking task. We found that laryngeal/pitch tracking activated Spt quite strongly, while tongue tracking activated ventral precentral and postcentral gyri (Isenberg, 2012).

It appears, then, that Spt, as it is typically localized in our speech experiments, is not responding predominantly to phonetic control, but rather to pitch control via the larynx (Hickok, 2017). This raised new problems. Specifically, if damage to Spt is part of an auditory-motor pitch control circuit, how does its damage cause the phonological problems in conduction aphasia? Similarly, while many of our speech Spt localizers included a strong prosodic component, not all of them did. For example, one study involved listening to and repeating four consonant-vowel (CV) syllables; it's not clear whether this task involved enough prosodic control to drive activation in Spt.

I think the answer to these questions is that Spt is a complex of auditory-motor interface systems rather than a homogeneous zone. You can get hints of this in the functional connectivity map in figure 10.11, in which Spt encompasses at least three subnetworks, one of which connects to portions of the IFG (green), while the other two relate to a more dorsal zone in the precentral gyrus, which traditionally hasn't been included in the language network.

Summary

For the first ten years of my career, I had no intention of ever studying speech production. But when the dual stream model predicted the existence of an auditory-motor interface serving speech production, I had to know if there actually was such a system. In retrospect, that was pretty easy to find initially, but it was a lot harder to confirm that it was indeed an auditory-motor interface system and then figure out how it relates to the psycholinguistic and motor control models. I don't think that we fully understand the system yet, but I do believe that it functions at the phonological level with an architecture homologous to other motor control systems. Future research will surely clarify the details of this network.

In the next chapter, we're going to tackle another mystery that was lurking in the results of functional imaging studies of speech processing since before the dual stream model was proposed (Price et al., 1996; Smith et al., 1996). Our focus will be on a region that showed up reliably in our Spt studies (Hickok et al., 2003), in studies investigating the role of the motor system in speech perception (Wilson et al., 2004), in an influential cortical parcellation using Human Connectome Project data (Glasser et al., 2016), in direct cortical stimulation mapping of speech (Lu et al., 2021), and in the Teghipco functional connectivity map. This region is the dorsolateral premotor area (blue outline, figure 10.12). The activation was consistent enough by 2004 to be included as part of the "articulatory speech codes" network in the dual stream model (Hickok & Poeppel, 2004), although I drew its location too far anteriorly in that iteration of the model. Clearly, it was related to speech, but no one knew exactly how.

11 Beyond Broca

> The integrity of the third frontal convolution (and perhaps of the second) seems indispensable to the exercise of the faculty of articulate language.
>
> —Paul Broca, 1861[1]

Broca famously claimed a role for the posterior part of the third frontal convolution in speech production. Not-so-famously, he also pondered a role for the posterior part of the *second* frontal convolution, today's middle frontal gyrus. His contemplation arose because he noticed that the posterior middle frontal gyrus was damaged in both of his first two cases. Commenting on his second case, Lelong, who had much more focal damage than his first case did, Broca writes, "Consequently, we can affirm that in our patient, the aphemia [his term for the speech disorder] was the result of a deep, but very clearly circumscribed lesion of the second and third frontal convolutions, in a part of their posterior third" (1861, translation courtesy of Jeremy Yeaton). Although the fact was not highlighted by Broca, his two cases also had damage to the precentral gyrus just posterior to both the inferior and middle frontal gyri (Dronkers et al., 2007).

Fast-forward to the age of functional brain imaging. Broca's area is firmly established as *the* frontal lobe language area, reinforced by early positron emission tomography (PET) and functional magnetic resonance imaging (fMRI) studies (Caplan et al., 2000; Embick et al., 2000; Hinke et al., 1993; Horwitz et al., 2003; Petersen et al., 1988; Price et al., 1996; Stromswold et al., 1996). Yet those same studies of speech and language processing, including work from my own lab, identified a remarkably consistent activation outside the classic language zone in the posterior middle frontal and adjacent precentral gyri (Hickok et al., 2003; Petersen et al., 1988; Petersen et al., 1989;

Price et al., 1996; Wilson et al., 2004). See figures 5.1, 5.5, 5.6, 7.3, 9.3, 10.3, 10.4, and 10.6 in previous chapters for examples of the ubiquity of this activation site in speech tasks, which is why the location found its way into the depictions of the anatomy of the dual stream model (Hickok & Poeppel, 2004, 2007). In fact, this more dorsal activation is the *most robust* frontal lobe area identified in a Neurosynth.org automated meta-analysis (642 studies) of the term "speech." There seems to some substance behind Broca's parenthetical reflection on the second frontal convolution.

The field at large, however, didn't recognize the region as being significant for speech and language until just recently. I largely ignored it myself, even though I saw it repeatedly in my own data. Several findings concerning this *dorsal precentral speech area,* as I'll call it, finally opened my eyes.

Area 55b

One was the rediscovery of an obscure cortical zone called *area 55b*. Area 55b was originally described in 1956 by Adolf Hupf (1923–2011), a German neuroanatomist who worked closely with the neuroscience power couple, Oskar (1870–1959) and Cécile (1875–1962) Vogt, and eventually took over directorship of their world-renowned Kaiser Wilhelm Institut für Hirnforschung, now the Max Planck Institute for Brain Research (Nieuwenhuys, 2013). Hupf studied anatomical variation in cortical myelination and used this method to develop parcellations of cortical zones similar to the cytoarchitectonic-based parcellations developed by Korbinian Brodmann (1868–1918) who also worked at the institute with the Vogts). One of the regions that Hupf identified was area 55b,[2] a lightly myelinated region on the border of the posterior middle frontal and precentral gyri (figure 9.3).

Area 55b received very little attention in the literature until 2016, when it was featured as the poster child for a new cortical parcellation study led by Matthew Glasser and David Van Essen based on Human Connectome Project data (Glasser et al., 2016). Rather than using just one feature, like myelination patterns, to generate a cortical parcellation, Glasser, Van Essen, and colleagues employed a multimodal approach including myelin maps, task fMRI, and functional connectivity. Area 55b popped out as a remarkable well delineated zone across the various imaging modes.

This got my attention for three reasons. First, area 55b looked like it partially overlapped the dorsal precentral speech area seen in many functional

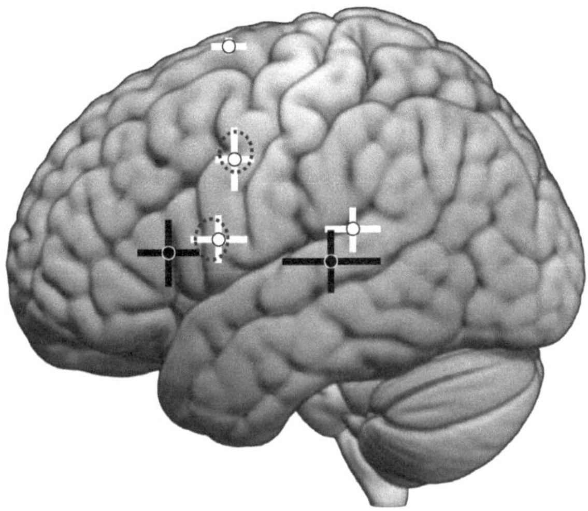

Figure 11.1
Centroid locations (circles) and standard deviations (bars) of direct cortical stimulation sites producing speech arrest (white) and anomia (black). Redrawn from Lu et al. (2021). The dotted black ovals indicate the approximate location of fMRI activation in a listen-silent repeat task (Rong et al., 2018); alignment reported in Hickok et al. (2023).

imaging studies, including those targeting the Spt-related phonological circuit. Second, and consistent with the first reason, the task fMRI study that helped delineate area 55b was the Human Connectome Project's language task, "story listening." Third, the functional connectivity map for area 55b looked basically identical to the Spt circuit, including parts of Broca's area, a region at the parietal-temporal boundary, and the superior temporal sulcus (STS; Glasser et al., 2016). Area 55b was apparently part of the network that I had been studying for years.

Direct Cortical Stimulation

Further evidence for the location and function of the dorsal precentral speech area comes from a large-scale, multisite, direct cortical stimulation mapping study of speech arrest and anomia in 598 participants (Lu et al., 2021). Speech arrest was defined as "the complete interruption of the ongoing number counting or continuous speaking, without apparent oral, facial, jaw, or tongue movements," whereas anomia was defined as an "inability to name an object or misnaming using the wrong word, while still being able to speak the initial words [i.e., 'This (picture) is a(n) . . .']" (Lu et al., 2021, p. 2542). The interpretation of these types of interruptions is that anomic errors reflect a breakdown of higher-level linguistic processing, while speech arrest reflects disruption at the level of motor *coordination*; that is, an intermediate process between linguistic planning and the lowest cortical level of motor control in primary cortex. Several important findings emerged from the massive study.

First, both types of interruption were noted in a *network* of areas that included both frontal and temporal regions. This is consistent with the long-standing observation dating back to Wernicke, that speech production planning involves both motor-related (frontal) and sensory-related (temporal) cortical zones (figure 11.1), as discussed in chapters 9 and 10.

Second, the two types of interruption implicate distinguishable hierarchies of planning. Speech arrest follows from stimulation of lower levels of the motor hierarchy on the precentral gyrus (roughly premotor cortex), the supplementary motor area, and, on the sensory side, the posterior superior temporal lobe near dorsal stream area Spt. Anomia implicates higher-level areas, Broca's area and classical Wernicke's area in the mid-to-posterior superior temporal gyrus (STG), including the *auditory phonological area* (APA), as discussed in chapter 5.

Third, the two speech arrest regions on the precentral gyrus align neatly with the dorsal and ventral sites that are activated in fMRI studies of the Spt circuit. Figure 11.1 shows the location of the geometric centers of the speech arrest zones (small solid circles, white bars) and the approximate location of fMRI activations in a syllable-level listen-repeat task (Rong et al., 2018) (dotted ovals). These are the *dorsal and ventral precentral speech areas* (Hickok et al., 2023). The implication for our attempt to understand the nature of these regions is that their location in premotor cortex (Hickok et al., 2023) points to these areas functioning at an intermediate level between primary motor cortex and higher levels of linguistic planning in the inferior and middle frontal gyri.

Speech arrest, though, is a very blunt behavioral response that could result from disruption to several different subsystems or subprocesses. In terms of subprocesses, one strong candidate is a neural process for *stopping*, the ability to halt midstream an operation such as a motor plan (Hannah & Aron, 2021). Recent evidence suggests that in premotor cortex, networks involved in speech planning and speech stopping live side by side (Zhao et al., 2023).[3]

In terms of subsystems, the fact that four different regions of cortex amount to epicenters for speech arrest strongly suggests a multifunctional network. We have a reasonably good idea of what some of these epicenters might be doing: the supplementary motor area (SMA) zone is likely involved in some process related to speech initiation or sequence processing (Bullock et al., 2023), and we argued in chapter 10 that the Spt-related epicenter is likely involved in some sort of auditory-motor transformation, linking motor plans with their auditory targets. The roles of the two premotor speech area will become clearer as we continue evaluating the recent evidence.

Spectrotemporal Receptive Fields in the Dorsal Premotor Cortex

A second reason why my attention finally turned toward the dorsal speech zone was that at around the same time that the Glasser parcellation was published, one of my graduate students, Jonathan Venezia, was working on mapping speech-driven spectrotemporal receptive field (STRF) properties of the auditory cortex using fMRI (see chapter 6). We identified a broad hierarchy of STRFs in the superior temporal lobe, consistent with previous work (Binder et al., 2000; Okada et al., 2010). We noticed something else: auditory STRFs could be identified outside of auditory cortex, prominently including a focus on the precentral gyrus, which seemed to overlap the posterior part of area 55b and the dorsal precentral speech area. In a follow-up study, we refined our analysis methods, which confirmed the existence of this precentral auditory responsive area and clarified the details of its location and spectrotemporal sensitivities (Venezia, Richards, & Hickok, 2021) (figure 11.2). This effort turned up three interesting facts.

First, the STRF tuning of this region looked a lot like what we saw in primary auditory cortex; that is, sensitivity to a broad range of spectrotemporal features including pitch (stippled area in the STRF map in figure 11.2). Second, its location indeed mapped quite nicely onto the dorsal precentral speech area (figure 11.2, circles). Third, the strongest functional connectivity of this precentral area was to core auditory areas in Heschl's gyrus and its immediate surroundings. While we did find mappable STRFs in the pars triangularis of Broca's area, this region only responded to intelligible speech (unlike the more dorsal area), was insensitive to pitch-related acoustic cues, and was functionally connected to higher-level areas in the STS and middle

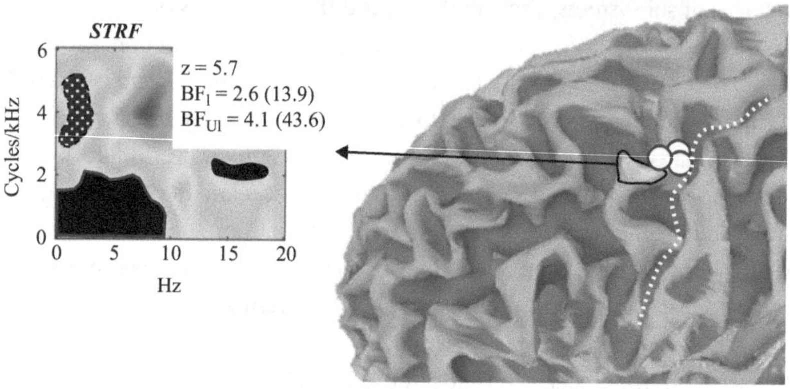

Figure 11.2
Speech-driven auditory STRFs on the precentral gyrus. Black outline on the brain map indicates the locations of regions containing auditory STRFs. Black-shaded/stippled zones on the STRF map indicate spectrotemporal regions of sensitivity; the stippled area is the region containing pitch cues. The white dotted line marks the central sulcus; white circles are the activation locations from three functional imaging studies of speech processing (Brown, Ngan, & Liotti, 2008; Buchsbaum et al., 2001; Wilson et al., 2004). Modified from Venezia et al. (2021), with permission.

temporal gyrus. As a collection of observations, these findings indicate that the auditory-responsive dorsal precentral gyrus is involved in processing fairly low-level acoustic features, including pitch, and does so uniquely among speech-related motor planning regions in the frontal cortex. This is a clear functional difference between the dorsal and ventral precentral speech areas.

Two Separate Laryngeal Motor Cortex Areas

The sensitivity of the dorsal precentral speech area to pitch features tied the region to another thread of research, the search for laryngeal motor cortex, which in turn provided additional clues to the functional role of the dorsal speech region. The larynx is the primary organ for producing modulations in voice pitch during speech and song (Collier, 1975; Roubeau, Chevrie-Muller, & Lacau Saint Guily, 1997; Titze, Luschei, & Hirano, 1989) and the neural circuitry behind its control has been a topic of much interest (Belkhir et al., 2021; Brown et al., 2008; Eichert et al., 2020; Kumar, Croxson, & Simonyan, 2016; Simonyan, 2014; Simonyan & Horwitz, 2011). One reason for this interest comes from comparative research on laryngeal control in humans versus nonhuman primates, which indicates that we are unique among primates in our ability for *voluntary* laryngeal control, an ability apparently enabled by direct connections between motor cortex and the brainstem centers that innervate the larynx (Fitch & Zuberbuhler, 2013; Jürgens, 2009; Simonyan & Jürgens, 2005).

Humans have two spatially separate motor areas involved in laryngeal control—one in ventral motor cortex (probably premotor cortex, to be precise) and one more dorsal in primary motor cortex—whereas nonhuman primates only have one, in ventral motor cortex (Belyk & Brown, 2017; Bouchard et al., 2013; Eichert et al., 2020; Eichert et al., 2021; Iwasaki et al., 2022; Jarvis, 2019; Simonyan, 2014). Thus, the dorsal laryngeal motor cortex is unique to humans and may be the cortical source of direct voluntary laryngeal control (Belyk & Brown, 2017); as such, it is likely a foundational circuit for the evolution of speech (Jürgens, 2002; Kuypers, 1958a, 1958b). With relevance for our purposes here, the dorsal laryngeal motor cortex appears to be localized to primary motor cortex in the central sulcus just posterior to the dorsal precentral speech area (Eichert et al., 2020), which itself is also responsive during voice pitch control, as we'll see next.

Voice Pitch Control Revealed by Direct Cortical Recordings and Stimulation

Benjamin Dichter, Edward Chang, and colleagues (2018) studied the cortical control of voice pitch in a direct cortical recording experiment involving patients undergoing neurosurgical treatment. While the research team monitored neural activity via electrode grids placed on the cortical surface, patients spoke a sentence repeatedly while putting a pitch accent on a different word each time—for example, *I never* SAID *she stole my money*, *I never said* SHE *stole my money*, *I never said she* STOLE *my money*, and so on. The goal was to identify cortical areas that showed a relation between the onset of the pitch accent and changes in neural activity. They identified such a region in the middle portion of the lateral precentral gyrus bilaterally, our dorsal precentral speech area. The team found that activity in this region increased as a function of voice pitch height during both speech and, in an additional experiment, during singing. It was also responsive during listening to speech, consistent with the auditory-motor response properties of this area measured using fMRI, as described previously. Finally, when stimulated, the same region could elicit involuntary vocalization—for example, an "aaah" sound—unlike more ventral areas, which did not.

Syllable Sequencing Associated with the Ventral, Not Dorsal Region

The studies reviewed here indicate that the dorsal precentral speech area is involved in the control of voice pitch. What's the ventral precentral speech area doing? One clue comes from an fMRI study led by Feng Rong and Lisette Isenberg in my lab (Rong et al., 2018). The study used sequences of four consonant-vowel (CV) syllables as the stimuli in a listen-and-silently-repeat Spt circuit localizer. Along with Spt and several other regions, we found the expected activations in the dorsal and ventral precentral speech areas.

After localizing the circuit, we carried out an additional study aimed at determining which nodes in the network were involved in coding sequences of syllables. We compared activation during two types of trials: (1) nonsequence trials, in which participants heard a syllable then repeated it, heard the next then repeated it, and so on, for a total of four syllables per trial; and (2) sequence trials, in which participants heard a sequence of four syllables and then repeated the sequence. We found that among Spt and the two precentral speech areas, only the ventral precentral speech area responded more during sequence repetition. The supplementary motor area also showed sequence sensitivity. This fit well with previous work showing sequencing-related responses in the ventral posterior frontal lobe (Bohland, Bullock, & Guenther, 2010). This study suggests a possible role for the ventral speech area in speech planning at the syllable and syllable sequence levels, rather than the voice pitch–related function of the dorsal precentral speech area.

Different Connectivity Patterns of the Dorsal and Ventral Precentral Speech Areas

Further clues to the differences between the dorsal and ventral precentral speech areas comes from their connectivity patterns. The first nugget of evidence that I encountered came from an analysis of white-matter connectivity predictors of repetition deficits in chronic stroke, a study carried out by Vatché Baboyan, a former graduate student in my lab (Baboyan

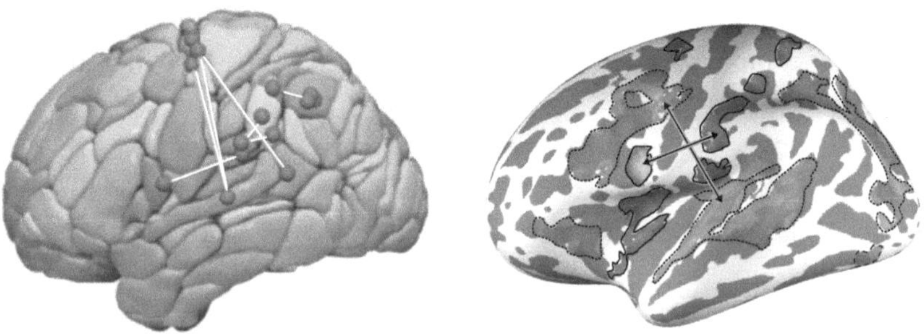

Figure 11.3
Connectivity patterns of the dorsal and ventral precentral speech areas. Left image: structural connectivity disruptions that predict speech repetition deficits in poststroke aphasia, showing dorsal frontal connectivity to auditory-related areas and ventral frontal connectivity to inferior parietal areas. Modified from Baboyan et al. (2021) under Creative Commons license. Right image: fMRI functional connectivity patterns that are stronger for seeds in the dorsal (dotted lines) versus ventral (solid lines) precentral speech areas. Arrows highlight different relative connectivity strengths between the dorsal precentral speech area and auditory-related areas, and between the ventral precentral speech area and somatosensory-related areas. Modified from Hickok et al. (2023), with permission.

et al., 2021). Repetition deficits were predicted by disruption of connectivity involving both the dorsal and ventral precentral speech areas, consistent with the existence of two separable speech planning networks. But the connectivity pattern differed for the two areas: the critical connectivity pattern of the dorsal precentral speech area was to auditory-related regions in the STG, while the critical connectivity pattern of the ventral precentral speech area was to somatosensory-related regions in the inferior parietal lobe (figure 11.3, left).

fMRI-based functional connectivity shows a similar pattern, with the dorsal precentral speech area more strongly connected to auditory-related areas in the superior temporal lobe and the ventral precentral speech area more strongly connected to somatosensory-related areas in the parietal lobe (Hickok et al., 2023; figure 11.3, right). We can see this too when we investigate different patterns of connectivity to define functionally distinct parcels of the posterior superior temporal lobe region. Glance back at figure 10.12 and notice the yellow/black regions, which corresponds to the dorsal precentral speech area (blue outline) and posterior auditory-related regions, including Spt (salmon outline). Notice, as well, the red network, which encompasses the ventral precentral speech area and the anterior parietal lobe.

Interestingly, a recent study showed that people who stutter have relatively weak connectivity of what the authors called the "ventral laryngeal motor cortex," but normal connectivity of the dorsal laryngeal network (Neef et al., 2021). This is interesting because it is well known that people who stutter can often sing without stuttering, a fact made famous by country singer Mel Tillis, presumably because of the spared dorsal network.

The Parallel Hierarchy Model of Speech Production

We've reviewed an impressive collection of evidence all pointing in the same direction. To recap, we started with the observation that there are two speech-related regions in the lateral precentral gyrus, but they differ in several ways:

- The dorsal region is part of area 55b, a well-delineated parcel of cortex in terms of its functional activation, myelination, and functional connectivity.
- The dorsal, but not the ventral, region is also uniquely auditory among premotor areas and exhibits STRFs that include tuning to pitch-related acoustic features.
- The dorsal speech area lies just anterior to dorsal laryngeal motor cortex, which is uniquely human and may be the cortical basis for voluntary laryngeal control, while the ventral speech area is closer to ventral laryngeal motor cortex.
- Activity in the dorsal, but not the ventral, precentral speech area is correlated with pitch height during vocalization and its stimulation can cause involuntary vocalization.
- Weak connectivity of the ventral speech system is associated with persistent developmental stuttering, with the dorsal speech connectivity system intact.

I don't think that anyone doubts that the classical ventral speech-related areas play an important role in motor speech coordination. But the evidence that we've surveyed strongly points to the view that the ventral premotor speech area is not coordinating vocal pitch for prosody and song, a function supported instead by the dorsal precentral speech area via the dorsal laryngeal motor cortex (Brown et al., 2008; Dichter et al., 2018). Jonathan Venezia,

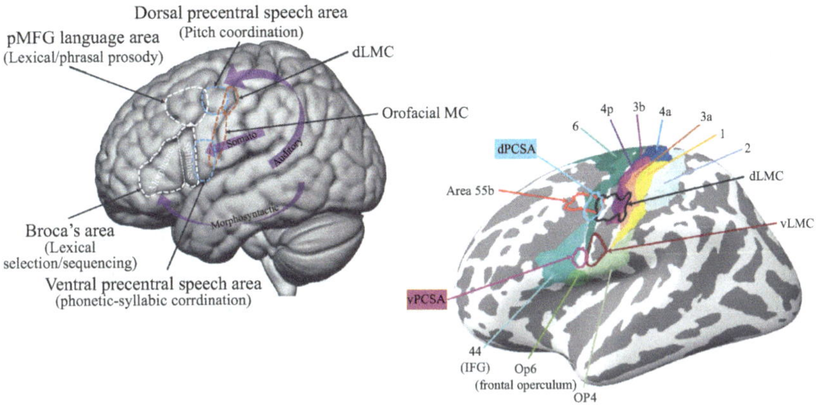

Figure 11.4
Left: The parallel hierarchy model of speech production represented schematically. The ventral hierarchy includes orofacial motor cortex (MC), including the ventral laryngeal motor cortex (vLMC), the ventral precentral speech area (vPCSA), and Broca's area, which has its own internal hierarchy. The dorsal hierarchy includes the dorsal laryngeal motor cortex (dLMC), the dorsal precentral speech area (dPCSA), and the posterior middle frontal gyrus (pMFG). Arrows indicate dominant functional connectivity inputs, not white-matter pathways. Right: Relation between functional areas of the hierarchy and cytoarchitectonic regions (numbered areas). Areas 1–3, primary somatosensory cortex; area 4, primary motor cortex; area 6, premotor cortex; OP/p, opercular areas; 44, posterior part of Broca's area roughly the pars opercularis of the inferior frontal gyrus (IFG). Area 6 was not fully mapped at the time that this map was generated, but it extends down to the vPCSA and vLMC. Here, dLMC is defined liberally to include both the sulcal (primary motor) and gyral (premotor) portions. Reprinted with permission from Hickok et al. (2023).

Alex Teghipco, and I have gone a step further, however, arguing that the dorsal precentral speech area is an intermediate level of processing in a distinct speech-planning *hierarchy* compared to the more traditional ventral hierarchy involving Broca's area, as shown in figure 11.4 (Hickok et al., 2023). Let me sketch the two hierarchies a bit more explicitly.

The ventral frontal speech production hierarchy is the familiar hierarchy that students of the neural organization of language have come to know and love. At the lowest level of the hierarchy is orofacial motor cortex in primary motor cortex (Breshears, Molinaro, & Chang, 2015; Simonyan et al., 2016), including the ventral laryngeal motor cortex (Eichert et al., 2020). The ventral precentral speech area is the next level up in the hierarchy, which we hypothesize coordinates the activity of the lower-level orofacial areas to produce speech gestures at the phonetic, syllabic, and syllable-sequence levels. The latter level may correspond to the concept of the *mental syllabary* from psycholinguistic traditions (Levelt & Wheeldon, 1994). Anterior to the ventral precentral speech area and higher in the planning hierarchy is Broca's area proper, which still resists a full understanding but is likely subdivided into at least two hierarchical linguistic levels: one involves the pars opercularis (area 44) at the level of phonological planning, perhaps playing a role in word-level phonological codes and syllabification processes (Indefrey, 2011; Indefrey & Levelt, 2004); and the next involves the pars triangularis (area 45), which appears to be involved in lexical selection (Conner et al., 2019) and morphosyntactic coding, as discussed previously (Hagoort, 2005; Matchin & Hickok, 2020).

The dorsal frontal speech production hierarchy requires a bit more motivation, as it has not figured prominently in models of speech planning until recently. At the lowest level of the hierarchy is the now-familiar dorsal laryngeal motor cortex. The next level up is the dorsal precentral speech area, which has been the topic of much of the preceding discussion. But what about the highest level? Previously, I mentioned that area 55b, with which the dorsal precentral speech area overlaps in its posterior portion (figure 11.4, right), is defined in part by its functional activation during a high-level language task (story listening). Take a look back at figure 9.3, the meta-analysis of functional imaging studies using the term "syntactic." I pointed out previously that both Broca's area and Wernicke's area are implicated, as expected, but so is the posterior part of the middle frontal gyrus. This region overlaps the *anterior* part of area 55b. Area 55b, then, appears to be two separable areas: an anterior, higher-level area that activates to story listening and "syntactic" tasks, and a posterior region that responds to lower-level speech tasks and contains STRF maps. You'll recall from chapter 9 that there is lesion evidence, too, for a higher-level language role of the posterior middle temporal gyrus, in that it is associated with expressive agrammatism (Matchin et al., 2020; Wilson et al., 2010).

My best guess regarding the syntax-related function carried out in the posterior middle frontal gyrus is that it is involved in phrasal-level prosodic planning (Hickok et al., 2023). This hypothesis makes sense in terms of the dorsal precentral gyrus's role in pitch coordination, and it could explain why syntactic deficits can emerge from damage to two systems. Specifically, the ventral frontal system (portions of Broca's area) may serve a linearization function in relation to morphosyntactic planning, as we discussed in chapter 9, while the more dorsal areas in the middle frontal gyrus play a role in prosodic planning, which some have argued serves as a planning frame for speech production generally (Shattuck-Hufnagel, 2015).[4] Broadly consistent with this idea, a meta-analysis of fMRI studies of the perception of linguistic and emotional prosody points to a bilateral network being involved in processing linguistic prosody, which includes the same dorsal and ventral frontal areas that are implicated in syntax (Belyk & Brown, 2014).

This parallel speech-planning hierarchy model differs from an alternative possibility—namely, that there is only one higher-level hierarchy, the traditional ventral one in Broca's area, which controls a collection of lower-level speech-related motor systems on the precentral gyrus (roughly premotor cortex) and in the precentral sulcus (primary motor cortex). In this sense, the new model goes "beyond Broca," in that it identifies two distinguishable streams of speech planning. The dorsal hierarchy, on this hypothesis, coordinates voice pitch for prosody and song via the larynx, while the ventral hierarchy coordinates phonetic articulation via the supralaryngeal vocal tract (and perhaps including nonpitch-related aspects of laryngeal control such as voicing).

Apraxia of Speech: An Intermediate Level Speech Planning Disorder

The present model was motivated in part by the observation of two separate speech areas in premotor cortex on the precentral gyrus. Within the parallel hierarchy model of speech planning, the dorsal and ventral precentral speech areas reside at an intermediate level of processing between lower-level primary motor cortex and higher-level planning systems, those traditionally referred to as "linguistic." Deficits associated with the lowest level of motor control are typically described as dysarthrias, "abnormalities in the strength, speed, range, steadiness, tone, or accuracy of movements required for breathing, phonatory, resonatory, articulatory, or prosodic aspects of speech production" (Duffy, 2019, p. 3). This results in slurred, slowed, and difficult-to-understand speech. Deficits associated with the highest levels of the system are traditional aphasic symptoms such as agrammatism, anomia, and various types of paraphasic errors. What deficits are associated with the premotor, intermediate level of processing?

A strong candidate is *apraxia of speech*, a disorder typically characterized, not as an aphasia proper, but a speech motor-planning disorder, hypothesized to occupy just this intermediate computational space between linguistic and lower-level motor planning. To quote a recent review chapter, "Apraxia of speech (AOS) is a motor speech disorder that . . . is conceptualized as disrupted translation of linguistic units into speech movement plans. The disorder does not reflect impairment at the linguistic or neuromuscular levels, where we respectively find the disorders of aphasia . . . and dysarthria. Distinct from problems with phonologic processing and motor execution, it is considered the quintessential disorder of speech motor planning/programming" (Jacks & Haley, 2021, p. 368).

People with apraxia of speech have difficulty with two main categories of speech production: articulation and prosody. Articulation symptoms include the production of distorted phonemes, phoneme substitutions and additions, and lengthened durations of individual speech sounds (segments) or the interval between them. Prosodic symptoms include slowed speech rate, a tendency to pause between syllables, and a tendency to apply equal stress on each syllable or to stress incorrect syllables. Some authors have suggested that the articulatory and prosodic features of apraxia of speech constitute two distinguishable subtypes of disorder (Mailend & Maas, 2020; Utianski et al., 2018), which in turn would suggest two distinguishable neural networks involved in articulatory versus prosodic coordination. A tempting conjecture is that these two subtypes of apraxic symptoms are related to damage to the ventral and dorsal precentral speech areas, respectively.

The neural correlates of stroke-related apraxia of speech, however, remain unclear despite decades of research. An early and influential study by Nina Dronkers reported an association between damage to the left precentral gyrus of the insula and apraxia of speech (Dronkers, 1996). Subsequent work, though, has questioned this association (Uddin et al., 2017) and instead points to involvement of portions of Broca's area (Hillis et al., 2004; Richardson et al., 2012), the precentral gyrus (Basilakos et al., 2015; Fridriksson et al., 2018; Graff-Radford et al., 2014; Hickok et al., 2014; New et al., 2015), posterior middle frontal gyrus/precentral gyrus (Chang et al., 2020; Itabashi et al., 2016), and/or the postcentral gyrus (Basilakos et al., 2015;

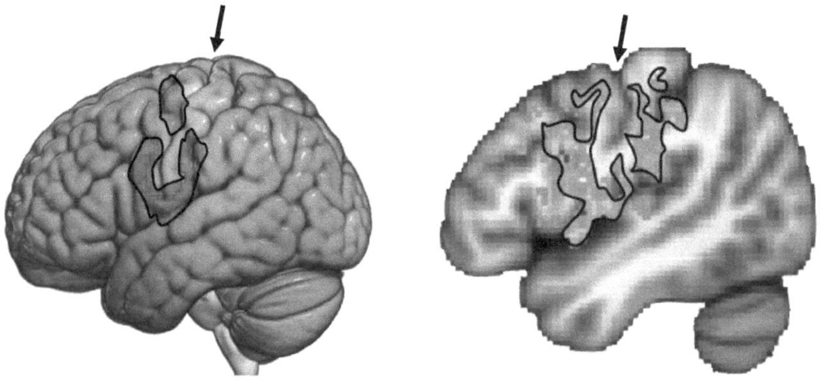

Figure 11.5
Left panel: Lesion pattern associated with clinically rated apraxia of speech in chronic stroke using a cross-validation lesion-symptom mapping analysis (Walker, Fridrikkson, Hickok, unpublished data). Right panel: Lesion/blood flow reduction pattern associated with clinically rated apraxia of speech in acute stroke. Arrows point to the central sulcus. Modified from Hickok et al. (2014) under the Creative Commons License.

Graff-Radford et al., 2014; Hickok et al., 2014). Research on a neurodegenerative form of apraxia of speech has additionally implicated the SMA and cerebellum (Utianski et al., 2018).

The left panel of figure 11.5 shows the lesion distribution associated with clinically rated apraxia of speech scores in a sample of 107 people with chronic left-hemisphere stroke in a cross-validated lesion-symptom mapping study (Walker, Fridriksson, Hickok, unpublished data). The right panel shows the distribution of disruption (frank damage or decreased blood flow) associated with clinically rated apraxia of speech in a sample of seventy-six people with acute stroke (Hickok et al., 2014). I find it quite interesting that both datasets point to a "ring around the central sulcus," apparently sparing the lowest levels of the cortical speech-planning hierarchy but involving the next level up on both the motor and somatosensory sides. I take this as further evidence for apraxia of speech involving an intermediate level of planning.

For our purposes, though, it is worth underlining that both of the precentral speech areas are within the distribution of regions implicated in causing apraxia of speech. Moreover, there is accumulating evidence for an association between pure apraxia of speech and the dorsal precentral speech area in particular, which is evident in acute stroke (Itabashi et al., 2016) and in a case of surgical resection (Chang et al., 2020). As for the possibility that distinguishable articulatory and prosodic subtypes of deficits in apraxia of speech are attributable to disruption of the ventral (articulatory) and dorsal (prosodic) precentral speech areas, we simply don't have sufficient data to know one way or another. More work is needed to test this hypothesis thoroughly.

Beyond Pitch in the Dorsal Speech Coordination Network

I've emphasized the centrality of pitch-related coding in the dorsal precentral speech area, as demonstrated by its correlated activity with vocal pitch height during speaking and singing (Dichter et al., 2018) and its responsivity to pitch-related acoustic features during perception (Berezutskaya et al., 2020; Venezia et al., 2021). But pitch isn't the only feature linked to dorsal precentral speech area activity. Recall that the STRF tuning was not *just* to pitch-related features, but also to a range of spectrotemporal modulation rates. It is also responsive to loudness (envelope) and to rhythmic patterns in perceived speech (Berezutskaya et al., 2020); that is, other important features of the prosody of language. In terms of motor-related activity, the dorsal precentral speech area responds not only during voice pitch modulation, but also during controlled breathing (Eichert et al., 2020), which is important for producing prosodic features such as loudness and prosodic phrasing. The dorsal precentral speech area is not a simple voice pitch control system. Instead, it seems to be using a range of acoustic features, and presumably higher-order, morphosyntactic-related, prosodic planning features, to guide the integration of laryngeal and respiratory coordination for the production of prosody. It seems to be doing something similar for song.

Another interesting property of the dorsal precentral speech area is its relation to attention to speech. In our study of STRF properties of the dorsal precentral speech area, we noticed an odd fact: the pitch-related responsivity to speech was mostly evident when speech was unintelligible, but not when it was intelligible (Venezia et al., 2021).[5] We reasoned that this may have something to do with the fact that pitch information is not necessary for speech perception under clear, intelligible listening conditions. When speech is noisy, however, tracking the voice pitch of the talker may be helpful for auditory object formation and suppression of acoustic noise that is not correlated with the talker's voice. Perhaps pitch-related responsivity in the dorsal precentral speech area ramped up in an unsuccessful effort to allocate attention to pitch cues during unintelligible speech.

Venezia and colleagues tested this hypothesis in a follow-up fMRI experiment that investigated speech perception in multitalker acoustic environments. The task was such that attention to pitch (male versus female voices) was critical to task performance, and under these conditions, the dorsal precentral speech area exhibited pitch-related acoustic responsivity for intelligible and unintelligible speech (Venezia et al., 2022). This result confirmed that the dorsal precentral speech area is concerned with pitch-related acoustic features, but only when the task requires it. It further suggests a role for the dorsal precentral speech area in task-dependent allocation of attention to auditory cues.

The Link between Voice Pitch Control and Orienting and Attention

I started thinking about a possible link between voice pitch control and orienting while pondering the question of why the dorsal premotor speech area and dorsal laryngeal motor cortex are located where they are, superior to the rest of the orofacial motor cortex, including the ventral laryngeal motor cortex (Brown, Yuan, & Belyk, 2021; Martins & Boeckx, 2014). I was particularly struck by the fact that the dorsal precentral speech area is sandwiched between the frontal eye field dorsally and the premotor eye field ventrally (Glasser et al., 2016). This is interesting from the standpoint of the types of sensory inputs that are useful for motor control of orienting responses, and it provides important clues to the question of why the dorsal precentral speech area is where it is.

When we think of orienting and attention, we often think about visual signals, such as detecting something moving in our peripheral visual field and then orienting ourselves toward it. But auditory signals are also used for orienting in space, such as orienting toward footsteps or a voice, or a crash coming from some part of the world around us. The most commonly studied acoustic cues for localizing sound sources in space are *interaural-level differences* and *interaural-time differences*. "Interaural" refers to the two ears, which receive sound waves that differ in *level* (loudness) and arrival *time* when the source of the sound is to the left or right of the head. Although the differences are physically small, they are powerful cues to the location of sounds in the horizontal plane. But interaural cues do not help with localization for sound sources that vary in the vertical plane. For *that,* we use spectral information, including pitch (Brown et al., 1982; Cabrera & Morimoto, 2007). There are also pitch-related environmental cues to location in the vertical plane, in that higher-frequency sounds tend to come from higher locations in space (e.g., singing birds) while lower-frequency sounds tend to come from locations closer to the ground (e.g., a babbling brook or the growl of a large animal) (Parise, Knorre, & Ernst, 2014).

Humans are highly sensitive to this environmental pitch gradient. Experiments have shown that higher-pitched sounds are systematically perceived as coming from higher positions in space relative to lower-pitched sounds presented at the same location: the *Pratt effect* (Cabrera & Morimoto, 2007; Pratt, 1930). This is not a learned association; newborns show the same tendency (Walker et al., 2018). Exploiting these cues must be adaptive because the shape of our outer ear, which acts as a kind of spectral filter, has evolved to amplify this vertical gradient (Parise et al., 2014).

The important point for us is that an orienting system needs access to relatively low-level auditory information, which it indeed receives according to direct cortical recordings from the frontal eye fields in humans and monkeys (Caruso et al., 2016; Kirchner et al., 2009). Why is this important? Because a laryngeal control system also needs access to relatively low-level auditory information (including pitch). What the preceding discussion indicates is that in the prevocal brain of our distant ancestor, the region of the cortex that was already wired for the right kind of auditory input is the eye (orienting) fields, making it a whole lot easier to evolve a new system for laryngeal control there, where it is situated in our own brains today. Comparative evidence from the evolution of birdsong reveals a similar history in which song control systems evolved in regions that were prewired for auditory input (Brenowitz, 1991).

There are some intriguing behavioral connections, as well, between voice pitch and orienting responses. Imagine a photo of a professional singer, head tilted up toward the sky, holding a note. Is that singer hitting a high or a low note? High, obviously. Such a posture is not just an artistic flourish or vocal trick to assist hitting those high notes, as head movement does not enhance pitch-related vocal quality (Knight & Austin, 2020). Rather, quantitative analyses show that when we raise and lower our voice pitch during singing and speaking, we tend to orient up and down with our head, eyes, and hands in sync (Burnham et al., 2022; Loehr, 2012; Munhall et al., 2004), reflecting some sort of relation between motor control of orienting and vocal pitch.

My collaborators and I suggested that this behavioral association between orienting and vocal pitch control is a remnant of the evolutionary association between the two systems, having evolved from a "common ancestor," so to speak. We suggested that the behavioral association resulted from "coactivation of related networks" (Hickok et al., 2023, p. 1784). We didn't say which networks exactly, although our thinking at the time was that the dorsal precentral speech area and the orienting fields had not fully differentiated over the course of their evolution, such that executing commands in one system drives the other system to some extent. Let's call this the *lack of differentiation hypothesis.*

There's an alternative explanation, however, that doesn't require one to assume lack of differentiation between orienting and vocal pitch control. Namely, the "coactivation of related networks" holds on the auditory side of these circuits in the temporal lobe rather than the motor side of the frontal lobe. Recall that auditory pitch information can serve as targets for action planning both for orienting in the vertical plane and for vocal pitch production. What this means is that when we activate an auditory pitch target for singing or speaking, it creates a simultaneous target for orienting, which then tends to drive a corresponding motor response in the orienting system. Let's call this the *shared auditory target hypothesis.*

Can we put these two hypotheses to a test? I think so, because they make different predictions regarding the conditions that will result in corresponding vocal pitch height/orienting actions. The *shared auditory target hypothesis* predicts a tendency to orient up or down when producing a higher or lower pitch because the activated target drives both actions. The *lack of differentiation hypothesis* makes this same prediction because activating the motor plan for generating a higher or lower pitch activates motor plans for orienting up or down. The two theories differ in their predictions for a situation in which a vertical plane-orienting action is planned in the absence of an auditory pitch target, such as if an orienting up/down action was cued by a visual signal or verbal instruction. There is no common auditory target, so no behavioral association should result according to the *shared auditory target hypothesis.* Contrastingly, if the orienting/vocal pitch association is grounded in shared (relatively undifferentiated) motor planning between the dorsal precentral speech area and the frontal orienting system, then activation of one of those motor-planning regions should active the other, leading to behavioral association even when orienting is driven by nonauditory signals.

The critical experiment, then, is to ask people to orient straight ahead, up, or down by using a visual target (or some other nonpitch cue) and then asking them to vocalize to see if their voice pitch rises and lowers in step with the orienting angle. Just such an experiment was published by Satoshi Iijima and colleagues, who asked participants to sing a note while fixating on a laser pointer–marked location that was at eye level, 35 degrees above eye level, or 35 degrees below eye level (Iijima, Ishimitsu, & Nakayama, 2019). Orienting to different locations had no effect on voice pitch, consistent with the *shared auditory target hypothesis*. However, participants were specifically asked to maintain a steady pitch, so this may not be a very strong test. We will have to await further study to determine which hypothesis is correct.

Relation to Spt

In chapter 10, I discussed the possibility that Spt isn't a monolithic functional zone and may instead be a composite system involved in processing pitch- and phonetic/syllabic information. We've noticed for years across many experiments that Spt is activated as part of a network that includes both the dorsal and ventral precentral speech areas (as I now call them). This pattern of coactivation and functional connectivity (figure 11.3) is consistent with the view that Spt participates in both circuits. More experiments are needed to firmly test this idea and map Spt's functional organization in more detail.

Rhythm

I've been talking about pitch coordination for prosody and song as the dominant function of the dorsal precentral speech area hierarchy. Rhythm is also a major component of prosody and song. So how does rhythm fit into this model? The answer is that we don't yet know, but new clues are starting to emerge. My personal hunch is that our rhythmic abilities stem from the necessity to synchronize the dorsal and ventral speech coordination systems (Hickok, 2024).[6] Let me unpack this idea.

Humans possess an extremely rare rhythmic proclivity. We can move our bodies to a beat—specifically an auditory beat—even if it is embedded in a complex rhythm. Researchers call this *beat perception and synchronization (BPS;*

see Patel, 2021). It is distinguished from the general ability to move rhythmically, which is common to most animals (think walking, flapping, chewing), and specifically refers to the ability to synchronize motorically to a heard beat. We are remarkably good at this basic ability. For example, whether presented with a simple metronome pulse or a more complex musical rhythm, we can tap to the beat quite precisely, within a few tens of milliseconds, and often predictively, tapping slightly ahead of the beat. This is true for a wide range of tempos, with intervals ranging from about 250 msec to 1 second (Patel et al., 2005). Nonhuman primates do not show this ability. They can be trained to tap to a metronome, but they tend to tap more reactively rather than predictively (Zarco et al., 2009).

As far as we can tell, the only other species that approaches our ability for BPS is parrots (Patel et al., 2009); search the web for "Snowball the Dancing Cockatoo" to see for yourself. This is interesting because humans and parrots possess another ability that is head and shoulders above the rest—the ability to learn and produce articulate speech. Researchers think that there is a connection. Cognitive neuroscientist Aniruddh Patel (2006, 2021), for example, has presented extensive arguments linking BPS and sophisticated vocal-learning ability. His argument is that (1) sophisticated vocal learning, like that seen in parrots and humans, requires strong auditory-motor connections capable of temporally precise, real-time control of vocalization; and (2) the evolution of these systems fortuitously granted the capacity for rhythmic movement to music (Patel, 2024).

I agree with Patel that the unique parallels between humans and parrots suggest a possible connection. I also agree that BPS emerged as a fortuitous trait (a *spandrel*) in parrots, where it remains so (it's not a behavior that is observed in the wild), as well as in humans, where it may or may not have undergone further selection (Patel, 2021). I further agree with Patel on two additional points: (1) what distinguishes parrots and humans from other vocal learners, like songbirds that don't exhibit BPS, is that parrots and humans have a more complex vocal learning system; and (2) this complexity is related to the added ability of parrots and humans to control supra-syringeal/laryngeal vocal organs, which seems to involve a dual neural control system in both species (Chakraborty & Jarvis, 2015; Hickok et al., 2023).

I disagree with Patel, though, in terms of *why* adding this more complex, dual vocal control system results in BPS. In the context of the human system, Patel argues that "the evolution of strong integration between auditory regions and vocal dorsal premotor regions in ancestral humans (via the laryngeal pitch control pathway) involved gene regulation changes which fortuitously enhanced the strength of neural connections between auditory and nonvocal dorsal premotor regions near the vocal dorsal premotor regions" (Patel, 2024, p. 8). But I think there's a deeper explanation.

My colleagues and I offered a brief conjecture, suggesting that "rhythmic synchronization, the ability to synchronize movement to an auditory beat . . . is a necessary function enabling the coordination of the two proposed [vocal control] streams" (Hickok, Venezia, & Teghipco, 2023, p. 1785). The idea is that precise temporal coordination of two separately controlled systems, like an ensemble of instruments in an orchestra, requires a "conductor"; rhythmic targets coded in the auditory system plus the ability to synchronize to that rhythm constitute the mechanism for this conducted coordination. To color in some of the details of this sketch of an idea, I'll now motivate the need for precise temporal coordination in speech, raise the problem of the quasi-rhythmic nature of speech, and then consider how the system might have developed to solve this problem.

The Need for Temporal Coordination and Prediction in Speech Planning

As we've discussed, modulation of pitch during speech, a component of prosody or intonation, occurs over different levels of linguistic organization, such as the phrase, word, and syllable. The temporal coordination of pitch features at all these levels has important effects on the communicative content of an utterance, cueing distinctions between questions and statements, phrasal boundaries, lexical contrasts (e.g., in tonal languages), and topic focus (e.g., pitch accents on different words signal different meanings, as in *I* THOUGHT *he left* versus *I thought* HE *left*). Thus, it is important to coordinate the timing of laryngeal pitch control with supralaryngeal phonetic/syllabic control to ensure that pitch modulations occur in the correct place in the sequence of syllables. Research on the timing of such coordination has indicated rather precise synchronization between syllable *boundaries* (in particular) and pitch features such as lexical tone or pitch accent (Ladd et al., 1999; Xu, 2002). Moreover, this work has shown that planning for coordinated articulation requires a form of prediction—an important feature of BPS—because generating changes in voice pitch during speech is not instantaneous; in fact, it is rather slow (on the order of 100 msec to traverse a 4-semitone pitch change)

(Xu & Sun, 2002). This means that to time a pitch peak or valley to a syllable boundary, pitch change toward that target needs to be initiated more than 100 msec earlier.

The Problem of Quasi-Rhythmic Speech

If speech rhythms pulsed at a regular, isochronous beat (i.e., more similar to musical rhythms), solving the dual system coordination problem might be relatively easy. After all, motor systems are adept at generating rhythmic movements. In principle, then, two motor control systems could simply align themselves separately to some kind of internal motor-centric metronome. But speech is only *quasi-periodic*. To be sure, speech does have an *average* rhythm, detectible in the amplitude envelope (basically tracking the syllable rate) of natural language (Poeppel & Assaneo, 2020), as discussed in chapter 6. But one cannot use that average rhythm to generate a precise prediction of the onset of the next syllable in any given utterance. To get an intuitive sense of the quasi-rhythmic speech problem, clap along with the syllables as you read the following sentence aloud, as I suggested in chapter 6: *Juan wrote a rough draft and then he edited it.* This is why we don't dance to conversational speech; the rhythm is not regular or predictable enough to the listener (Dalla Bella, Bialunska, & Sowinski, 2013).

So the fact that the motor system is adept at generating rhythmic movements is not particularly helpful in coordinating laryngeal and supralaryngeal speech effectors for the purpose of generating natural speech. What is needed is a means to coordinate these separate subsystems so they align to an utterance-dependent, quasi-rhythm that *breaks* the natural tendency of the motor system to generate rhythmic regularity.

Toward a Quasi-Rhythmic Beat Perception and Synchronization System

While there is extensive research regarding the perception of, and synchronization to, predictable rhythmic patterns, relatively little attention has been paid to how less predictable rhythms are processed. This is not surprising, at least for overt motor synchronization (like tapping to a beat), because syncing to an unpredictable beat equates to a simple reaction time task. The perceptual side is potentially informative, however, because it can answer the question of whether unpredictable rhythms can be learned and remembered efficiently. If the answer is yes, then we have a candidate basis for synchronizing independent motor systems to an *irregular* rhythm (i.e., a neural code for the target rhythm that can be used to plan the timing of motor commands in the two systems). Put differently, a rhythm can be predictable in more than one way. It can follow a regular pattern (incoming beats follow the pattern of previous beats), or it can follow a previously learned template for an irregular (or quasi-rhythmic) pattern.

Can humans learn irregular rhythms efficiently? Apparently, yes. Kang and colleagues (2017) presented listeners with temporal patterns comprised of random sequences of irregularly spaced clicks. They report that listeners were able to learn these patterns rapidly after only a few exposures, and did so over a broad range of interclick intervals, from a sparse five clicks per second to a dense fifty clicks per second. Moreover, learning was also shown to occur implicitly over the course of the session for sequences that were repeated occasionally, interspersed among other unique stimuli during testing. This ability does *not* appear to be unique to humans (Kang et al., 2021), arguing that quasi-rhythmic BPS in humans does not result from a specialized auditory system, but rather a unique sensorimotor connectivity pattern (Patel, 2024).

It seems, then, that we have a possible foundation for supporting quasi-rhythmic-BPS: the auditory system is capable of learning and storing irregular temporal patterns that could serve as the targets for coordinated action across multiple motor subsystems, assuming appropriate connectivity. This is a testable hypothesis, in that it predicts that tapping to a learned, *random* beat (or speechlike quasi-rhythm) would show similar characteristics to what is reported for BPS experiments with regular rhythms in humans. [7]

But how would this basic ability translate to speech? Surely, the rhythmic patterns that we generate during speaking are not called up from a great store of previously learned patterns covering all possible sequences of syllables, stress patterns, and rates of speech. No—like other aspects of linguistic structure, prosodic planning and the rhythms that it contains are part of a generative system that is integrated with the generation of segmental/phonemic and morphosyntactic patterns (Buring, 2013). One consequence of such planning (i.e., a level of representation generated) may be a quasi-rhythm pattern coded at some level in the auditory system that could serve as the target for motor coordination.

Although this idea is speculative so far, there is some evidence to support the existence of an auditory rhythmic code playing a role is speech production. It comes from research on delayed auditory feedback, which disrupts speech fluency and tends to do so most dramatically at a delay interval of 200 msec. The standard assumption is that the delay causes interference at the segmental/syllabic phonological level: the system expects to hear syllable *S* in the feedback but hears *S-1* (the previous syllable) instead. Howell et al. (1983) proposed a *rhythm-based* alternative account to the standard *phonemic content* model, noting that 200 msec is approximately the duration of a syllable, and speech delayed at this interval would cause mismatches in expected versus perceived rhythmic stress patterns. Kaspar and Rübeling (2011) tested this hypothesis directly by asking participants to read repeated sequences of syllable pairs that varied in two dimensions, phonemic content (same syllable: *tata tata tata* . . . versus different syllables: *tali tali tali* . . .) and stress pattern (uniform stress: *tata tata tata* . . . versus accented stress: TA*ta* TA*ta* TA*ta*) under conditions of delated auditory feedback. Note that for the different syllable and accented stress stimuli, 200-msec feedback delays would result in a phase shift of phonemic content (*tali→lita*) or stress pattern (TA*ta* →*ta*TA). They report that phonemic content mismatches (*tali* relative to *tata*) had no effect on reading times, whereas rhythmic mismatches did (TA*ta*/TA*li* relative to *tata*/*tali*). This demonstrates the influence of some sort of auditory rhythmic code during speech production planning and fits nicely with recent proposals that prosodic features provide a planning frame for speech production (Shattuck-Hufnagel, 2015). Interestingly, delayed auditory feedback seems to drive activity in the dorsal, laryngeal pitch-related pathway more strongly and faster than the ventral pathway (Ozker et al., 2022).

Together, these arguments at least establish the feasibility of auditory-related rhythmic/prosodic representations serving as targets for coordinating multiple motor systems for speech. To summarize, the problem is how to temporally coordinate pitch features with phonetic/syllabic boundaries during vocalization, which has been shown to be quite precise even though the timing is only quasi-rhythmic and varies from utterance to utterance. The solution that I am proposing is that speech planning involves the integration of prosodic features and phonetic/syllabic sequences that generates an auditory-based representation of the quasi-rhythmic target, which is used as a temporal reference to synchronize articulatory planning of separately controlled effectors. Once the system has the ability to generate an auditory-based target rhythm and, crucially, to wire it up to motor systems for the purpose of synchronizing to it, an externally provided and *predictable* rhythm would function as a simpler target for motor synchronization. BPS emerges for free.

Why Does BPS Involve Nonvocal Systems?

Patel (2024) suggests that BPS spreads to nonvocal motor systems simply as an unexplained, fortuitous consequence of auditory-laryngeal system strengthening, implemented anatomically via an indirect pathway from auditory cortex to the angular gyrus to nonvocal dorsal premotor areas. The idea that BPS results from the need to coordinate independent vocal articulators gets us one step closer to an explanation because synchronizing more than one effector system to a quasi-rhythm *is* the computational problem that parrot- and human-level vocal learning solved. In short, some degree of quasi-rhythmic BPS spreading is a computational requirement for speech production. But it doesn't get us all the way there because, logically, quasi-rhythmic BPS didn't have to spread to nonvocal systems in order for speech to emerge.

I don't think we have an answer to all this yet, but I can see at least one avenue to explore in the search for a solution. At the most general level, perhaps it's easier to spread quasi-rhythmic BPS to the entire motor system than to just two motor-planning pathways. It may be relevant that the human dorsal laryngeal pathway is more strongly auditory weighted than the ventral phonetic/syllabic pathway, hinting that the neural architecture of quasi-rhythmic BPS spreading did not involve the evolution of a duplicate, parallel auditory-motor circuit into the ventral phonetic/syllabic pathway, which isn't consistent with pan-motor quasi-rhythmic BPS spreading anyway. It seems more likely that multieffector synchronization was achieved by patching into a system that has broader communication with multiple motor subsystems, such as circuits involving the basal ganglia, supplementary motor complex, and cerebellum, all of which have been implicated in BPS, as Patel (2024) notes. Perhaps it's relevant that a particular zone of the cerebellum has been implicated in ataxic dysarthria, a motor speech disorder involving dyscoordination during the articulation of individual speech gestures (i.e., consonants and vowels), as well as a tendency to produce isochronously timed syllables rather than the normal quasi-rhythmic timing (Ackermann et al., 2007).

Summary

This chapter sketched a more detailed functional architecture of the dorsal (temporal-parietal-frontal) network supporting speech production. We saw that there are two separate hierarchies in the frontal lobe involved in speech planning. The ventral frontal hierarchy is the one that we are familiar with, including Broca's area in the IFG (pars triangularis and pars opercularis) and premotor and primary motor regions just posterior to Broca's area. The ventral frontal hierarchy is concerned with speech and language planning at the morphosyntactic level (pars triangularis), phonological planning at the syllable sequence level (pars opercularis), phonological planning at the phoneme sequence or syllable level (the ventral precentral speech area), and motor control of the orofacial articulators (ventral primary motor cortex).

The dorsal frontal hierarchy has not figured into standard models of speech and language organization until recently, but substantial evidence points to a network that includes the posterior middle frontal gyrus and premotor and primary motor areas just posterior to it. We argued here that the dorsal frontal hierarchy is the phylogenetically older circuit (in terms of learned vocal communication ability) and is concerned with planning pitch-related vocalizations for both speech and song. The highest levels of planning in the middle frontal gyrus (the anterior part of area 55b) may be related to prosody at the phrasal and lexical levels and can influence syntactic ability via the proposed prosodic scaffolding for sentence planning. The intermediate level of the precentral gyrus (the dorsal precentral speech area or posterior area 55b), with its strong auditory inputs, appears to coordinate vocal pitch (and perhaps respiration) planning, while the lowest level (dorsal laryngeal motor cortex) provides more direct descending control of the larynx for pitch. These systems necessarily interact to produce speech. How they interact is an open question, but we suggested that rhythmic synchronization may be the cornerstone of this ability.

Our focus in this chapter on the two lateral frontal cortical hierarchies for speech and language planning left much undiscussed regarding the motor control of speech. Although we touched on other core subregions, such as the cerebellum and the supplementary motor area, we did not take a deep dive into these systems. Nor did we discuss the role of an important subcortical network for motor control, the basal ganglia, reward systems, or the many details of the organization of primary motor cortex. I have little to add on these topics, so I will refer the reader to other sources for excellent recent reviews (e.g., Guenther, 2016; Simonyan et al., 2016).

V The Big Picture

12 The Neural Architecture of Language

> As of now, there is no clear mapping between the processing components of comprehension and production and their neural substrates in the brain.
> —Michael Gazzaniga, Richard Ivry, and George Mangun, 1998

We've come a long way in the last quarter-century since Gazzaniga, Ivry, and Mangun (1998) surveyed the field for their first edition of the now-classic text *Cognitive Neuroscience: The Biology of the Mind*. Today, we have a much better handle on the mapping between brain and language, not only for the most basic of abilities, like comprehension and production, but also for some finer-grained psycholinguistic components of language processing. In the preceding chapters, we traced the neural architecture of language from the early stages of speech perception up to semantics and syntax, and then back down to speech planning and coordination. In this chapter, I'll try to pull all of this work together and provide an overview of the architecture of the language system.

An Integrated Linguistic-Sensorimotor Model of the Neural Architecture of Language

Figure 12.1 represents my attempt to depict the architecture of the language system in graphic form. It is motivated by the following principles, which I hope have come to light in the previous chapters:

- *Sensory and motor processes are hierarchically organized both generally and within the language system.* These are not new ideas, and as far as I can tell, they are generally accepted in the field.

- *All levels of linguistic processing are organized into a sensorimotor-like neurocomputational architecture.* We discussed evidence for this neurocomputational architecture applying at both the phonological and morphosyntactic levels (chapters 9 and 10). This is a rather controversial principle and is easily misunderstood, as I emphasized in chapter 10. To repeat, this principle does not imply that linguistic processing can be reduced to more general sensorimotor systems or that the structure of language as elucidated by linguistic research over the decades is invalid or unduly rich. What it does mean, though, is that the architecture of language processing is *homologous* to the architecture of nonlinguistic sensorimotor systems such as visuomotor control, and furthermore employs homologous computations such as feedback control. Thus, this principle is evolutionarily grounded: language systems and nonlinguistic sensorimotor systems share a common neuro-architectural/computational ancestor; they are *homologous*. To draw a rough analogy, language processing is a species of sensorimotor processing in the same relational sense that humans are a species of ape. This does not imply, of course, that we are "reducible" to chimps or that humans are not unique in many ways. It just means that we are related evolutionarily by a common ancestor and thus share many homologous traits. The same holds for the sensorimotor relation between language and nonlinguistic sensorimotor systems.
- *Different portions of sensorimotor architectures are engaged in a task-dependent fashion.* This is the principle that is reflected in modern dual stream models of perception and action (Milner & Goodale, 1995; Rauschecker & Tian, 2000), including dual stream models of language processing (Hickok & Poeppel, 2000, 2004, 2007; Rauschecker & Scott, 2009), which observed that perception for understanding (one type of task) and perception for action (another type of task) involve distinguishable ventral and dorsal circuits, respectively (chapter 2). It is also the principle that undergirds the classical diagnostic tasks that together differentiate classical aphasia syndromes: production fluency (disrupted by dorsal stream damage), comprehension (disrupted by ventral stream damage), and repetition (when occurring in relative isolation, disrupted by the dorsal-ventral interface). Linking these observations to the context of our own discussion, this principle is also the basis of the *expressive-receptive asymmetry* documented in chapters 9 and 10, as speaking and understanding are different linguistic tasks that recruit different parts of the sensorimotor

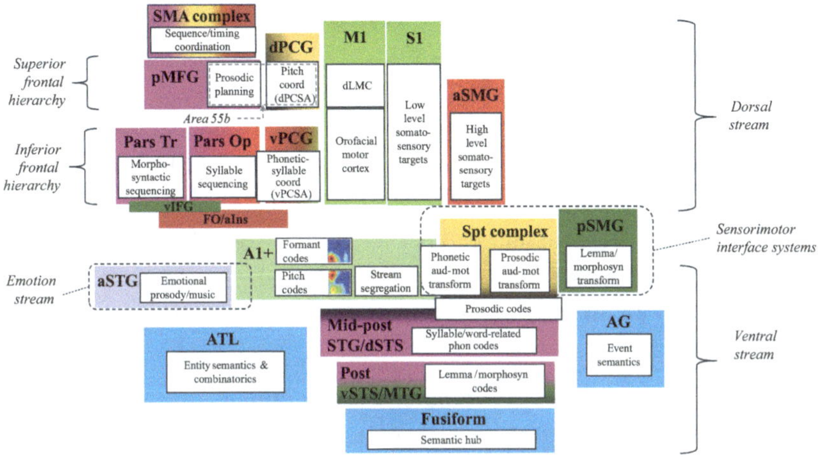

Figure 12.1

The integrated linguistic sensorimotor model (LSM) of the neural architecture of language. A1+, primary auditory cortex and surrounding regions; STG, superior temporal gyrus; aSTG, anterior superior temporal gyrus; dSTS, dorsal bank superior temporal sulcus; vSTS; ventral bank of the superior temporal sulcus; MTG, middle temporal gyrus; ATL, anterior temporal lobe; AG, angular gyrus; Spt, Sylvian parietal-temporal; pSMG, posterior supramarginal gyrus; aSMG, anterior supramarginal gyrus; S1, primary somatosensory cortex; M1, primary motor cortex; vPCG, ventral precentral gyrus, dPCG, dorsal precentral gyrus, Pars Op, pars opercularis portion of Broca's area; FO, frontal operculum; aIns, anterior insula; Pars Tr, pars triangular gyrus portion of Broca's area; pMFG, posterior middle frontal gyrus; aud-mot, auditory-motor.

architecture: the entire system for speaking, and just the ventral portion for understanding.

- *Network components vary in their degree of laterality at the population and individual level.* This principle refers to the observation that some linguistic abilities tend to more or less lateralized than others in general (e.g., speech production is more strongly left dominant than speech comprehension), and that the strength of the lateralization for a given linguistic ability can vary from person to person.

With these principles in hand, we are almost ready to make sense of figure 12.1. First, though, I need to mention some important features and caveats. The spatial layout of the components of the diagram are roughly organized according to neural topography. Box size is not to scale. Functional labels do not imply selectivity or exhaustivity of function of a region. Some nodes or functions are better supported by evidence than others, which I will discuss shortly. Now we are ready to work through the diagram.

The hierarchy of the system is color-coded, with like colors representing similar levels of the hierarchy. In fact, the colors are derived from the Teghipco functional connectivity map in figure 10.12. This map is relatively coarse-grained, but it was tuned to identify connectivity patterns associated with the temporal-parietal junction, which, broadly speaking, is a major sensorimotor hub for language. As such, it suffices for our purposes, with only a minor modification or two, as we'll discuss. Here are the hierarchies:

- Light green—Low-level sensorimotor systems: primary motor cortex (M1), primary somatosensory cortex (S1), primary auditory cortex and surrounds (secondary auditory cortex or belt regions, A1+). This level corresponds functionally to low-level sensorimotor control and clinically is associated with dysarthria (when sensorimotor cortex is involved) and word deafness (when bilateral auditory cortex is involved).
- Red—The ventral frontoparietal portions of the premotor level in the sensorimotor hierarchy: ventral precentral gyrus (vPCG), anterior supramarginal gyrus (aSMG), frontal operculum (FO), anterior insula (aIns), and portions of the supplementary motor area (SMA) complex. This level corresponds functionally to coordination of speech articulators and is clinically associated with motor speech coordination deficits.
- Yellow/black—The dorsal fronto-parietal-temporal portions of the premotor level in the sensorimotor hierarchy: dorsal precentral gyrus (dPCG), Sylvian parietal temporal area (Spt), and portions of the SMA complex. This level corresponds functionally to coordination of voice pitch and respiration and is clinically associated with motor speech coordination deficits (anterior damage) or conduction aphasia (posterior damage).
- Pink—The linguistic levels of the hierarchy: pars opercularis (Pars Op) and pars triangularis (Pars Tr), posterior middle frontal gyrus (pMFG), middle-to-posterior superior temporal gyrus (STG), superior temporal sulcus (STS), middle temporal gyrus (MTG), and portions of the SMA complex. Although the pink level of the Teghipco map lumps the Pars Tr/Op and STG/MTG structures together, I assume from other evidence discussed in this book that further functional and hierarchical divisions can be made within these traditional language areas. Specifically, phonological

processes are more strongly associated with STG/dorsal STS, while lemma/morphosyntactic processes are more strongly associated with ventral STS/MTG. This level of processing is associated clinically with typical aphasic symptoms, including paraphasic errors, agrammatism and paragrammatism, repetition impairment, and comprehension difficulty.

- Dark green—A possible circuit associated with lemma/morphosyntactic processing. The Teghipco functional connectivity map does not indicate the involvement of the middle temporal gyrus in this network, but other whole brain functional connectivity maps do (Yeo et al., 2011). For this reason, I've filled in the vSTS/MTG box with two-tone colors.
- Blue—The semantic network, as discussed in chapter 8.
- Lavender—The auditory-emotional stream (which will be discussed shortly).

In addition to the hierarchical organization of the system, the dual stream architecture is evident in the model, with the regions in the bottom portion of the diagram corresponding to the ventral, auditory to conceptual stream; the top portion of the diagram corresponding to the dorsal, auditory to motor stream; and some of the nodes in the middle corresponding to the interface between the two. Also evident in the model are the parallel inferior and superior motor speech-planning hierarchies, as discussed in chapter 11. Now let's walk through the network components.

Comprehending Language

The mapping from sound to meaning involves several cortical levels implicating the ventral stream in dual stream models. The dorsal stream can be recruited during some receptive tasks if working memory demand is high or if the task requires access to information that is not naturally encoded in ventral stream systems.

The spectrotemporal level. This level involves primary auditory cortex and immediately surrounding belt regions (A1+), where speech-relevant acoustic features are coded. These include, but are not limited to, pitch- and formant-related codes. I've highlighted these two types of codes in figure 12.1 to reflect the hypothesized distinction between them in terms of their engagement with different circuits for motor speech control and their differing association with phoneme-related versus prosody-related processes. This system is bilateral, but with possible mildly different distributions of pitch versus phonetic sensitivities. See part II of this book for details.

Stream segregation. This refers to the process of sorting a noisy acoustic environment into distinct source streams. Auditory cortex in the anterior planum temporale is the likely location for this mechanism. This system is bilateral (Smith et al., 2010). See chapter 10.

The phonological level. This level involves the lateral mid/posterior STG and the dorsal bank of the STS. Two subcomponents are hypothesized, with largely overlapping distributions: one for representing syllable/word-related phonological codes and one for linguistic prosodic codes. The latter may have a more posterior distribution (Turker et al., 2023; Venezia et al., 2019). This system is bilateral in most individuals (> 50 percent) but can be left-hemisphere dominant to varying degrees in some people. See part II for details.

The lemma/morphosyntactic level. This level involves the posterior ventral bank of the STG and the MTG. This system is a bit more left dominant than lower levels. See chapter 9 for details.

The conceptual semantic level. This level involves portions of the ATL, AG, and fusiform gyrus. See chapter 8 for details.

Producing Language

Natural language production involves the entire network, starting with a semantic representation and then activating each of the levels of the sensorimotor-like hierarchy. Again, the sensory-related components of each level of the hierarchy serve as targets for planning in the motor-related components. In general, the production network is more strongly left-hemisphere dominant than the receptive network. Several levels of planning are involved, as described next.

Lemma/morphosyntactic planning. The highest level of linguistic production planning is the lemma/morphosyntactic level (chapter 9). The sensory-related component of the ventral STS and MTG codes lemmas/morphemes and the hierarchical relationships between them during sentence planning. This representation does not code the linear sequence of a sentence, only the hierarchical representation. The motor component in the inferior frontal regions (Pars Tr) codes the linearized sequences of morphemes that correspond to the hierarchy coded in the vSTS/MTG. Reasoning from homology to nonlinguistic sensorimotor architectures, there should be a region in the vicinity of the temporal-parietal junction that serves as an interface or coordinate transform between the morphosyntactic hierarchies in the posterior temporal lobe and the morphosyntactic sequences in the inferior frontal lobe. A candidate region is the posterior portion of the SMG, which is implicated in language dysfunction following stroke (Baldo & Dronkers, 2006; Baldo, Katseff, & Dronkers, 2012; Buchsbaum et al., 2011) and is more sensitive to lexical (as opposed to sublexical) manipulations in functional imaging measurements (Oberhuber et al., 2016). On the Teghipco functional connectivity map, this pSMG region is connected to the ventral portion of the IFG, a zone that some authors linked to syntactic function (Schell, Zaccarella, & Friederici, 2017). As mentioned earlier in this chapter, this pSMG region is functionally connected to the MTG (Yeo et al., 2011). See chapter 9 for further discussion.

Phonological planning. The phonological level likely has several subparts, although the evidence for the neural correlates of the various sublevels is weak. A major distinction is between phonetic and prosodic codes. By "phonetic," I mean to include phonological codes for phonemes, sequences of phonemes, syllables, words, and even chunked/idiomatic phrases (such as, "How ya doing?"). By "prosodic," I mean primarily linguistic prosody, including stress patterns and intonation at the syllable, lexical, and phrasal levels, prosodic phrasing, and the like. It is reasonable to assume distinguishable circuits for phonetic versus prosodic planning, given that the two systems are distinct in phonological theory (Ladd, 2008) and are supported by distinguishable motor speech coordination hierarchies in frontal cortex (see Hickok et al., 2023, and chapter 11 of this book).

Another major distinction concerns subhierarchies within phonetic and prosodic systems (as I've defined them). Within the phonetic system, it seems reasonable to suppose that there are different neural circuits involved in coding, such as the sound of a particular vowel or syllable complex compared to the code for a multisyllabic word or phrase. For example, altered auditory feedback of the first formant (F1) leads to low-level motor articulatory compensation to vocal tract configurations, implicating quite a low level of planning, and this occurs whether the F1 shift crosses a vowel category or not, indicating a correspondingly low level of auditory coding for speech planning (Niziolek & Guenther, 2013). Assuming that vowel formant information and multisyllable word information are relevant for different hierarchical levels of speech motor planning, it seems likely that they would be coded in different levels of the auditory and motor cortical hierarchy. A similar hierarchical gradient holds within the prosodic system, spanning relatively low-level voice pitch information, as revealed in F0 altered auditory feedback experiments, to lexical-level stress patterns, to phrasal level contours. See chapter 10 for a discussion of altered auditory feedback experiments.

All of this is to say that the phonological processing level is surely much more complex than the model in figure 12.1 indicates (which is surely true of the entire diagram as well). Nonetheless, we can provide a general sketch of the architecture as follows: Auditory-phonological targets for phonetic and prosodic planning involve overlapping but nonidentical distributions in the mid-to-posterior STG and the dorsal bank of the STS, likely organized into some kind of hierarchical gradient within the phonetic and prosodic subsystems. The organization of the frontal motor component of the phonological planning circuit reflects the phonetic/prosodic distinction, involving the ventral and dorsal frontal hierarchies, respectively, with the hierarchical levels on both the auditory and motor sides appropriately aligned. The auditory-motor interface at the phonological level involves the Spt complex, which may have separable subregions also reflecting the phonetic/prosodic distinction, and which may be hierarchically organized as well. Finer-grained functional dissection of this system is needed.

The sequencing system. Although sequencing is not traditionally singled out as a linguistic level in its own right, it is incorporated into multiple levels of linguistic planning as a linearization process involving phoneme-level gestures within a syllable, sequences of syllables in words and phrases, sequences of morphemes in syntactic structures, and sequences of prosodic/intonational patterns. The SMA complex has been argued to be a key structure involved in sequencing across multiple domains, including language (Cona & Semenza, 2017). The Teghipco map suggests connectivity with both the dorsal and ventral precentral speech areas, as well as higher-level speech/language planning networks in the inferior and middle frontal gyri (pink areas, figure 10.12; note that the SMA complex is not shown).

Low-level motor articulation. The lowest level of the process of producing language predominantly involves the speech-related portion of primary motor cortex (M1) and primary somatosensory cortex (S1) but it also may involve inputs from primary auditory cortex and immediate surrounds (A1+). Following sensorimotor control architectures generally, sensory systems code the targets (here, somatosensory and/or auditory) for speech gesture planning that motor plans aim to hit. Within M1, the orofacial motor cortex and the dorsal laryngeal motor cortex (dLMC) are the lowest levels of distinct phonetic versus pitch control systems. The ventral laryngeal motor cortex is argued to be in premotor cortex, but it may be lumped in with the function of orofacial motor cortex (e.g., for the control of voicing as opposed to pitch). The cerebellum likely plays an important role at this level of control, serving as a somatosensory-motor transform system ((Hickok, 2012a), chapter 10).

I have not discussed the role of the basal ganglia in these networks, but surely this system plays a role in speech and language production planning similar to its role across motor and cognitive domains generally. These include the selection and facilitation of action plans, including sequences, reinforcement learning, and stopping (Lanciego, Luquin, & Obeso, 2012; Schroll & Hamker, 2013).

The "Third Stream": Auditory-Emotion

Dual stream models developed from the sensory perspective emphasize two systems that sensory information must interface with: ventral conceptual semantic systems and dorsal sensorimotor systems. But, of course, there is a third interface, to emotion systems, which provides the associated valence of a sensory signal. In the auditory domain, music is the most highly studied emotional signal and is well known to activate emotion and reward circuits (Zatorre, 2015). A number of subregions of the auditory cortex respond to musical stimuli, but the dorsal aspect of the anterior STG (planum polare) appears to be particularly important for music perception, including emotional aspects (Angulo-Perkins et al., 2014; Koelsch, 2014; Koelsch et al., 2013; LaCroix, Diaz, & Rogalsky, 2015; Rogalsky et al., 2011; Zatorre, 2015). A study of emotional valence in sounds more generally (not just music) has also implicated the anterior STG and its connectivity with the amygdala (Viinikainen, Katsyri, & Sams, 2012). A full treatment of this issue is beyond the scope of this book, but the evidence is sufficient to warrant the hypothesis of a dorsal anterior temporal auditory-emotion stream.

Dynamics

So far in this chapter, and the book as a whole, we have focused mostly on the neural *architecture* of language. The dynamics within the network architecture is critical too, of course. I want to highlight just a few features in addition to the discussion on neural oscillations in chapter 6.

As the twentieth century approached the new millennium, a major debate in psycholinguistics, indeed psychology at large, was whether information processing was strictly serial, progressing sequentially from one stage to the next, or was interactive with information flowing between levels in the network interactively and in parallel (Dell, 1986, 1995; Fodor, 1983; Levelt et al., 1999; Levelt, 1989; Marslen-Wilson & Tyler, 1980; Swinney, 1979). As is often the case with opposing theories taking the extreme positions, the truth ends up somewhere in the middle.

My read of the evidence—particularly from intracranial recording data where we have high temporal and spatial resolution—is that both positions find support. At a coarse level, activation of various hierarchical language production networks proceeds in a rough temporal order, such that during speech production the *onset* or *peaks* of activation at higher levels generally precede lower levels, consistent with serial processing models. However, the duration of activation of a given level is substantially longer than the phase delay between the levels, resulting in the entire network being active simultaneously for a substantial amount of time and therefore interacting, consistent with parallel interactive models (Conner et al., 2019; Flinker et al., 2015). More important, though, analyses of the interaction between coactive neural regions and levels of processing indicate a serial progression of cross-regional neural network *states*, defined as patterns of interaction within the network (Conner et al., 2019; Giahi Saravani et al., 2019). In other words, even when multiple areas and levels in the language network are simultaneously active and interacting (supporting parallel interactive models), a serial progression is evident through the various interactive states (supporting serial models). Although the research that I've cited here concerns word production, it also holds for receptive language (Gwilliams et al., 2024). I suspect that this is how complex cortical networks work generally and for the language system more specifically: information is processed largely at the network level in a series of staged network states reflecting different patterns of interaction.

White-Matter Correlates

The preceding summary of the language network discusses functional but not structural white-matter pathway connectivity. This would seem to be a glaring omission, given that white-matter pathways, and the arcuate fasciculus in particular, have figured prominently in neural models of language since the 1870s, and given that advances in neuroimaging (Catani et al., 2002) and neurosurgical investigations (Duffau et al., 2002) have enabled the identification of a number of language-related white-matter pathways (Chang, Raygor, & Berger, 2015; Duffau, 2016; Friederici, 2015). This is important and relevant work, particularly in the clinic, which we can align somewhat with parts of the linguistic sensorimotor model (LSM).

For example, the superior longitudinal fasciculus segment III (SLF III) is a strong candidate for the white-matter pathway supporting the "red" network involving inferior parietal and frontal regions (Baboyan et al., 2021), while the arcuate fasciculus is a candidate for connectivity of the "pink" network linking traditional Wernicke's area and Broca's area. Tempting candidates for the "green" network linking the MTG, posterior SMG, and ventral IFG are the posterior and anterior subsegments of the arcuate fasciculus as identified by Catani and colleagues (2005). In addition, white-matter connectivity to dorsal premotor regions (the "yellow/black" circuit) has been identified, such as portions of the SLF II (reviewed in Chang et al., 2015) or in our more direct identification of white-matter connections predictive of repetition deficits in aphasia (Baboyan et al., 2021; also see chapter 11).

Yet it is far from clear whether these named tracts, sub-tracts, and sub-branches hold a one-to-one correspondence to functionally relevant cortical networks. It seems likely to me, and I suspect that most researchers would agree, that a given tract such as the SLF III is not connecting just two of our identified cortical areas, nor are the cortical areas getting input only from this tract. To illustrate this point, consider that none of these major tracts connect primary auditory cortex with motor planning areas, and yet we see speech-related *functional* connectivity between Heschl's gyrus and the dorsal precentral speech area (Venezia et al., 2021). This could be because the functional connectivity is mediated by a third area, or perhaps there are direct connections that so far have gone undetected. In short, white-matter tract tracing is still a rather blunt instrument, and as such, it is not as useful as it could be (yet) for working out the functional architecture of language at the level of detail that we have now achieved. This is why I have not spent a great deal of time in this volume discussing structural connectivity.

Relation to Other Large-Scale Models

There are a number of large-scale models of the functional neuroanatomy of language in the literature. It is worth our time to consider how some of the major ones compare to the LSM. I'll consider them in three separate groups according to the different approaches that this collection of authors have taken: the auditory-grounded approach, the psycholinguistic-grounded approach, and the motor control–grounded approach.

The Auditory-Grounded Approach

The auditory-grounded approach is adopted both by Hickok and Poeppel (2000, 2004, 2007) and Rauschecker and Scott (2009). It is no accident that both teams propose a dual stream model of speech processing because, from the perspective of the auditory system, processing of speech can proceed along two very different pathways—a ventral comprehension stream and a dorsal auditory-motor stream. The dual stream architecture is built into the LSM, as we saw earlier—evident from the perspective of the auditory system—but it has been elaborated on quite a bit compared to the earlier architectures. This elaboration includes the delineation of the two frontal motor speech planning hierarchies, the addition of somatosensory areas as part of the system (also common to motor control approaches), a shift in the thinking about Spt from a unified area to a complex of interfaces, the inclusion of a possible higher-level lemma/morphosyntactic interface in the posterior SMG, and the inclusion of prosody-related systems.

Although the LSM incorporates the auditory-centric dual stream architecture, I do not think it is helpful to call it a "dual stream model" because, again, the dual streams make sense only from the auditory perspective. From the perspective of volitional speech production, the system is not dual stream at all. I hope that this will alleviate some confusion surrounding attempts to interpret dual stream models as general models of the neural organization of language.

Both the dual stream models and the LSM take seriously the concept of feedback control in speech planning. One difference with the LSM, though, is that it takes feedback control seriously at all levels of linguistic planning, from low-level articulation to phonological planning to morphosyntactic planning. All these hierarchical levels are based on (i.e., homologous to) a sensorimotor control architecture, as I've said again and again.

Other differences can also be pointed out. One is the role of the anterior temporal lobe, which Hickok and Poeppel (2007) hypothesized to be a "combinatorial network" and which Rauschecker and Scott (2009) argue is the target of the ventral stream. I think both concepts are wrong to some extent. By "combinatorial network," Poeppel and I had in mind something related to syntax, as that area was showing fairly selective sentence responses in functional magnetic resonance imaging (fMRI) at the time. Subsequent evidence indicated that the ATL region in question had more to do with conceptual combination than a syntactic function, as reviewed in chapter 9. I suppose that Poeppel and I hedged enough to be partially right.

Rauschecker and Scott's view is that the ventral stream progresses anteriorly (not posteriorly) from Heschl's gyrus, which they call the "antero ventral stream." Evidence reviewed in chapter 5, however, indicates that the mid-to-posterior temporal lobe is the critical zone and thus is reflected in the LSM. Rauschecker and Scott (2009, p. 722) also proposed "a common computational structure for space processing and speech control in the postero-dorsal auditory stream." This is based on evidence from the monkey suggesting an auditory spatial function for the dorsal stream. We reviewed evidence in chapter 10, however, showing that auditory spatial information in the planum temporal region is distinct from auditory-motor speech processing. Accordingly, the LSM distinguishes these by indicating a space-related auditory stream segregation function to the anterior planum temporale and an auditory-motor transform function to the posterior planum temporale. That said, I think that Rauchecker and Scott were on to something in terms of a "common computational structure" being related to dorsal stream sensorimotor function: both spatial and speech information can inform action planning (via different subnetworks, as discussed in chapter 11), and that is the functional role of the dorsal stream.

The Psycholinguistic Approach

The psycholinguistic approach has been adopted by a number of researchers following a tradition dating back to Roman Jakobson (chapter 9). The major difference between the psycholinguistic approach and the LSM is that the latter adopts a sensorimotor control-like architecture even for linguistic levels of processing. They share, however, a commitment to taking psycholinguistic theory seriously, and all involve some kind of mapping between linguistic levels of processing and neural circuits. The psycholinguistic approach models do this more straightforwardly, generally aligning components of a particular psycholinguistic theory with particular brain areas.

Indefrey and Levelt's neural model (Indefrey, 2011; Indefrey & Levelt, 2004) adopts Levelt and colleagues' psycholinguistic model of word production (Levelt et al., 1999; Levelt, 1989). Each component of the processing model—lemma retrieval, phonological retrieval, syllabification, articulation, and self-monitoring—is mapped onto circumscribed cortical regions, as discussed in chapter 10. Indefrey and Levelt's model shares quite a bit with the LSM, including the distinction between lemma retrieval and phonological retrieval, a role for aspects of phonological processing in both the inferior frontal lobe (syllabification, a component of sequencing) and the posterior temporal lobe (phonological retrieval), the localization of lemma retrieval in the middle temporal gyrus, and self-monitoring (feedback control) functions. There are notable differences, however, most of which are related to the more detailed functional anatomy of the LSM and its sensorimotor-like architecture applied to each linguistic level. Another major difference is Indefrey and Levelt's assumption of strictly serial processing, which is not a feature of the LSM. On balance, I find the two models to be generally consistent, given the differences in scope, granularity, and perspective.

Angela Friederici's neural model places emphasis on syntactic function and adopts Noam Chomsky's syntactic framework, notably featuring the operation, Merge (Chomsky, 1995). Merge is argued to be a very basic operation, that of combining two linguistic elements (words or morphemes), which, when applied recursively, combine to build hierarchical phrasal units. Friederici localizes Merge to posterior Broca's area (BA 44, pars opercularis) and the frontal operculum and assumes its application to sentence processing both in production and comprehension (Friederici, 2017). While Merge might be reframed as a kind of sequencing, a general function of the IFG in the LSM, morphosyntactic sequencing is argued to be a function of *anterior* Broca's area (BA 45, pars triangularis) in the LSM. Further, the LSM assumes that the sequencing function of inferior frontal areas is engaged only during production (or high-load tasks that benefit from motor recoding, as discussed in chapter 9).

Friederici argues that the pMTG is the storage site of lexical semantic information and is integrated with syntactic information in the posterior IFG via the pSTG/STS and anterior Broca's area (BA 45, pars triangularis). Two points of convergence between the two models can be found here. One is that there is an alignment between lexical information and pMTG, which is engaged both during comprehension and production; the other is that anterior Broca's area plays a role in sentence-level processing. Beyond that, the models diverge.

Receptive phonological processing, in Friederici's model, is argued to involve regions in and around Heschl's gyrus (lower-level processing) and both anterior and posterior portions of the superior temporal cortex (higher-level phonological processing). This is broadly consistent with the LSM in terms of the role of the mid/posterior STG in phonological processing.

In my view, a major challenge faces Friederici's model, stemming from the assumption that the core function supporting syntactic processing, Merge, operates in the left posterior inferior frontal cortex. The problem is that this predicts severe syntactic deficits, both receptive and expressive, following damage to that system. This prediction is not supported by the evidence, as discussed in chapter 9. Instead, syntactic expressive deficits are associated with more anterior and dorsal lesions (Matchin et al., 2020) and do not cooccur with receptive syntactic deficits (Matchin et al., 2023), while posterior damage consistently disrupts receptive syntactic ability, including acceptability judgments (Fahey et al., 2024), and cooccurs with syntactic expressive deficits (Matchin et al., 2023).

Like Friederici, Peter Hagoort adopts a linguistic framework, that of Jackendoff (2007), and attempts to map its components to neural areas. The broad components are memory, unification, and control (MUC), where *memory* refers to stored knowledge of phonological, morphological, and syntactic building blocks of language; *unification* refers to the combinatorial part of language that assembles the building blocks into larger structures at each level; and *control* refers to the social and executive components. Hagoort (2013) localizes *memory* to the lateral temporal lobe, *unification* to the IFG, and *control* to prefrontal and cingulate cortex. Within the broader *memory* and *unification* systems can be found subdivisions corresponding to the traditional distinctions between phonology, syntax, and semantics appropriately connected to their corresponding level across the broader memory and unification zones.

The MUC model shares some important features with the LSM, notably its linguistic hierarchy and sensorimotor-like architecture, with each level involving both a frontal and temporal lobe component, although Hagoort (2013) does not describe the MUC in these terms. The two models also share the assumption that there is syntactic information coded in the temporal lobe. In the MUC, this involves "syntactic templates" associated with verbs, nouns, and adjectives. Such templates, "treelets" (see Matchin & Hickok, 2020 for a discussion), or something like them, are also assumed to hold for temporal lobe structures in the LSM. The models differ at the syntactic level in that the MUC assumes that the frontal-based *unification* component is required for building phrase-level hierarchies and for both comprehension and production, whereas the LSM assumes that the temporal lobe is responsible for building hierarchies for both receptive and expressive functions, while frontal syntax-related regions are involved in linearization (sequencing) for speech production (chapter 9). Thus, like Friederici's model, the MUC suffers from lack of ability to explain the expressive-receptive asymmetry as it predicts significant receptive syntactic deficits following IFG lesions.

The Motor Control–Grounded Approach

The motor control–grounded approach generates models of the neural organization of language that appear to be birds of a different feather, particularly compared to the psycholinguistic approach. But, as I argued in chapter 10, this is more apparent than real, and it is not that hard to roughly align the various components of the two sorts of models, at least at the phonological level.

Consider Frank Guenther's DIVA model (Guenther, 2016). The *auditory target/error/state maps* are all localized to the pSTG and can be conceptualized as coding phonological information and processes at various levels, while the *speech sound map* and *feedback control map* are localized to ventral premotor cortex and the posterior IFG, also coding motor-related gestures at various phonological levels. Although the terminology is starkly different, DIVA agrees with models in other traditions that propose phonological-level processes linked to the pSTG and inferior IFG. DIVA's inclusion of somatosensory-motor circuits also aligns with Indefrey and Levelt's *articulation* level—a level that the other theorists would certainly acknowledge, even if it is not part of their models.

The two major differences between DIVA and the psycholinguistic models are the scope of the models and the architecture. In terms of scope, DIVA focuses on lower-level processes, from the phonological level to primary motor areas, whereas psycholinguistic models range up to the lemma/morphosyntax and conceptual levels. In terms of architecture, DIVA and psycholinguistic models diverge, with DIVA following a sensorimotor-like architecture, similar to the auditory-inspired models and, of course, the LSM. But, as noted earlier, even the psycholinguistically inspired theorists acknowledge both frontal and temporal involvement in processing linguistic information at a given level.

The LSM model differs from DIVA in important respects. One computational difference is that LSM assumes internal feedback control, whereas DIVA does not, as discussed in Guenther and Hickok (2015). Other differences include the identification of a distinct pitch/prosodic system in the LSM model, its inclusion of higher-level linguistic systems, and its explicit attempt to integrate motor control and psycholinguistic approaches.

Conclusion

As should be clear, the LSM model is not a dramatic departure from any of the existing large-scale models of the neural architecture of language; it shares several features with each one. This is on purpose, as I've been inspired by and tried to retain the best components of the various approaches while abandoning assumptions that aren't strongly supported by a broad assessment of the evidence from all sources relevant to understanding the neural architecture of language.[1] As such, the model includes linguistic representations and processes that are familiar and well established from the psycholinguistic tradition, as well as familiar and well-established principles from research on motor control and dual stream sensory processing architectures. There is much work left to do, but my own assessment is that the LSM model represents a viable set of hypotheses for an integrated large-scale model of the language network that accounts for both functional activation and neurolinguistic data based on an evolutionarily grounded sensorimotor framework.

I think, too, that we've been able to go beyond the glaring problem noted by Gazzaniga, Ivry, and Mangun a quarter-century ago. Not only do we have a reasonable degree of basic understanding of the relation between the processing components of comprehension and production and their neural architecture in the brain, but we now understand something about the relation between the architecture of language and other systems in the brain. They are distinguishable but share a homologous neurocomputational architecture.

Appendix: The Anatomical Foundations of Dorsal and Ventral Processing Streams

Why does the "how" system set up shop in the dorsal fronto-parietal portions of the primate brain and the "what" system occupy the ventral occipital-temporal region? It has to do with the networks with which the two systems must interact, given their computational goals and a basic principle of neural connectivity: minimize distance between connected networks. Let's start with the latter because that principle is easy to motivate.

Why should brain organization aim to minimize connectivity distances? Because, as Jon Kaas (2004, p. 1155) puts it, "distance is time in neuronal computations," and time is critical. Speed of response is obviously a life-or-death matter on the predator-prey battlefield, but even in mundane moments, integration of signals from multiple senses and coordination of motor control across multiple muscle groups and effectors require rapid neural conduction with precise timing. Evolution has taken this problem very seriously, independently evolving myelin-insulated conduction fibers four times, showing up in diverse animals ranging from humans to earthworms to shrimp (Castelfranco & Hartline, 2016). If it is worth the effort to evolve elaborate mechanisms to save neural computation time, it is worth the effort to reduce neural conduction distance as efficiently as possible (Bullmore & Sporns, 2012; Cherniak, 1994; Cherniak et al., 2004; Chklovskii & Koulakov, 2004; Chklovskii, Schikorski, & Stevens, 2002).

The "How" Interface—Primacy of the Body Sense

It is an interesting fact that some of our distant, small-brained mammalian cousins, such as the opossum, lack a differentiable motor cortex (Beck, Pospichal, & Kaas, 1996). It's not that their cortex is undifferentiated in general. No—they possess a very well differentiated complement of sensory cortices, including five distinct somatosensory fields. It's just that they lack a zone that has the anatomy and functional properties of motor cortex. Instead, what cortical motor function that exists in the opossum can be found in somatosensory cortex: electrical stimulation produces movement, albeit at much elevated thresholds (Beck et al., 1996). Even in species that have a differentiated motor cortex, one finds direct neuronal projections from somatosensory fields into the corticospinal tract (Galea & Darian-Smith, 1994; Nudo & Masterton, 1990). More tellingly, research on the cortical control of the mouse whisker system has uncovered evidence of direct control of certain whisking actions by neurons in primary somatosensory cortex (Matyas et al., 2010). Among the cortical sensory systems, these motor-related properties are unique to somatosensation (Galea & Darian-Smith, 1994), indicating a uniquely intimate relation between the lowest levels of motor control and body sense.

The dominance of somatosensory information at the lowest levels of cortical motor control is also revealed by comparing the influence of information from different senses during a motor task. That is, we can ask the question: Which sensory system has the strongest influence over the motor system? One study that directly compared artificially altered somatosensory versus visual feedback in a hand-movement task found that behavior was almost entirely predicted by somatosensory feedback (Crevecoeur, Munoz, & Scott, 2016). A particularly dramatic example comes from the case of Ian Waterman, who lost somatosensory input from the neck down as a result of a viral infection. Despite fully intact motor, visual, and auditory systems, he was left unable to move purposefully. Eventually, and with great conscious effort, he taught himself to move again—walk, drive, gesture, and so on—using visual feedback (Cole & Sedgwick, 1992). But his case was unusual; most people with similar conditions remain wheelchair bound for the rest of their days. Control of the body depends on somatosensory input.

This all makes sense, at least from the perspective of an armchair theorist: body position information is critical for controlling body movements because you have to know where the body is now(ish) to know how to move it next, and that information is most reliably given by somatosensation. How could it *not* be the case that somatosensation dominates the other senses for arm movements, locomotion, and similar actions? Now, says the armchair theorist, if we were talking about controlling a part of the body that was intimately tied to vision (such as ocular control) or hearing (such as speech), we would find visual or auditory dominance in motor control. This is what I thought, but astonishingly, it's not the case.

Consider the case of speech, the quintessential auditory action in the sense that we get precise acoustic feedback on the sensory consequences of our speech gestures. We know that auditory information is absolutely critical in enabling speech motor control, as profound early-onset deafness all but precludes the ability to speak. (It certainly doesn't preclude language, though, as we know from the signed languages of the Deaf.) However, when auditory and somatosensory cues in speech motor control are pitted against one other in artificially altered sensory feedback experiments, somatosensory cues are processed more rapidly (suggesting more direct, lower-level control) and are weighted more strongly, at least for some aspects of speech motor control (Larson et al., 2008). Another clue is provided by the differential effects of blocking somatosensory versus auditory feedback during speech production. It is a common experience that speech is difficult to articulate after a mouth-numbing dental visit. Controlled experiments have confirmed this informal observation by selectively anesthetizing the lingual nerve, which carries sensory information from the tongue, and documenting the deleterious effects on speech (Niemi et al., 2002, 2004, 2006).

Blocking auditory feedback does not produce a similar difficulty with speech, at least not unless it is blocked chronically. You can get an intuitive sense of this by plugging your ears and whispering to a friend (the whispering prevents you from hearing yourself via bone conduction). It should feel like your speech is perfectly clear and easy to produce under these conditions. Ask your friend to confirm this, or record it and listen to it yourself. Experimentally, this has been studied by asking participants to speak while a masking noise is played to them over headphones while their speech is recorded noise-free. No decrements in speech articulation have been reported under these conditions. In fact, quite the opposite—talkers actually generate *more* intelligible speech under noise-masked conditions (Dreher & O'Neill, 1958; Summers et al., 1988), presumably in an effort, conscious or not, to help their listeners. But the fact that they can increase intelligibility when acoustic feedback is severely limited indicates that, unlike in the somatosensory case, overt auditory feedback is not necessary for low-level control over motor speech ability. This is not to say that auditory information is unimportant, but the effects are at a higher level (chapter 10).

What about vision? If speech is the quintessential auditory action, ocular control—looking from one place to the next, tracking a moving target with the eyes—would seem to be the quintessential *visually* guided motor act. Certainly, one might muse, somatosensory information from the muscles around the eyes plays a back-seat role, if any at all. Remarkably, however, in experiments that pit the retinal position of a target against extraretinal eye position information from muscle receptors in a point-to-target task, the latter trumps the former. Observers systematically mispoint to targets that are accurately recorded by retinal position if their sensed eye position is artificially displaced (Rine & Skavenski, 1997).

To be clear, I'm not necessarily suggesting that somatosensory is king in all circumstances and at all levels of the sensorimotor hierarchy. I'm talking here primarily about sensorimotor circuits in the cerebral cortex, where much evidence points to the view that the somatosensory system has a particularly tight and low-level relationship with the cortical motor system. To recapitulate the highlights, (1) ancestrally in our mammal line, the cortical somatosensory and motor systems were one and the same, and distinct from visual and auditory systems; (2) the somatosensory system is the only sensory modality with direct input to the corticospinal tract, preserving its ancestral relation; and (3) somatosensory information often takes precedence over other sensory information in rapid guidance of motor acts. We can think of somatosensation as the system holding the reins of the motor system, rendering the most direct and lowest level of control, with vision and hearing seated behind it in the metaphorical carriage providing information about what is beyond its immediate ability to detect.

Why bother to traverse this territory of the relation between the senses in the context of motor control? Recall that we are trying to understand why the "how" stream is organized in dorsal-anterior brain regions relative to the other (nonchemical) senses. The answer that I'm working toward is that "how" networks in general will ultimately need to talk to the cortex's gatekeeper to the motor system, the lowest level of cortical control. I've argued that the gatekeeper is somatosensory, which means that "how" streams will have to interface with this network.

Where does the somatosensory gatekeeper live in the cerebral cortex? In the dorsal part of the brain. Specifically, primary somatosensory cortex (S1) in primates is located in and immediately posterior to the central sulcus, which separates the frontal and parietal lobes. Primary motor cortex (M1) is located in and immediately anterior to the central sulcus. These facts dictate that visual information emanating from the far posterior end of the brain in the occipital lobe, as well as auditory information emanating from the dorsal surface of the temporal lobe, must make its way toward somatosensory cortex in the parietal lobe if it is to play a role in guiding action. There are strong genetic constraints on this organizational recipe, in that the relative locations of the different sensory fields are remarkably conserved in mammals (Kaas, 2004; Krubitzer, 2009; Manns & Eichenbaum, 2006). To be sure, the precise relationships between the sensory fields have drifted around a bit, plate tectonic–like, during the evolution of divergent mammal lines, but the overall similarity from platypus to primate suggests evolutionary roots of the dorsal "how" stream can be traced to at least 200 million years ago.

The "What" Interface

We now understand why the "how" system is consistently found in dorsal brain regions. What about "what"? We adopt the same strategy and look to the system that a "what" stream must interface with and see if we can derive its location.

The macaque "what" visual stream interfaces with three cortical systems: the medial temporal lobe memory system, which is important for forming long-term memories; the orbital frontal cortex, which is important for building (or dismantling) stimulus-reward associations; and the ventral-lateral prefrontal cortex (VLPFC), which plays a role in linking stimulus information with the task goals at hand (Kravitz et al., 2013). To illustrate, upon encountering a colorful object dangling from a tree limb, the ventral visual stream would do the job of extracting a representation of the object's identity, both its semantic category and how that particular object relates to other instances of that category (Hung et al., 2005), which would have been established and honed by past experiences with similar objects via the interface with the medial temporal lobe memory system. The potential reward value of the object—how tasty it will be compared to other types of food or past experiences with similar objects—would be determined via the interface with the orbital frontal cortex. If the reward is deemed worthy of the effort, a plan for retrieving the object, including the temporary memory of it should the plan require briefly losing sight of the morsel, would be enabled by the interface with VLPFC. Given the location of the cortical networks that the visual "what" stream interfaces with (i.e., the medial temporal and inferior/lateral frontal lobes), it is no surprise to find it coursing toward the temporal pole, which sits roughly between its interface networks.

In humans, cortical networks near the temporal pole have a distinctly semantic function (chapter 8), which is broadly consistent with the picture that has emerged from monkey visual "what" stream research. A neurodegenerative condition called *semantic dementia* (also called the *semantic variant of primary progressive aphasia*) causes a gradual but ultimately severe decline in semantic knowledge, which has been linked to cell loss in the anterior temporal lobes (Acosta-Cabronero et al., 2011; Davies et al., 2005; Hodges & Patterson, 2007; Mummery et al., 2000; Mummery et al., 1999). Affected individuals have difficulty recognizing and categorizing objects by sight or through language, suggesting that the problem is not tied to just the visual modality but instead is more general. This and other research, discussed in more depth in chapter 8, has led to the conclusion that the anterior temporal lobe (ATL) is an important hub for representing object- or entity-based semantic information (Binder & Desai, 2011; Binder et al., 2009; Mesulam et al., 2015; Patterson et al., 2007). Such a network would seem to be an important target for "what" streams, in that sensory modality-specific object information could be linked with a more abstract, semantic memory. It is unlikely, however, that such a network served as a phylogenetically preexisting anatomical target for an evolving "what" stream, as a homologue in even our primate relatives has not been identified (Kravitz et al., 2013). More likely, in my view, an abstract semantic hub in the ATL was a consequence, not a cause, of ventral "what" stream evolution.

So far, our discussion of the "what" stream has focused mostly on vision. What about the other nonchemical senses? Let's start with audition. We've known for more than a century that deficits in the ability to comprehend the "what" of speech are associated with temporal lobe lesions (Wernicke, 1874/1969; also see chapter 5). Research on the "what" stream in macaque monkeys has also found a similar temporal lobe organization (Heffner & Heffner, 1984; Ortiz-Rios et al., 2015; Rauschecker et al., 1995). In that species, all evidence points toward an anterior-going stream in the superior part of the temporal lobe that parallels the ventral visual "what" stream. There also appears to be some similarities in the targets of the auditory and visual "what" streams, including the orbital frontal cortex, VLPFC, and the medial temporal lobe (Romanski et al., 1999; Suzuki & Amaral, 1994). The ventral stream organization in humans is a bit more controversial, but most agree that the temporal lobe is implicated and an important target is the anterior temporal lobe and beyond.

It is uncontroversial, then, that both vision and hearing have "what" streams. What about body sense? Is there a somatosensory "what" stream? Functionally, it not hard to imagine. We can readily recognize by touch the distinction between our phone and lip balm in the dark of night and do so independent of their position or orientation on the bedside table or whether we are exploring with our left or right hand. We can even recognize objects by touch when they are passively impressed on our bodies. This provides a conceptual argument for the existence of a functional division of labor, even within the somatosensory system. Not surprisingly then, the neuropsychology backs it up, providing evidence for dissociations between the ability to recognize objects by touch versus generating exploratory movements based on touch information (Caselli, 1991; Dijkerman & de Haan, 2007). Consider the following cases.

E.C. was a sixty-five-year-old woman who suffered a left inferior parietal lobe stroke and complained of difficulty recognizing objects in her pocketbook by touch with her right hand (Reed & Caselli, 1994; Reed, Caselli, & Farah, 1996). Formal testing revealed that her basic somatosensory function was intact in both hands, including the ability to detect light touch, pinpricks, temperature, vibration, and other sensations. She could also correctly discriminate weight, size, and texture differences among objects. Visual object recognition was flawless. However, when asked to recognize a set of common objects by touch, her tactile agnosia, as it is called, became plainly obvious. Using her right hand, E.C. failed to recognize common objects by touch alone roughly half the time, whereas using her left hand, she performed at 90 percent correct or better. Observation of her exploratory movements with her agnostic hand suggested no abnormalities, and formal testing confirmed that this was not the problem: directed exploration or passive impression of objects on her hand did not resolve her deficit (Reed et al., 1996). E.C.'s difficulty had to do with recovering the "what" of objects that she touched with her affected hand.

A rather different clinical picture characterized a twenty-eight-year-old woman who also complained of difficulty recognizing objects by touch, this time with the left hand following a large right-hemisphere stroke (Valenza et al., 2001). Testing revealed that, like E.C., basic somatosensory function was well preserved in both hands in this person, but tactile recognition of common objects was quite poor in her affected left hand (three out of fourteen items correct) but quite good with her right hand (thirteen of fourteen correct). She had similar trouble recognizing geometric forms (letters, digits, and shapes) by touch with the left hand (five of forty) versus the right hand (thirty-seven of forty). But unlike E.C., her ability to recognize these forms improved dramatically when they were passively impressed on her left hand (thirty of forty). It appeared that her deficit was not one of recognition but of exploration: she simply wasn't able to produce the hand movements necessary to properly explore the shape. The authors describe the difference in exploratory behavior in the right versus left hand:

> When using her right hand, the patient first wrapped the object with her hand, searched for its salient features (e.g. the hairs of a toothbrush) and named it very quickly. In contrast, exploration by the left hand was characterized by slow and iterative rubbing movements with the index and the thumb, often limited to non-discriminative features of the object. The patient hardly ever explored the whole object and never appeared to search for its salient features. (Valenza et al., 2001, p. 2289)

Is this because of some basic dexterity deficit involving the left hand, such that she was incapable of even executing the necessary exploratory movements? No. When the exploration of a shape was guided by verbal instruction ("Move your finger down, now to the right . . .") she was able both to perform the movements and recognize the objects (ten out of ten trials). Her deficit involved the use of tactile object information to guide appropriate movements aimed at exploring that object. Similar cases have been reported by other authors (e.g., Pause et al., 1989).

Thus, conceptual and neuropsychological arguments align with the existence of dissociable "what" and "how" streams in the cortical somatosensory system. What is their neuroanatomical organization? Here, things get a little fuzzy. It's fairly clear from functional imaging in humans and electrophysiology in monkeys that the somatosensory hierarchy involves a pathway that flows from primary areas to regions in the parietal operculum, the underbelly of the parietal lobe, which corresponds to the higher-level somatosensory cortex (Dijkerman & de Haan, 2007; Sathian, 2016; Sathian et al., 2011; Yau et al., 2016). This region is sometimes referred to simply as "S2" for the secondary somatosensory cortex, but in fact, there are at least two separate regions (probably more), termed "S2" and "PV" (the parietal ventral area) (Disbrow et al., 2003; Disbrow, Roberts, & Krubitzer, 2000; Krubitzer et al., 1995).

A third region just posterior to these in the parietal operculum, area 7b, has sensory responses that are predominantly somatosensory, but also with a mix of motor-related responses (Andersen et al., 1990). Of course, there are a whole host of regions on the convexity of the parietal lobe, some of which exhibit somatosensory response properties, but these are less somatocentric and are generally considered multisensory and sensorimotor, including strong visual-related response properties—that is, what many consider the dorsal visual stream.

Returning to the somatosensory regions in the parietal operculum, there is functional imaging evidence for the region playing a role in object recognition in humans and evidence from the monkey that it connects via the posterior insula with limbic structures related to emotion and memory (Friedman et al., 1986). The insular cortex, an islandlike region buried deep inside the Sylvian fissure, is potentially interesting in that it appears to contain an extension of the somatosensory cortex that has been likened to the ventral visual stream (Schneider, Friedman, & Mishkin, 1993). If true, a pathway from S1 to S2/PV to the insula to medial temporal lobe memory/emotion systems would be a strong candidate for the somatosensory ventral stream. Functional imaging studies of haptic object perception reliably report activation in the parietal operculum and insula consistent with this hypothesis, but evidence is not clear-cut: a range of other areas also activate during haptic object perception, including those involved in sensorimotor function (Savini et al., 2010). Further complicating the picture, the connectivity pattern of S2/PV in monkeys includes areas with sensorimotor function such as 7b (connections with both S2 and PV), the intraparietal sulcus (connections with PV), and the premotor cortex (connections with PV), leading some authors to seriously question a division of labor in the somatosensory system (e.g., Disbrow et al., 2003).

I wouldn't go that far. My own evaluation of the state of play with respect to dual somatosensory streams is that conceptual arguments for the "what"/"how" distinction and the corroborating neuropsychological dissociations make a pretty solid case for a functional-anatomic distinction. What's unclear is the neuroanatomy. I think that there are three reasons why. One is that much less research has been conducted on somatosensation compared to vision or audition, so we simply have less information, leaving much room for ambiguity. Another is that the functional distinction between "what" and "how" processes in somatosensation is less clear-cut, given that in many cases, "how" is required to determine "what": objects must be actively explored—a "how" function—in order to build up a sense of "what" it is. This does not imply that the two are one and the same, only that they are often confounded when looking at neural activation and connectivity patterns.

A third reason is that the anatomy of the two proposed somatosensory streams is crowded in the parietal lobe. Recall the argument that a "how" stream for vision or audition will need to make contact with the sensorimotor gatekeeper, somatosensation. This leads the "how" stream for these modalities away from their primary areas and toward the somatosensory cortex. But by the same logic, a somatosensory "how" stream will need to flow toward *itself,* limiting the anatomical differentiation among the streams. For these reasons, the anatomy of the two somatosensory streams remains sketchy.

Let us tentatively propose the following organization for somatosensory cortical streams based on the available evidence plus general organizational principles. The haptic "what" stream emanates from S1 and is elaborated in S2 and then the posterior insula, which connects to the medial temporal lobe. Haptic object-related information also makes its way into the ventral occipital-temporal lobe, activating object-related visual areas (lateral occipital complex, LOC) (Amedi et al., 2001; Tal & Amedi, 2009), but lesion evidence (bilateral LOC lesions) suggests that this area is not necessary for haptic object recognition (Snow, Goodale, & Culham, 2015), so it may serve to augment visual object recognition. The somatosensory "how" system is also hierarchically organized. Direct connections between S1 and M1 constitute the lowest level of control, representing the gatekeeper function discussed earlier. S2, with its dense connectivity from and to S1, likely represents a pathway for higher-order somatosensory guidance of action of the sort that was impaired in the young women described here. Thus S2, or subsections thereof, may serve both "what" and "how" functions, perhaps informed by tactile object-based information via reciprocal inputs from the insula. A further extension of the somatosensory "how" system involves its integration with information from the other senses in the posterior parietal cortex via S2, perhaps PV in particular (Andersen et al., 1990; Borra et al., 2008; Dijkerman & de Haan, 2007; Disbrow et al., 2003). Connectivity between somatic and auditory cortical processing is also well established (Hackett et al., 2007a, 2007b; Schroeder et al., 2003)—they are immediate neighbors in the parietal operculum.

Notes

Chapter 1

1. I was lucky enough to meet Dr. Sacks when I was a postdoc working at the Salk Institute. I had been working with Ursula Bellugi on neuromorphological correlates of auditory and music ability in Williams syndrome, a topic that Sacks was researching for his book *Musicophilia* (Sacks, 2008). He liked my work, which led to a minor publication in *Science* (Hickok, Bellugi, & Jones, 1995) as a follow-on to something that Sacks himself had published there. We corresponded enough that he agreed to write a letter of recommendation for me for my job application at the University of California, Irvine, where I've been a faculty member ever since.

2. The recent development of large language models has led to new debates about how language is learned ("Language models and linguistic theories beyond words," 2023). I do not think that these impressive feats of computer/data science engineering change the way that we think about the biology of language. But that's a topic for another book.

Chapter 2

1. In fact, conscious awareness of a highly practiced automatic behavior might actually interfere with successful execution. Athletes and performers refer to this as the state of being "in your head." The remedy for this problem is "Stop thinking about it," "Let it go," "Your body knows what to do," "Use your muscle memory"; that is, let the automatic action do its thing without conscious interference.

2. Neither was Rauchecker's. In fact, Deutsch and Roll proposed separate "what" and "where" mechanisms in humans based on an interesting auditory illusion. Listeners were presented with different five-tone sequences in each ear. The tone was either 400 Hz (Low) or 800 Hz (High) and followed this pattern:

Sequence 1: H H H L L
Sequence 2: L L L H H

If both sequences are presented to both ears, one hears a sequence of five buzzy tones with a pitch equal to the low tone, as expected. But when the sequences are presented separately to the two ears, most people perceive either HHHLL or LLLHH, indicating that the pattern or auditory object presented to one ear is passing through to consciousness while the pattern in the other ear is suppressed. However, when listeners were asked which ear contained the pattern that they heard, most reported that the high tones were in one ear and the low tones were in the other, in contrast to what was actually presented to them. Listeners' auditory systems appeared to be tracking objects (patterns) in one way and locations in another, presumably a consequence of "separate 'what' and 'where' decision mechanisms," as the paper's title indicated.

An even earlier historical precedent to a dual stream model of audition goes back to the work of Stjepan Poljak, who in 1926 discussed the various subdivisions in "the connections of the acoustic nerve" and came to the conclusion that foreshadowed the modern "what/how" system by the better part of a century. Poljak (1926, p. 468) wrote, "The constituent parts of the central auditory system have mostly a double function—viz. to conduct the peripheral auditory sensations to the prosencephalon [forebrain] on the one hand, and on the other, to establish a reflex path for the cochlear stimuli to the motor mechanisms of the brain stem."

3. Some readers will assert that cupness includes sensorimotor information regarding how we interact with cups. But this can't be right, or at least it contributes very little to the idea of cupness for the reasons that we've already considered: "how" depends on the peculiarities of a given cup (many "hows" to the same "what") and isn't particularly correlated with "what" (the same "how" can be applied to many "whats"). To make the point concrete, consider what a grasp of your morning coffee cup contributes to your concept of "cup." The particular trajectory of a particular reach can't contribute anything useful to the concept, nor can the precise grasp position on the cup, the grip forces applied given the current liquid level, whether you grasp with your left or right hand, or other factors. If there is a sensorimotor contribution to cupness, it must be more abstract. Perhaps it is the type of grip. But this can't be right, as a coffee mug can be grasped differently than a teacup or an espresso cup and the same grip type would apply to a small flowerpot, trophy, or candle. Perhaps it is the act of grasping to bring the item to the mouth and drink. But what about when you grasp a cup and hand it to a friend or move it aside or dump out its contents? Is it less of a cup in those cases? Do people who are unable to drink from cups fail to extract the full concept of a "cup"? If so, what precisely is missing? If what is missing is simply the knowledge of "how" to interact with cups in general, then we have simply stipulated that "how" knowledge is part of the concept, as proof that "how" knowledge contributes to the "what" of the concept.

4. I'm ignoring the role of frontal lobe networks in the present discussion, but I have no doubt that they play critical roles in both streams, which target different frontal areas, and provide another arena for the interaction between the streams.

5. Indeed, the idea of the dual stream model of speech processing that David Poeppel and I proposed in 2000 was directly inspired by Milner and Goodale's research into visual-processing streams in monkeys and humans.

Chapter 3

1. The use of the English term "clang" likely derives from earlier descriptions by the German authors, who used the term *wortklänge*. This would appear to translate directly to the English "word-clang" but is better translated as "word sound." Something about the word "clang," though, captures the cacophony of what speech sounds like when one can't make sense of it, so I use it here interchangeably with more traditional terms.

2. This surely has to do with Geschwind's primary interest in describing the existence of neurological disorders that occur as a result of the disconnection of otherwise-intact brain areas, as the title of his article, "Disconnexion Syndromes in Animals and Man," attests. Pure word deafness wouldn't have made the cut in the article had he not analyzed it as a disconnection syndrome.

3. I made an attempt to quantify Goodglass's puzzle and concluded—by admittedly very rough approximation—that assuming that Geschwind's analysis is right, the likelihood of observing the frequency of bilateral versus unilateral cases of word deafness that are reported in the literature is about 1 in a million (Hickok, 2018). Goodglass had a good point.

4. If one counts unrelated errors as a severe speech sound perception problem, then the rate of semantic errors (15 percent) is slightly less than the rate of presemantic perceptual errors (12.2 + 4.3 = 16.5 percent). However, the authors argued from the pattern in their data that unrelated errors were often a result of combined phonemic and semantic errors: "bear" is misheard as "pear" (phonemic error), which is misunderstood as "grapes" (semantic error). Counting unrelated errors as *both* phonemic and semantic changes the probability of occurrence estimate to .166 for phonemic errors and .194 for semantic ones. No matter how you tally the results, though, the overall deficit is a mix of relatively mild phonemic and semantic problems.

5. A follow-up to this study carried out by Sheila Blumstein's group in 2012 claimed to have provided evidence for a very strong correlation between discrimination (measured differently) and comprehension in Wernicke's aphasia (Robson et al. 2012). However, their comprehension measure was the combined comprehension scale from the Boston Diagnostic Aphasia Examination, which includes a semantically demanding word-to-picture matching test, a test of comprehending complex multistep commands, and a yes/no test involving comprehending complex ideational material (e.g., "Will a cork sink in water?"). The association between discrimination and comprehension, therefore, cannot be tied to misperceiving sounds because a comparative comprehension test was not used.

6. Continuing the earlier quote from Goodglass's puzzle, "It is possible that this disorder [word deafness] can appear only when the Wernicke's area homologue in the right hemisphere is totally unable to process word meanings from auditory input. Such a total absence of rudimentary language comprehension in the right hemisphere may be exceptional and account in part for the rarity of severe word-deafness" (Goodglass, 1993, p. 125).

Chapter 4

1. Petersen et al. (1989, p. 163): "This result has identified the left supramarginal gyrus with phonological as distinct from auditory processing. The activation of the left temporoparietal focus during passive auditory word presentation, but not for auditory clicks or tones, makes this area a good candidate for phonological encoding of words."

Zatorre et al. (1992, p. 1765): "It is consistent with our data to suggest that phonological processing is accomplished through a network including the left posterior temporal and parietal regions as well as Broca's area, although each region's precise contribution remains to be elucidated."

Demonet et al. (1992, p. 1765): "The neural correlates of phonological processes are located primarily in the left associative auditory cortex and spread to Broca's area."

2. In task subtraction approaches to studying speech perception, this result appears to be the norm: bilateral but more extensive on the left. It is natural to assume that this is a functional/computational difference, and indeed it may be. But it still *is* an assumption. There are many noncomputational reasons why one might see differences in activation from one region to the next. There could be differences in vasculature, processing efficiency (perhaps the right is *more* efficient at speech perception than the left, and therefore activates less), or individual variability, which would affect the way that activations average out in a group. So far as I am aware, no one has attempted to assess these potential sources for the left-right differences in such studies.

3. The human face is a good analogy that illustrates this point. Every face is different, but the basic layout is the same. They are more similar to each other than they are different.

4. Many studies of hemispheric asymmetries use an asymmetry index calculated by a formula such as L-R/(R+L), but the simple difference score that I have used is easier to interpret in this case.

Chapter 5

1. Sampling at one of our research sites included both people with and without aphasia. If only cases with aphasia are included, the left-hemisphere stroke group's average percent correct is 96 percent.

2. Discriminability is traditionally measured using a statistic called "d-prime," notated as *d'*, which is based on signal detection theory (Green & Swets, 1966). The advantage of this statistic is that it takes into account an individual's potential response bias, a tendency to default to "same" or "different" in cases of uncertainty. A score near zero would indicate complete guessing and a score of 1 is the conventional threshold boundary for the ability to discriminate the pairs above chance. Integer units correspond to standard deviations so that the best (about 5) and the worst (about 1) performers are four standard deviations apart in discrimination performance (that's a big difference).

3. This is not to say that these tasks are useless—just that they are confounded by their reliance on an additional process beyond speech recognition, phonological working memory, which shows up in functional activation and lesion deficit patterns. Some researchers have been resistant to this claim over the years, and maybe you have doubts as well. I've found the following informal experiment helpful to get a more intuitive sense of how different comprehension and sublexical tasks are. I use it as a demo in my lectures. I ask my audience to listen very carefully to the next statement that I'm going to make. I then speak a sentence, usually "Sublexical tasks depend on processes that are not used during normal speech comprehension." I then immediately ask whether there is a */b/* sound in the sentence. Most people have no idea. Some people try to replay the sentence in their heads to check for a */b/*, and some can do this, but *that's* using working memory. I then ask them to listen for a */b/* as I repeat the sentence. Now most people can hear the */b/*. The point here is that we don't hear speech sounds when we converse; we hear words or word meanings. We *can* listen to sounds consciously, but that requires some additional process or effort. It is this additional process that drives activation in the left frontoparietal circuit, and that can be disrupted following left dorsal stream damage.

4. This is debatable but irrelevant here.

5. TMS is a noninvasive method in which a coil is placed on the scalp over a chosen region. The coil delivers a strong and localized magnetic field through the skull and into the brain tissue below it, which can interfere with neural activity, producing a kind of temporary and fully reversible "virtual lesion."

6. Dewitt and Rauschecker (2012) also looked at activations associated with phrasal-level processing and found even more anterior (and posterior) involvement, including the zone reported by Scott et al. (2000). This suggests to me that the anterior temporal activation is more involved in higher-level sentence or meaning-related processing than speech perception or word recognition. We will examine the neural architecture at these higher levels in subsequent chapters.

Chapter 6

1. For a visual demonstration, search YouTube for "The Basilar Membrane—AT&T Archives," and you should find a video of a model built in the 1960s by AT&T engineers using von Békésy's measurements.

2. Mathematically, these spectrotemporal modulation filters are typically modeled using wavelet analysis resulting in a four-dimensional representation including frequency, time, spectral modulation, and temporal modulation. We don't need to delve into those details here.

Chapter 7

1. *Lemma* is a term borrowed from mathematics and logic that refers to a stepping-stone theorem enabling a larger result.

2. A recent study of the semantic variant of primary progressive aphasia (sv-PPA) claims to have identified cases with dissociations between word comprehension (impaired) and nonverbal semantic ability (spared) (Mesulam et al., 2019). However, their word comprehension task is semantically much more complicated than the bear-pear-moose-grapes task from chapter 3. In the authors' own words: "Although the PPVT-IV is a word–picture matching task, fewer than half of the items represent concrete objects. The majority of the remaining items (e.g., salutation, perplexed, culinary) require extensive conceptual interpretation (i.e., comprehension) of the words" (Mesulam et al., 2019, p. e225). This task, therefore, is likely tapping into conceptual knowledge systems rather than a more basic lexical access process (Matchin et al., 2022).

Chapter 8

1. You may be familiar with Noam Chomsky's interesting claim that language did not evolve for communication but for thought (Chomsky, 2015). This is different from what I'm arguing here in a couple of senses. First, by "language," Chomsky means roughly syntax, the ability for recursive, hierarchical processing. Second, rather than augmentative, he sees this system as an integral part of our species' intellectual ability. Many people find the idea odd at best—that "language" didn't evolve for communication—since communication appears to be what language is mostly used for. I'm skeptical of Chomsky's idea too, but I don't think it is crazy. One has to remember that *language* in his framework has a restricted scope, one that doesn't include phonology or linearization. For him, language is the part of the sound-meaning-sound transformation that is the closest to the conceptual system. It very well may have been that some of the core computational abilities of syntax (operations like Merge) had their start in enhancing reasoning ability, which later became linked (*externalized,* in Chomsky's terminology) with vocal and auditory abilities that evolved for more basic communication. This is a fascinating topic, but one that takes us too far afield from the goals in this book, so I'll leave it at that.

Chapter 9

1. In those days, the term "grammatical structure" was used to refer to morphological markings, which are rich in some languages, such as German, and impoverished in

others, such as English, which used "syntax" (i.e., word order) to convey the same information.

2. Nowadays, we associate agrammatism primarily with nonfluent Broca's aphasia and the tendency to omit grammatical morphemes. Kussmaul's grammatical acataphasia also describes cases of reduced grammatical morphology, but he does not associate this symptom with nonfluent aphasia. The example that he gave from a "feeble-minded girl" was likely a fluent utterance, and he compared her speech pattern with that of young children who have not yet mastered grammatical morphology. Moreover, Kussmaul specifically noted that "*dysgrammatical disorders or pathological agrammatism*" (he proposed alternative terms for grammatical faults of speech) are frequently "a complication of amnestic aphasia and paraphasia" (1877, p. 793), which implicates the fluent aphasias.

3. Norman Geschwind, recall, also worked at the Boston VA, and with Edith Kaplan, he founded the Boston University Aphasia Research Center in the early 1960s. This group, initially directed by Geschwind and later by Goodglass, was a major hub of aphasia research through the 1960s, 1970s, and 1980s.

4. The closest activation to Broca's area in many of these studies was a frontal opercular site that was quite posterior to Broca's area.

5. While working as a postdoctoral scholar in my lab, Meropi Topalidou showed that a phonological sequence can be coded elegantly in terms of connection strength differences between each element in the sequence and a higher-level representation such as a lemma node (Topalidou, Neftci, & Hickok, 2018).

Chapter 10

1. From https://abilitymagazine.com/aphasia-the-movie-a-film-beyond-words/.

2. The existence of separable input versus output subregions of the left auditory-related cortex wouldn't explain why damage to the entirety of the posterior temporal lobe failed to yield significant speech perception problems nor how the isolated right hemisphere could perceive speech so well, but such evidence was not as strong then as it is now, so doubts remained.

3. In the same year, Richard Wise and colleagues independently published a PET study that identified what appeared to be the same posterior medial planum temporale area that activated during overt and silently articulated speech, but not during imagined speech. They interpreted the region as "an interface between posterior temporal cortex and motor cortex for speech" (Wise et al., 2001, p. 93).

4. We're talking here about a basic ability to hear a short melody, retain it, and be able to hum it back recognizably, not trained singing.

5. Truth be told, the acronym was reverse engineered to be an anatomically reasonable approximation to the lab nickname for our new favorite area: *The Spot.*

6. Since this is a rather long-distance projection from the frontal to the temporal lobe areas, this signal probably involves excitatory activation from motor-related systems to interneurons in the auditory system, which then provide the local inhibitory signal.

7. We included simulations to ensure that the fit improvement was not simply due to the added free parameters.

8. Looking at this image, you may wonder why I didn't use the *planning <> interface <> target* architecture at the word (lemma) level, as I proposed in chapter 9. I had suspected that it did follow a sensorimotor architecture, but it would take several more years to work out the details with William Matchin.

Chapter 11

1. Translated in Dronkers et al. (2007).

2. This appears to be a finer-grained parcellation of area 55 described by Oskar Vogt (1910) based on his myelination study.

3. Speech motor planning may incorporate a stopping mechanism as an integral part of the planning process rather than simply a top-down system to halting an ongoing plan. For example, in chapter 10, we discussed internal feedback control as a mechanism for detecting and correcting production planning errors. But this process takes time, which varies from one item to another, necessitating a mechanism to prevent execution before errors are corrected. Planning systems, therefore, may have built-in inhibitory subcircuits that monitor activity in the network and release inhibition only when, for example, a response conflict signal drops below a certain level, reflecting a convergence on an appropriate plan (Nozari, Dell, & Schwartz, 2011).

4. For example, the probability of a speech-sequencing error is related to the position of the element within the larger prosodic phrase: more errors occur the deeper into the phrase one gets, presumably reflecting a decay in the strength of the prosodic planning frame (Choe & Redford, 2012).

5. This does not imply that the dorsal precentral speech area does not respond to intelligible speech. All of the functional activation recordings reviewed in this chapter show that it does respond very well to intelligible speech, in fact, and we found the same in our STRF study. What was missing from the dorsal precentral speech area's activation pattern to intelligible speech was a robust sensitivity to pitch-related regions of spectrotemporal modulation space, which is not necessary for intelligibility of single talkers.

6. The following discussion is from the cited comment (Hickok, 2024) under the Creative Commons license.

7. I say "similar" because in general, we might expect better performance on regular rhythms compared to irregular rhythms simply as a function of task difficulty. This was true in part for the Kang et al. (2021) experiment.

Chapter 12

1. The relative paucity of inclusion of the large body of electroencephalography (EEG) and magnetoencephalography (MEG) data reflects their relative weakness in spatial resolution, which is critical for informing models of the neural architecture.

References

Ackermann, H., Mathiak, K., & Riecker, A. (2007). The contribution of the cerebellum to speech production and speech perception: Clinical and functional imaging data. *Cerebellum, 6*(3), 202–213.

Acosta-Cabronero, J., Patterson, K., Fryer, T. D., Hodges, J. R., Pengas, G., Williams, G. B., & Nestor, P. J. (2011). Atrophy, hypometabolism and white matter abnormalities in semantic dementia tell a coherent story. *Brain, 134*(Pt 7), 2025–2035.

Ahissar, E., Nagarajan, S., Ahissar, M., Protopapas, A., Mahncke, H., & Merzenich, M. M. (2001). Speech comprehension is correlated with temporal response patterns recorded from auditory cortex. *Proceedings of the National Academy of Sciences, 98*(23), 13367–13372.

Albouy, P., Benjamin, L., Morillon, B., & Zatorre, R. J. (2020). Distinct sensitivity to spectrotemporal modulation supports brain asymmetry for speech and melody. *Science, 367*(6481), 1043–1047.

Aliu, S. O., Houde, J. F., & Nagarajan, S. S. (2009). Motor-induced suppression of the auditory cortex. *Journal of Cognitive Neuroscience, 21*(4), 791–802.

Amedi, A., Malach, R., Hendler, T., Peled, S., & Zohary, E. (2001). Visuo-haptic object-related activation in the ventral visual pathway. *Nature Neuroscience, 4*(3), 324–330.

Andersen, R. (1997). Multimodal integration for the representation of space in the posterior parietal cortex. *Philosophical Transactions of the Royal Society of London B Biological Sciences, 352*(1360), 1421–1428.

Andersen, R. A., Asanuma, C., Essick, G., & Siegel, R. M. (1990). Corticocortical connections of anatomically and physiologically defined subdivisions within the inferior parietal lobule. *Journal of Comparative Neurology, 296*(1), 65–113.

Anderson, J. M., Gilmore, R., Roper, S., Crosson, B., Bauer, R. M., Nadeau, S., Beversdorf, D. Q., Cibula, J., Rogash, M. III, Kortencamp, S., . . . , Heilman, K. M. (1999). Conduction aphasia and the arcuate fasciculus: A reexamination of the Wernicke-Geschwind model. *Brain and Language, 70*(1), 1–12.

Angulo-Perkins, A., Aube, W., Peretz, I., Barrios, F. A., Armony, J. L., & Concha, L. (2014). Music listening engages specific cortical regions within the temporal lobes: Differences between musicians and non-musicians. *Cortex, 59*, 126–137.

Archibald, S. J., Mateer, C. A., & Kerns, K. A. (2001). Utilization behavior: Clinical manifestations and neurological mechanisms. *Neuropsychology Review, 11*(3), 117–130.

Archibald, Y. M., Wepman, J. M., & Jones, L. V. (1967). Nonverbal cognitive performance in aphasic and nonaphasic brain-damaged patients. *Cortex, 3*, 275–294.

Assaneo, M. F., Rimmele, J. M., Orpella, J., Ripolles, P., de Diego-Balaguer, R., & Poeppel, D. (2019). The lateralization of speech-brain coupling is differentially modulated by intrinsic auditory and top-down mechanisms. *Frontiers in Integrative Neuroscience, 13*, 28.

Avrutin, S., & Hickok, G. (1992). *Operator/variable relations, referentiality, and agrammatic comprehension.* Unpublished manuscript, Massachusetts Institute of Technology (MIT).

Awh, E., Jonides, J., Smith, E. E., Schumacher, E. H., Koeppe, R. A., & Katz, S. (1996). Dissociation of storage and rehersal in working memory: PET evidence. *Psychological Science, 7*(1), 25–31.

Azadpour, M., & Balaban, E. (2008). Phonological representations are unconsciously used when processing complex, non-speech signals. *PLoS One, 3*(4), e1966.

Baars, B. J., Motley, M. T., & MacKay, D. G. (1975). Output editing for lexical status in artificially elicited slips of the tongue. *Journal of Verbal Learning and Verbal Behavior, 14*(4), 382–391.

Baboyan, V., Basilakos, A., Yourganov, G., Rorden, C., Bonilha, L., Fridriksson, J., & Hickok, G. (2021). Isolating the white matter circuitry of the dorsal language stream: Connectome-symptom mapping in stroke induced aphasia. *Human Brain Mapping, 42*(17), 5689–5702.

Baddeley, A. (1981). The role of subvocalisation in reading. *Quarterly Journal of Experimental Psychology, 33*(4), 439–454.

Baddeley, A. D. (1992). Working memory. *Science, 255*(5044), 556–559.

Baddeley, A., Thomson, N., & Buchanan, M. (1975). Word length and the structure of short-term memory. *Journal of Verbal Learning and Verbal Behavior, 14*(6), 575–589.

Baker, E., Blumsteim, S. E., & Goodglass, H. (1981). Interaction between phonological and semantic factors in auditory comprehension. *Neuropsychologia, 19*(1), 1–15.

Baldo, J. V., Arevalo, A., Patterson, J. P., & Dronkers, N. F. (2013). Grey and white matter correlates of picture naming: evidence from a voxel-based lesion analysis of the Boston Naming Test. *Cortex, 49*(3), 658–667.

Baldo, J. V., Bunge, S. A., Wilson, S. M., & Dronkers, N. F. (2010). Is relational reasoning dependent on language? A voxel-based lesion symptom mapping study. *Brain and Language, 113*(2), 59–64.

Baldo, J. V., & Dronkers, N. F. (2006). The role of inferior parietal and inferior frontal cortex in working memory. *Neuropsychology, 20*(5), 529–538.

Baldo, J. V., Dronkers, N. F., Wilkins, D., Ludy, C., Raskin, P., & Kim, J. (2005). Is problem solving dependent on language? *Brain and Language, 92*(3), 240–250.

Baldo, J. V., Katseff, S., & Dronkers, N. F. (2012). Brain regions underlying repetition and auditory-verbal short-term memory deficits in aphasia: Evidence from voxel-based lesion symptom mapping. *Aphasiology, 26*(3–4), 338–354.

Baldo, J. V., Paulraj, S. R., Curran, B. C., & Dronkers, N. F. (2015). Impaired reasoning and problem-solving in individuals with language impairment due to aphasia or language delay. *Frontiers in Psychology, 6*, 1523.

Barton, B., Venezia, J. H., Saberi, K., Hickok, G., & Brewer, A. A. (2012). Orthogonal acoustic dimensions define auditory field maps in human cortex. *Proceedings of the National Academy of Sciences, 109*(50), 20738–20743.

Basilakos, A., Rorden, C., Bonilha, L., Moser, D., & Fridriksson, J. (2015). Patterns of poststroke brain damage that predict speech production errors in apraxia of speech and aphasia dissociate. *Stroke, 46*(6), 1561–1566.

Basso, A., Casati, G., & Vignolo, L. A. (1977). Phonemic identification defects in aphasia. *Cortex, 13*(1), 84–95.

Bates, E., Friederici, A., & Wulfeck, B. (1987a). Comprehension in aphasia: A cross-linguistic study. *Brain and Language,* 32(1), 19–67.

Bates, E., Friederici, A., & Wulfeck, B. (1987b). Grammatical morphology in aphasia: Evidence from three languages. *Cortex, 23*(4), 545–574.

Bates, E., Wilson, S. M., Saygin, A. P., Dick, F., Sereno, M. I., Knight, R. T., & Dronkers, N. F. (2003). Voxel-based lesion-symptom mapping. *Nature Neuroscience, 6*(5), 448–450.

Bates, E., Wulfeck, B., & MacWhinney, B. (1991). Cross-linguistic research in aphasia: An overview. *Brain and Language, 41*(2), 123–148.

Beauchamp, M. S., Lee, K. E., Argall, B. D., & Martin, A. (2004). Integration of auditory and visual information about objects in superior temporal sulcus. *Neuron, 41*(5), 809–823.

Beck, P. D., Pospichal, M. W., & Kaas, J. H. (1996). Topography, architecture, and connections of somatosensory cortex in opossums: Evidence for five somatosensory areas. *Journal of Computational Neurology, 366*(1), 109–133.

Beeman, M. (1993). Semantic processing in the right hemisphere may contribute to drawing inferences from discourse. *Brain and Language, 44*(1), 80–120.

Belin, P., Zatorre, R. J., & Ahad, P. (2002). Human temporal-lobe response to vocal sounds. *Cognitive Brain Research, 13*(1), 17–26.

Belin, P., Zatorre, R. J., Lafaille, P., Ahad, P., & Pike, B. (2000). Voice-selective areas in human auditory cortex. *Nature, 403*(6767), 309–312.

Belkhir, J. R., Fitch, W. T., Garcea, F. E., Chernoff, B. L., Sims, M. H., Navarrete, E., Haber, S., Paul, D. A., Smith, S. O., Pilcher, W. H., & Mahon, B. Z. (2021). Direct electrical stimulation evidence for a dorsal motor area with control of the larynx. *Brain Stimulation, 14*(1), 110–112.

Belyk, M., & Brown, S. (2014). Perception of affective and linguistic prosody: An ALE meta-analysis of neuroimaging studies. *Social Cognitive and Affective Neuroscience, 9*(9), 1395–1403.

Belyk, M., & Brown, S. (2017). The origins of the vocal brain in humans. *Neuroscience & Biobehavioral Reviews, 77*, 177–193.

Bemis, D. K., & Pylkkanen, L. (2011). Simple composition: A magnetoencephalography investigation into the comprehension of minimal linguistic phrases. *Journal of Neuroscience, 31*(8), 2801–2814.

Bemis, D. K., & Pylkkanen, L. (2013). Basic linguistic composition recruits the left anterior temporal lobe and left angular gyrus during both listening and reading. *Cerebral Cortex, 23*(8), 1859–1873.

Benton, A. (1991). Aphasia: Historical perspectives. In M. T. Sarno (Ed.), *Acquired aphasia* (2nd ed., pp. 1–26). San Diego: Academic Press.

Berezutskaya, J., Baratin, C., Freudenburg, Z. V., & Ramsey, N. F. (2020). High-density intracranial recordings reveal a distinct site in anterior dorsal precentral cortex that tracks perceived speech. *Human Brain Mapping, 41*(16), 4587–4609.

Berker, E. A., Berker, A. H., & Smith, A. (1986). Translation of Broca's 1865 report: Localization of speech in the third left frontal convolution. *Archives of Neurology, 43*(10), 1065–1072.

Berko, J. (1958). The child's learning of English morphology. *Word, 14*(2–3), 150–177.

Berthier, M. (1999). *Transcortical aphasias*. East Sussex, UK: Psychology Press.

Berwick, R. C., & Chomsky, N. (2015). *Why only us: Language and evolution*. Cambridge, MA: MIT Press.

Bever, T. G., & Chiarello, R. J. (1974). Cerebral dominance in musicians and nonmusicians. *Science, 185*(4150), 537–539.

Binder, J. R. (2016). In defense of abstract conceptual representations. *Psychonomic Bulletin & Review, 23*(4), 1096–1108.

Binder, J. R., & Desai, R. H. (2011). The neurobiology of semantic memory. *Trends in Cognitive Sciences, 15*(11), 527–536.

Binder, J. R., Desai, R. H., Graves, W. W., & Conant, L. L. (2009). Where is the semantic system? A critical review and meta-analysis of 120 functional neuroimaging studies. *Cerebral Cortex, 19*(12), 2767–2796.

Binder, J. R., Frost, J. A., Hammeke, T. A., Bellgowan, P. S., Springer, J. A., Kaufman, J. N., & Possing, E. T. (2000). Human temporal lobe activation by speech and nonspeech sounds. *Cerebral Cortex, 10*(5), 512–528.

Binder, J. R., Frost, J. A., Hammeke, T. A., Rao, S. M., & Cox, R. W. (1996). Function of the left planum temporale in auditory and linguistic processing. *Brain, 119*(4), 1239–1247.

Biran, I., & Chatterjee, A. (2004). Alien hand syndrome. *Archives of Neurology, 61*(2), 292–294.

Bishop, D. V., Brown, B. B., & Robson, J. (1990). The relationship between phoneme discrimination, speech production, and language comprehension in cerebral-palsied individuals. *Journal of Speech and Hearing Research, 33*(2), 210–219.

Blackmer, E. R., & Mitton, J. L. (1991). Theories of monitoring and the timing of repairs in spontaneous speech. *Cognition, 39*(3), 173–194.

Blakemore, S. J., Wolpert, D., & Frith, C. (2000). Why can't you tickle yourself? *Neuroreport, 11*(11), R11–R16.

Blank, I., Balewski, Z., Mahowald, K., & Fedorenko, E. (2016). Syntactic processing is distributed across the language system. *NeuroImage, 127*, 307–323.

Blumstein, S. E., Baker, E., & Goodglass, H. (1977). Phonological factors in auditory comprehension in aphasia. *Neuropsychologia, 15*(1), 19–30.

Blumstein, S., & Cooper, W. E. (1974). Hemispheric processing of intonation contours. *Cortex, 10*(2), 146–158.

Boatman, D., Gordon, B., Hart, J., Selnes, O., Miglioretti, D., & Lenz, F. (2000). Transcortical sensory aphasia: Revisited and revised. *Brain, 123*(Pt 8), 1634–1642.

Boatman, D., Hart, J. J., Lesser, R. P., Honeycutt, N., Anderson, N. B., Miglioretti, D., & Gordon, B. (1998). Right hemisphere speech perception revealed by amobarbital injection and electrical interference. *Neurology, 51*(2), 458–464.

Bogen, J. E. (1969). The other side of the brain. II. An appositional mind. *Bulletin of the Los Angeles Neurological Society, 34*(3), 135–162.

Bogen, J. E., Fisher, E. D., & Vogel, P. J. (1965). Cerebral commissurotomy. A second case report. *JAMA, 194*(12), 1328–1329.

Bohland, J. W., Bullock, D., & Guenther, F. H. (2010). Neural representations and mechanisms for the performance of simple speech sequences. *Journal of Cognitive Neuroscience, 22*(7), 1504–1529.

Booij, G. (2010). *Construction morphology*. Oxford: Oxford University Press.

Bornkessel-Schlesewsky, I., & Schlesewsky, M. (2013). Reconciling time, space and function: A new dorsal-ventral stream model of sentence comprehension. *Brain and Language, 125*(1), 60–76.

Borra, E., Belmalih, A., Calzavara, R., Gerbella, M., Murata, A., Rozzi, S., & Luppino, G. (2008). Cortical connections of the macaque anterior intraparietal (AIP) area. *Cerebral Cortex, 18*(5), 1094–1111.

Bouchard, K. E., Mesgarani, N., Johnson, K., & Chang, E. F. (2013). Functional organization of human sensorimotor cortex for speech articulation. *Nature, 495*(7441), 327–332.

Brais, B. (1992). The third left frontal convolution plays no role in language: Pierre Marie and the Paris debate on aphasia (1906–1908). *Neurology, 42*(3 Pt 1), 690–695.

Brazier, M. A. (1963). The history of the electrical activity of the brain as a method for localizing sensory function. *Medical History, 7*(3), 199–211.

Bregman, A. S. (1990). *Auditory scene analysis: The perceptual organization of sound.* Cambridge, MA: MIT Press.

Bremmer, F., Kubischik, M., Hoffmann, K. P., & Krekelberg, B. (2009). Neural dynamics of saccadic suppression. *Journal of Neuroscience, 29*(40), 12374–12383.

Brenowitz, E. A. (1991). Evolution of the vocal control system in the avian brain. *Seminars in Neuroscience, 3*(5), 399–407.

Breshears, J. D., Molinaro, A. M., & Chang, E. F. (2015). A probabilistic map of the human ventral sensorimotor cortex using electrical stimulation. *Journal of Neurosurgery, 123*(2), 340–349.

Breska, A., & Deouell, L. Y. (2017). Neural mechanisms of rhythm-based temporal prediction: Delta phase-locking reflects temporal predictability but not rhythmic entrainment. *PLoS Biology, 15*(2), e2001665.

Broadbent, D. E. (1954). The role of auditory localization in attention and memory span. *Journal of Experimental Psychology, 47*(3), 191–196.

Broca, P. (1861). Nouvelle observation d'aphémie produite par une lésion de la troisième circonvolution frontale. *Bulletins de la Société d'anatomie (Paris), 2e serie, 6,* 398–407.

Broderick, M. P., Anderson, A. J., & Lalor, E. C. (2019). Semantic context enhances the early auditory encoding of natural speech. *Journal of Neuroscience, 39*(38), 7564–7575.

Brown, C. H., Schessler, T., Moody, D., & Stebbins, W. (1982). Vertical and horizontal sound localization in primates. *Journal of the Acoustical Society of America, 72*(6), 1804–1811.

Brown, S., Ngan, E., & Liotti, M. (2008). A larynx area in the human motor cortex. *Cerebral Cortex, 18*(4), 837–845.

Brown, S., Yuan, Y., & Belyk, M. (2021). Evolution of the speech-ready brain: The voice/jaw connection in the human motor cortex. *Journal of Comparative Neurology, 529*(5), 1018–1028.

Brysbaert, M., Stevens, M., Mandera, P., & Keuleers, E. (2016). How many words do we know? Practical estimates of vocabulary size dependent on word definition, the degree of language input and the participant's age. *Frontiers in Psychology, 7,* 1116.

Buchman, A. S., Garron, D. C., Trost-Cardamone, J. E., Wichter, M. D., & Schwartz, M. (1986). Word deafness: One hundred years later. *Journal of Neurology, Neurosurgery, and Psychiatry, 49*(5), 489–499.

Buchsbaum, B. R., Baldo, J., Okada, K., Berman, K. F., Dronkers, N., D'Esposito, M., & Hickok, G. (2011). Conduction aphasia, sensory-motor integration, and phonological short-term memory—An aggregate analysis of lesion and fMRI data. *Brain and Language, 119*(3), 119–128.

Buchsbaum, B. R., & D'Esposito, M. (2008). The search for the phonological store: From loop to convolution. *Journal of Cognitive Neuroscience, 20*(5), 762–778.

Buchsbaum, B., Hickok, G., & Humphries, C. (2001). Role of left posterior superior temporal gyrus in phonological processing for speech perception and production. *Cognitive Science, 25*(5), 663–678.

Buchsbaum, B. R., Olsen, R. K., Koch, P., & Berman, K. F. (2005). Human dorsal and ventral auditory streams subserve rehearsal-based and echoic processes during verbal working memory. *Neuron, 48*(4), 687–697.

Buckingham, H. W. (1992). Phonological production deficits in conduction aphasia. In S. E. Kohn (Ed.), *Conduction aphasia* (pp. 77–116). Hillsdale, NJ: Lawrence Erlbaum Associates.

Bullmore, E., & Sporns, O. (2012). The economy of brain network organization. *Nature Reviews Neuroscience, 13*(5), 336–349.

Bullock, L., Forseth, K. J., Woolnough, O., Rollo, P. S., & Tandon, N. (2023). Supplementary motor area in speech initiation: a large-scale intracranial EEG evaluation of stereotyped word articulation [Electronic Version]. *bioRxiv* (2023).

Buring, D. (2013). Syntax, information structure, and prosody. In M. den Dikken (Ed.), *The Cambridge Handbook of Generative Syntax* (pp. 860–896). Cambridge: Cambridge University Press.

Burnett, T. A., Freedland, M. B., Larson, C. R., & Hain, T. C. (1998). Voice F0 responses to manipulations in pitch feedback. *Journal of the Acoustical Society of America, 103*(6), 3153–3161.

Burnett, T. A., Senner, J. E., & Larson, C. R. (1997). Voice F0 responses to pitch-shifted auditory feedback: A preliminary study. *Journal of Voice, 11*(2), 202–211.

Burnham, D., Vatikiotis-Bateson, E., Vilela Barbosa, A., Menezes, J. V., Yehia, H. C., Morris, R. H., Vignali, G., & Reynolds, J. (2022). Seeing lexical tone: Head and face motion in production and perception of cantonese lexical tones. *Speech Communication, 141*, 40–55.

Burton, M. W., Small, S., & Blumstein, S. E. (2000). The role of segmentation in phonological processing: An fMRI investigation. *Journal of Cognitive Neuroscience, 12*(4), 679–690.

Butterworth, B., & Howard, D. (1987). Paragrammatisms. *Cognition, 26*(1), 1–37.

Buzsáki, G. (2006). *Rhythms of the brain*. Oxford and New York: Oxford University Press.

Cabrera, D., & Morimoto, M. (2007). Influence of fundamental frequency and source elevation on the vertical localization of complex tones and complex tone pairs. *Journal of the Acoustical Society of America, 122*(1), 478–488.

Calvert, G. A., Bullmore, E. T., Brammer, M. J., Campbell, R., Williams, S. C., McGuire, P. K., Woodruff, P. W., Iversen, S. D., & David, A.S. (1997). Activation of auditory cortex during silent lipreading. *Science, 276*(5312), 593–596.

Cancelliere, A. E., & Kertesz, A. (1990). Lesion localization in acquired deficits of emotional expression and comprehension. *Brain and Cognition, 13*(2), 133–147.

Caplan, D., Alpert, N., Waters, G., & Olivieri, A. (2000). Activation of Broca's area by syntactic processing under conditions of concurrent articulation. *Human Brain Mapping, 9*(2), 65–71.

Caplan, D., Gow, D., & Makris, N. (1995). Analysis of lesions by MRI in stroke patients with acoustic-phonetic processing deficits. *Neurology, 45*(2), 293–298.

Caplan, D., & Waters, G. (1999). Verbal working memory and sentence comprehension. *Behavioral and Brain Sciences, 22*(1), 114–126.

Caramazza, A., & Zurif, E. B. (1976). Dissociation of algorithmic and heuristic processes in sentence comprehension: Evidence from aphasia. *Brain and Language, 3*(4), 572–582.

Carey, D. P., Harvey, M., & Milner, A. D. (1996). Visuomotor sensitivity for shape and orientation in a patient with visual form agnosia. *Neuropsychologia, 34*(5), 329–337.

Caruso, V. C., Pages, D. S., Sommer, M. A., & Groh, J. M. (2016). Similar prevalence and magnitude of auditory-evoked and visually evoked activity in the frontal eye fields: Implications for multisensory motor control. *Journal of Neurophysiology, 115*(6), 3162–3173.

Caselli, R. J. (1991). Rediscovering tactile agnosia. *Mayo Clinic Proceeding, 66*(2), 129–142.

Castelfranco, A. M., & Hartline, D. K. (2016). Evolution of rapid nerve conduction. *Brain Research, 1641*(Pt A), 11–33.

Catani, M., Howard, R. J., Pajevic, S., & Jones, D. K. (2002). Virtual in vivo interactive dissection of white matter fasciculi in the human brain. *NeuroImage, 17*(1), 77–94.

Catani, M., Jones, D. K., & Ffytche, D. H. (2005). Perisylvian language networks of the human brain. *Annals of Neurology, 57*(1), 8–16.

(1, R. (1877). IV.—Interim report on investigation of the electric currents of the brain. *British Medical Journal, Suppl. 1*, 62.

Caton, R. (1971). IV.—Interim report on investigation of the electric currents of the brain. *American Journal of EEG Technology, 11*(1), 23–24.

Cavanna, A. E., Critchley, H. D., Orth, M., Stern, J. S., Young, M. B., & Robertson, M. M. (2011). Dissecting the Gilles de la Tourette spectrum: A factor analytic study on 639 patients. *Journal of Neurology, Neurosurgery & Psychiatry, 82*(12), 1320–1323.

Chakraborty, M., & Jarvis, E. D. (2015). Brain evolution by brain pathway duplication. *Philosophical Transactions of the Royal Society B: Biological Sciences, 370*: 20150056. http://doi.org/10.1098/rstb.2015.0056.

Chang, E. F., Kurteff, G., Andrews, J. P., Briggs, R. G., Conner, A. K., Battiste, J. D., & Sughrue, M. E. (2020). Pure apraxia of speech after resection based in the posterior middle frontal gyrus. *Neurosurgery, 87*(3), E383–E389.

Chang, E. F., Kurteff, G., & Wilson, S. M. (2018). Selective interference with syntactic encoding during sentence production by direct electrocortical stimulation of the inferior frontal gyrus. *Journal of Cognitive Neuroscience, 30*(3), 411–420.

Chang, E. F., Raygor, K. P., & Berger, M. S. (2015). Contemporary model of language organization: An overview for neurosurgeons. *Journal of Neurosurgery, 122*(2), 250–261.

Cherniak, C. (1994). Component placement optimization in the brain. *Journal of Neurosurgery, 14*(4), 2418–2427.

Cherniak, C., Mokhtarzada, Z., Rodriguez-Esteban, R., & Changizi, K. (2004). Global optimization of cerebral cortex layout. *Proceedings of the National Academy of Sciences USA, 101*(4), 1081–1086.

Chi, T., Ru, P., & Shamma, S. A. (2005). Multiresolution spectrotemporal analysis of complex sounds. *Journal of the Acoustical Society of America, 118*(2), 887–906.

Chklovskii, D. B., & Koulakov, A. A. (2004). Maps in the brain: What can we learn from them? *Annual Review of Neuroscience, 27*, 369–392.

Chklovskii, D. B., Schikorski, T., & Stevens, C. F. (2002). Wiring optimization in cortical circuits. *Neuron, 34*(3), 341–347.

Choe, W. K., & Redford, M. A. (2012). The distribution of speech errors in multi-word prosodic units. *Laboratory Phonology, 3*(1), 5–26.

Chomsky, N. (1956). Three models for the description of language. *IRE Transactions on Information Theory, 2*(3), 113–124.

Chomsky, N. (1957). *Syntactic structures*. The Hague: Mouton & Co.

Chomsky, N. (1959). Verbal behavior by B. F. Skinner. *Language, 35*(1), 26–58.

Chomsky, N. (1995). *The minimalist program*. Cambridge, MA: MIT Press.

Chomsky, N. (2010). Some simple evo devo theses: How true might they be for language? In R. K. Larson, V. Deprez, & H. Yamakido (Eds.), *The evolution of human language* (pp. 45–62). Cambridge: Cambridge University Press.

Chomsky, N. (2015). *What kind of creatures are we?* New York: Columbia University Press.

Clerget, E., Winderickx, A., Fadiga, L., & Olivier, E. (2009). Role of Broca's area in encoding sequential human actions: A virtual lesion study. *Neuroreport, 20*(16), 1496–1499.

Cohen, B. D., Noblin, C. D., Silverman, A. J., & Penick, S. B. (1968). Functional asymmetry of the human brain. *Science, 162*(3852), 475–477.

Cole, J. (2015). Prosody in context: A review. *Language, Cognition, and Neuroscience, 30*(1-2), 1–31.

Cole, J. D., & Sedgwick, E. M. (1992). The perceptions of force and of movement in a man without large myelinated sensory afferents below the neck. *Journal of Physiology, 449*(1), 503–515.

Collier, R. (1975). Physiological correlates of intonation patterns. *Journal of the Acoustical Society of America, 58*(1), 249–256.

Cona, G., & Semenza, C. (2017). Supplementary motor area as key structure for domain-general sequence processing: A unified account. *Neuroscience Biobehavioral Reviews, 72*, 28–42.

Conner, C. R., Kadipasaoglu, C. M., Shouval, H. Z., Hickok, G., & Tandon, N. (2019). Network dynamics of Broca's area during word selection. *PLoS One, 14*(12), e0225756.

Corballis, M. C. (1980). Laterality and myth. *American Psychologist, 35*(3), 284–295.

Corkin, S. (2013). *Permanent present tense: The unforgettable life of the amnesic patient, H. M.* New York: Basic Books.

Creutzfeldt, O., Hellweg, F. C., & Schreiner, C. (1980). Thalamocortical transformation of responses to complex auditory stimuli. *Experimental Brain Research, 39*(1), 87–104.

Crevecoeur, F., Munoz, D. P., & Scott, S. H. (2016). Dynamic multisensory integration: Somatosensory speed trumps visual accuracy during feedback control. *Journal of Neuroscience, 36*(33), 8598–8611.

Crick, F., & Koch, C. (1990). Towards a neurobiological theory of consciousness. *Neurosciences, 2*, 263–275.

Curtiss, S. (2014). The case of Chelsea: The effects of late age at exposure to langauge on language performance and evidence for the modularity of language and mind. *UCLA Working Papers in Linguistics, 18*, 115–146.

Cutler, A. (2005). Lexical stress. In D. B. Pisoni & R. E. Remez (Eds.), *The handbook of speech perception* (pp. 264–289). Oxford, UK: Blackwell.

Cutler, A., Dahan, D., & van Donselaar, W. (1997). Prosody in the comprehension of spoken language: A literature review. *Lang Speech, 40*(Pt 2), 141–201.

Cutler, A., & Norris, D. (1988). The role of strong syllables in segmentation for lexical access. *Journal of Experimental Psychology: Human Perception and Performance, 14*(1), 113–121.

Dahl, G., E, Yu, D., Deng, L., & Acero, A. (2012). Context-dependnent pre-trained deep neural networks for large-vocabulary speech recognition. *IEEE Transactions on Audio, Speech, and Language Processing, 20*(1), 30–42.

Dalla Bella, S., Bialunska, A., & Sowinski, J. (2013). Why movement is captured by music, but less by speech: Role of temporal regularity. *PLoS One, 8*(8), e71945.

Damasio, A. R. (1992). Aphasia. *New England Journal of Medicine, 326*(8), 531–539.

Damasio, A., Bellugi, U., Damasio, H., Poizner, H., & Van Gilder, J. (1986). Sign language aphasia during left-hemisphere Amytal injection. *Nature, 322*, 363–365.

Damasio, H. (1991). Neuroanatomical correlates of the aphasias. In M. Sarno (Ed.), *Acquired aphasia* (2nd ed., pp. 45–71). San Diego: Academic Press.

Damasio, H., & Damasio, A. R. (1980). The anatomical basis of conduction aphasia. *Brain, 103*(2), 337–350.

Darwin, C. (1871). *The descent of man and selection in relation to sex*. London: John Murray.

D'Ausilio, A., Pulvermuller, F., Salmas, P., Bufalari, I., Begliomini, C., & Fadiga, L. (2009). The motor somatotopy of speech perception. *Current Biology, 19*(5), 381–385.

Davies, R. R., Hodges, J. R., Kril, J. J., Patterson, K., Halliday, G. M., & Xuereb, J. H. (2005). The pathological basis of semantic dementia. *Brain, 128*(Pt 9), 1984–1995.

de Boysson-Bardies, B., Bacri, N., Sagart, L., & Poizat, M. (1981). Timing in late babbling. *Journal of Child Language, 8*(3), 525–539.

Dehaene, S., & Cohen, L. (2011). The unique role of the visual word form area in reading. *Trends in Cognitive Sciences, 15*(6), 254–262.

De La Tourette, G. (1998). Study of a nervous affliction characterized by motor incoordination with echolaha and coprolalia (Jumping, Latah, Myriachit). *History of Psychiatry, 9*(33), 102–120.

Dell, G. S. (1986). A spreading-activation theory of retrieval in sentence production. *Psychological Review, 93*(3), 283–321.

Dell, G. S. (1995). Speaking and misspeaking. In L. R. Glietman & M. Liberman (Eds.), *An invitation to cognitive science: Language* (2nd ed., Vol. 1, pp. 183–208). Cambridge, MA: MIT Press.

Dell, G. S., & O'Seaghdha, P. G. (1992). Stages of lexical access in language production. *Cognition, 42*(1–3), 287–314.

Dell, G. S., Schwartz, M. F., Martin, N., Saffran, E. M., & Gagnon, D. A. (1997). Lexical access in aphasic and nonaphasic speakers. *Psychological Review, 104*(4), 801–838.

Dell, G. S., Schwartz, M. F., Nozari, N., Faseyitan, O., & Branch Coslett, H. (2013). Voxel-based lesion-parameter mapping: Identifying the neural correlates of a computational model of word production. *Cognition, 128*(3), 380–396.

Demonet, J.-F., Chollet, F., Ramsay, S., Cardebat, D., Nespoulous, J.-L., Wise, R., Rascol, A., & Frackowiak, R. (1992). The anatomy of phonological and semantic processing in normal subjects. *Brain, 115*(6), 1753–1768.

den Ouden, D. B., Malyutina, S., Basilakos, A., Bonilha, L., Gleichgerrcht, E., Yourganov, G., Hillis, A. E., Hickok, G., Rorden, C., & Fridriksson, J. (2019). Cortical and structural-connectivity damage correlated with impaired syntactic processing in aphasia. *Human Brain Mapping, 40*(7), 2153–2173.

Deutsch, D., Henthorn, T., & Lapidis, R. (2011). Illusory transformation from speech to song. *Journal of the Acoustical Society of America, 129*(4), 2245–2252.

Deutsch, D., & Roll, P. L. (1976). Separate "what" and "where" decision mechanisms in processing a dichotic tonal sequence. *Journal of Experimental Psychology: Human Perception & Performance*, 2(1), 23–29.

DeWitt, I., & Rauschecker, J. P. (2012). Phoneme and word recognition in the auditory ventral stream. *Proceedings of the National Academy of Sciences USA, 109*(8), E505–E514.

DeWitt, I., & Rauschecker, J. P. (2016). Convergent evidence for the causal involvement of anterior superior temporal gyrus in auditory single-word comprehension. *Cortex, 77*, 164–166.

Dichter, B. K., Breshears, J. D., Leonard, M. K., & Chang, E. F. (2018). The control of vocal pitch in human laryngeal motor cortex. *Cell, 174*(1), 21–31 e29.

Diehl, R. L., & Kluender, K. R. (1989). On the objects of speech perception. *Ecological Psychology, 1*(2), 121–144.

Dijkerman, H. C., & de Haan, E. H. (2007). Somatosensory processes subserving perception and action. *Behavioral and Brain Sciences, 30*(2), 189–201; discussion 201–239.

di Pellegrino, G., Fadiga, L., Fogassi, L., Gallese, V., & Rizzolatti, G. (1992). Understanding motor events: A neurophysiological study. *Experimental Brain Research, 91*(1), 176–180.

Disbrow, E., Litinas, E., Recanzone, G. H., Padberg, J., & Krubitzer, L. (2003). Cortical connections of the second somatosensory area and the parietal ventral area in macaque monkeys. *Journal of Comparative Neurology, 462*(4), 382–399.

Disbrow, E., Roberts, T., & Krubitzer, L. (2000). Somatotopic organization of cortical fields in the lateral sulcus of Homo sapiens: Evidence for SII and PV. *Journal of Comparative Neurology, 418*(1), 1–21.

Doelling, K. B., Arnal, L. H., Ghitza, O., & Poeppel, D. (2014). Acoustic landmarks drive delta-theta oscillations to enable speech comprehension by facilitating perceptual parsing. *NeuroImage, 85*(Pt 2), 761–768.

Doupe, A. J., & Kuhl, P. K. (1999). Birdsong and human speech: Common themes and mechanisms. *Annual Review of Neuroscience, 22*(1), 567–631.

Dreher, J. J., & O'Neill, J. J. (1958). Effects of ambient noise on speaker intelligibility of words and phrases. *Laryngoscope, 68*(3), 539–548.

Dronkers, N. F. (1996). A new brain region for coordinating speech articulation. *Nature, 384*, 159–161.

Dronkers, N. F., Plaisant, O., Iba-Zizen, M. T., & Cabanis, E. A. (2007). Paul Broca's historic cases: High resolution MR imaging of the brains of Leborgne and Lelong. *Brain, 130*(Pt 5), 1432–1441.

Dronkers, N. F., Wilkins, D. P., Van Valin, R. D., Jr., Redfern, B. B., & Jaeger, J. J. (2004). Lesion analysis of the brain areas involved in language comprehension. *Cognition, 92*(1–2), 145–177.

Duffau, H. (2016). White matter pathways in the human. In G. Hickok & S. L. Small (Eds.), *Neurobiology of language* (pp. 129–137). New York: Academic Press.

Duffau, H., Capelle, L., Sichez, N., Denvil, D., Lopes, M., Sichez, J.-P., Bitar, A., & Fohanno, D. (2002). Intraoperative mapping of the subcortical language pathways using direct stimulations. An anatomo-functional study. *Brain, 125*(Pt 1), 199–214.

Duffy, J. R. (2013). *Motor speech disorders: Substrates, differential diagnosis, and management*. St. Louis: Elsevier.

Duffy, J. R. (2019). *Motor speech disorders: Substrates, differential diagnosis, and management* (4th ed.). New York: Elsevier.

Durfee, A. Z., Sheppard, S. M., Blake, M. L., & Hillis, A. E. (2021). Lesion loci of impaired affective prosody: A systematic review of evidence from stroke. *Brain Cognition, 152,* 105759.

Eichert, N., Papp, D., Mars, R. B., & Watkins, K. E. (2020). Mapping human laryngeal motor cortex during vocalization. *Cerebral Cortex, 30*(12), 6254–6269.

Eichert, N., Watkins, K. E., Mars, R. B., & Petrides, M. (2021). Morphological and functional variability in central and subcentral motor cortex of the human brain. *Brain Structure and Function, 226*(1), 263–279.

Eimas, P. D., Siqueland, E. R., Jusczyk, P., & Vigorito, J. (1971). Speech perception in infants. *Science, 171*(968), 303–306.

Elli, G. V., Lane, C., & Bedny, M. (2019). A double dissociation in sensitivity to verb and noun semantics across cortical networks. *Cerebral Cortex, 29*(11), 4803–4817.

Embick, D., Marantz, A., Miyashita, Y., O'Neil, W., & Sakai, K. L. (2000). A syntactic specialization for Broca's area. *Proceedings of the National Academy of Sciences, 97*(11), 6150–6154.

Engel, A. K., Fries, P., & Singer, W. (2001). Dynamic predictions: Oscillations and synchrony in top-down processing. *Nature Reviews Neuroscience, 2*(10), 704–716.

Erber, N. P. (1969). Interaction of audition and vision in the recognition of oral speech stimuli. *Journal of Speech and Hearing Research, 12*(2), 423–425.

Fahey, D., Fridriksson, J., Hickok, G., & Matchin, W. (2024). Lesion-symptom mapping of acceptability judgments in chronic poststroke aphasia reveals the neurobiological underpinnings of receptive syntax. *Journal of Cognitive Neuroscience, 36*(6), 1141–1155.

Fairbanks, G. (1955). Selective vocal effects of delayed auditory feedback. *Journal of Speech and Hearing Disorders, 20*(4), 333–346.

Farahbod, H., Saberi, K., & Hickok, G. (2020). The rhythm of attention: Perceptual modulation via rhythmic entrainment is lowpass and attention mediated. *Attention, Perception, & Psychophysics, 82*(7), 3558–3570.

Fedorenko, E., Behr, M. K., & Kanwisher, N. (2011). Functional specificity for high-level linguistic processing in the human brain. *Proceedings of the National Academy of Sciences USA, 108*(39), 16428–16433.

Fedorenko, E., Blank, I. A., Siegelman, M., & Mineroff, Z. (2020). Lack of selectivity for syntax relative to word meanings throughout the language network. *Cognition, 203,* 104348.

Fedorenko, E., Duncan, J., & Kanwisher, N. (2012). Language-selective and domain-general regions lie side by side within Broca's area. *Current Biology, 22*(21), 2059–2062.

Fedorenko, E., Fillmore, P., Smith, K., Bonilha, L., & Fridriksson, J. (2015). The superior precentral gyrus of the insula does not appear to be functionally specialized for articulation. *Journal of Neurophysiology, 113*(7), 2376–2382.

Fedorenko, E., Piantadosi, S. T., & Gibson, E. A. F. (2024). Language is primarily a tool for communication rather than thought. *Nature, 630*(8017), 575–586.

Fernandino, L., Binder, J. R., Desai, R. H., Pendl, S. L., Humphries, C. J., Gross, W. L., Conant, L. L., & Seidenberg, M. S. (2016). Concept representation reflects multimodal abstraction: A framework for embodied semantics. *Cerebral Cortex, 26*(5), 2018–2034.

Fernandino, L., Tong, J. Q., Conant, L. L., Humphries, C. J., & Binder, J. R. (2022). Decoding the information structure underlying the neural representation of concepts. *Proceedings of the National Academy of Sciences USA, 119*(6), e2108091119.

Ferreira, C. T., Ceccaldi, M., Giusiano, B., & Poncet, M. (1998). Separate visual pathways for perception of actions and objects: Evidence from a case of apperceptive agnosia. *Journal of Neurology, Neurosurgery & Psychiatry, 65*(3), 382–385.

Fiez, J. A., Raichle, M. E., Miezin, F. M., Petersen, S. E., Tallal, P., & Katz, W. F. (1995). PET studies of auditory and phonological processing: Effects of stimulus characteristics and task demands. *Journal of Cognitive Neuroscience, 7*(3), 357–375.

Fisher, C. M. (1988). Neurologic fragments. I. Clinical observations in demented patients. *Neurology, 38*(12), 1868–1873.

Fitch, W. T. (2010). *The evolution of language*. Cambridge: Cambridge University Press.

Fitch, W. T. (2017). Empirical approaches to the study of language evolution. *Psychonomic Bulletin & Review, 24*(1), 3–33.

Fitch, W. T. (2018). The biology and evolution of speech: A comparative analysis. *Annual Review of Linguistics, 4*, 255–279.

Fitch, W. T. (2019). Sequence and hierarchy in vocal rhythms and phonology. *Annals of the New York Academy of Sciences, 1453*(1), 29–46.

Fitch, W. T., de Boer, B., Mathur, N., & Ghazanfar, A. A. (2016). Monkey vocal tracts are speech-ready. *Science Advances, 2*(12), e1600723.

Fitch, W. T., Hauser, M. D., & Chomsky, N. (2005). The evolution of the language faculty: Clarifications and implications. *Cognition, 97*(2), 179–210; discussion 211–225.

Fitch, W. T., & Zuberbuhler, K. (2013). Primate precursors to human language: Beyond discontinuity. In E. Altenmuller, S. Schmidt, & E. Zimmermann (Eds.), *Evolution of emotional communication: From sound in nonhuman mammals to speech and music in man.* (pp. 26–48). Oxford: Oxford University Press.

Flinker, A., Doyle, W. K., Mehta, A. D., Devinsky, O., & Poeppel, D. (2019). Spectrotemporal modulation provides a unifying framework for auditory cortical asymmetries. *Nature Human Behaviour, 3*(4), 393–405.

Flinker, A., Korzeniewska, A., Shestyuk, A. Y., Franaszczuk, P. J., Dronkers, N. F., Knight, R. T., & Crone, N. E. (2015). Redefining the role of Broca's area in speech. *Proceedings of the National Academy of Sciences USA, 112*(9), 2871–2875.

Fodor, J. A. (1983). *The modularity of mind.* Cambridge, MA: MIT Press.

Fonseca, J., Raposo, A., & Martins, I. P. (2018). Cognitive performance and aphasia recovery. *Topics in Stroke Rehabilitation, 25*(2), 131–136.

Formisano, E., De Martino, F., Bonte, M., & Goebel, R. (2008). "Who" is saying "what"? Brain-based decoding of human voice and speech. *Science, 322*(5903), 970–973.

Forseth, K. J., Hickok, G., Rollo, P. S., & Tandon, N. (2020). Language prediction mechanisms in human auditory cortex. *Nature Communications, 11*(1), 5240.

Forseth, K. J., Kadipasaoglu, C. M., Conner, C. R., Hickok, G., Knight, R. T., & Tandon, N. (2018). A lexical semantic hub for heteromodal naming in middle fusiform gyrus. *Brain, 141*(7), 2112–2126.

Foulis, D. (1879). A case in which there was destruction of the third left frontal convolution without aphasia. *British Medical Journal, 1*(950), 383–384.

Foygel, D., & Dell, G. S. (2000). Models of impaired lexical access in speech production. *Journal of Memory and Language, 43*(2), 182–216.

Freed, D. B., Marshall, R. C., & Chuhlantseff, E. A. (1996). Picture naming variability: A methodological consideration of inconsistent naming responses in fluent and nonfluent aphasia. *Clinical Aphasiology, 24,* 193–205.

Fridriksson, J., den Ouden, D. B., Hillis, A. E., Hickok, G., Rorden, C., Basilakos, A., Yourganov, G., & Boniha, L. (2018). Anatomy of aphasia revisited. *Brain, 141*(3), 848–862.

Fridriksson, J., Fillmore, P., Guo, D., & Rorden, C. (2015). Chronic Broca's aphasia is caused by damage to Broca's and Wernicke's areas. *Cerebral Cortex, 25*(12), 4689–4696.

Friederici, A. (2006). The neural basis of sentence processing: Inferior frontal and temporal contributions. In Y. Grodzinsky & K. Amunts (Eds.), *Broca's region* (pp. 196–217). Oxford: Oxford University Press.

Friederici, A. D. (2015). White-matter pathways for speech and language processing. *Handbook of Clinical Neurology, 129,* 177–186.

Friederici, A. (2017). *Language in our brain: The origins of a uniquely human capacity.* Cambridge, MA: MIT Press.

Friederici, A. D., Makuuchi, M., & Bahlmann, J. (2009). The role of the posterior superior temporal cortex in sentence comprehension. *Neuroreport, 20*(6), 563–568.

Friederici, A. D., Meyer, M., & von Cramon, D. Y. (2000). Auditory languge comprehension: An event-related fMRI study on the processing of syntactic and lexical information. *Brain and Language, 74*(2), 289–300.

Friedman, D. P., Murray, E. A., O'Neill, J. B., & Mishkin, M. (1986). Cortical connections of the somatosensory fields of the lateral sulcus of macaques: Evidence for a corticolimbic pathway for touch. *Journal of Comparative Neurology, 252*(3), 323–347.

Friston, K. J., Price, C. J., Fletcher, P., Moore, C., Frackowiak, R. S., & Dolan, R. J. (1996). The trouble with cognitive subtraction. *NeuroImage, 4*(2), 97–104.

Fritsch, G., & Hitzig, E. (2009/1870). Electric excitability of the cerebrum (Uber die elektrische Erregbarkeit des Grosshirns). *Epilepsy Behavior, 15*(2), 123–130.

Fromkin, V. (1971). The non-anomalous nature of anomalous utterances. *Language 47*(1), 27–52.

Gagnepain, P., Chetelat, G., Landeau, B., Dayan, J., Eustache, F., & Lebreton, K. (2008). Spoken word memory traces within the human auditory cortex revealed by repetition priming and functional magnetic resonance imaging. *Journal of Neuroscience, 28*(20), 5281–5289.

Gainotti, G., Caltagirone, C., & Ibba, A. (1975). Semantic and phonemic aspects of auditory language comprehension in aphasia. *Linguistics, 13*(154–155), 15–30.

Gajardo-Vidal, A., Lorca-Puls, D. L., Team, P., Warner, H., Pshdary, B., Crinion, J. T., Leff, A. P., Hope, T. M. H., Geva, S., Seghier, M. L., . . . , & Price, C. J. (2021). Damage to Broca's area does not contribute to long-term speech production outcome after stroke. *Brain, 144*(3), 817–832.

Galaburda, A. M. (1982). Histology, architectonics, and asymmetry of language areas. In M. A. Arbib, D. Caplan, & J. C. Marshall (Eds.), *Neural models of language processes* (pp. 435–445). San Diego: Academic Press.

Galaburda, A., & Sanides, F. (1980). Cytoarchitectonic organization of the human auditory cortex. *Journal of Comparative Neurology, 190*(3), 597–610.

Galantucci, B., Fowler, C. A., & Turvey, M. T. (2006). The motor theory of speech perception reviewed. *Psychonomic Bulletin & Review, 13*(3), 361–377.

Galea, M. P., & Darian-Smith, I. (1994). Multiple corticospinal neuron populations in the macaque monkey are specified by their unique cortical origins, spinal terminations, and connections. *Cerebral Cortex, 4*(2), 166–194.

Gall, F. J. (1835). *On the functions of the brain and of each of its parts: With observations on the possibility of determining the instincts, propensities, and talents or the moral and intellectual dispositions of men and animals, by the configuration of the brain and head.* Translated by W. Lewis. Boston: Marsh, Capen, & Lyon.

Gallese, V., Fadiga, L., Fogassi, L., & Rizzolatti, G. (1996). Action recognition in the premotor cortex. *Brain, 119*(Pt 2), 593–609.

Ganong, W. F. III. (1980). Phonetic categorization in auditory word perception. *Journal of Experimental Psychology: Human perception and Performance, 6*(1), 110–125.

Ganos, C., Ogrzal, T., Schnitzler, A., & Munchau, A. (2012). The pathophysiology of echopraxia/echolalia: Relevance to Gilles de la Tourette syndrome. *Movement Disorders, 27*(10), 1222–1229.

Gardner, H. (1987). *The mind's new science: A history of the cognitive revolution.* New York: Basic Books.

Garrett, M. F. (1975). The analysis of sentence production. In G. H. Bower (Ed.), *The psychology of learning and motivation* (pp. 133–177). New York: Academic Press.

Gauvin, H. S., & Hartsuiker, R. J. (2020). Towards a new model of verbal monitoring. *Journal of Cognition, 3*(1), 17.

Gazzaniga, M. S. (1967). The split brain in man. *Scientific American, 217*(2), 24–29.

Gazzaniga, M. S. (2015). *Tales from both sides of the brain: A life in neuroscience.* New York: Ecco/HarperCollins Publishers.

Gazzaniga, M. S., Bogen, J. E., & Sperry, R. W. (1962). Some functional effects of sectioning the cerebral commissures in man. *Proceedings of the National Academy of Sciences USA, 48*(10), 1765–1769.

Gazzaniga, M. S., Ivry, R., & Mangun, G. (1998). *Cognitive neuroscience: The biology of the mind.* New York: Norton.

Gazzaniga, M. S., Smylie, C. S., Baynes, K., Hirst, W., & McCleary, C. (1984). Profiles of right hemisphere language and speech following brain bisection. *Brain and Language, 22*(2), 206–220.

Gazzaniga, M. S., & Sperry, R. W. (1967). Language after section of the cerebral commissures. *Brain, 90*(1), 131–148.

Geschwind, N. (1965). Disconnexion syndromes in animals and man. *Brain, 88*(2), 237–294, 585–644.

Geschwind, N. (1970). The organization of language and the brain. *Science, 170*(3961), 940–944.

Geschwind, N., & Levitsky, W. (1968). Human brain: Left-right asymmetries in temporal speech region. *Science, 161*(3837), 186–187.

Geschwind, N., Quadfasel, F. A., & Segarra, J. M. (1968). Isolation of the speech area. *Neuropsychologia, 6*(4), 327–340.

Ghazanfar, A. A., Takahashi, D. Y., Mathur, N., & Fitch, W. T. (2012). Cineradiography of monkey lip-smacking reveals putative precursors of speech dynamics. *Current Biology, 22*(13), 1176–1182.

Giahi Saravani, A., Forseth, K. J., Tandon, N., & Pitkow, X. (2019). Dynamic brain interactions during picture naming. *eNeuro, 6*(4), ENEURO.0472-18.2019.

Gibson, E. (1991). *A computational theory of human linguistic processing: Memory limitations and processing breakdown.* Unpublished PhD thesis, Carnegie Mellon University, Pittsburgh.

Gibson, E. (1998). Linguistic complexity: Locality of syntactic depedencies. *Cognition, 68*(1), 1–76.

Gibson, E., & Fedorenko, E. (2013). The need for quantitative methods in syntax and semantics research. *Language and Cognitive Processes, 28*(1–2), 88–124.

Giraud, A. L., Lorenzi, C., Ashburner, J., Wable, J., Johnsrude, I., Frackowiak, R., & Kleinschmidt, A. (2000). Representation of the temporal envelope of sounds in the human brain. *Journal of Neurophysiology, 84*(3), 1588–1598.

Giraud, A. L., & Poeppel, D. (2012). Cortical oscillations and speech processing: Emerging computational principles and operations. *Nature Neuroscience, 15*(4), 511–517.

Glasser, M. F., Coalson, T. S., Robinson, E. C., Hacker, C. D., Harwell, J., Yacoub, E., Ugurbil, K., Andersson, J., Beckermann, C. F., Jenkinson, M., . . . , & Van Essen, D. O. (2016). A multi-modal parcellation of human cerebral cortex. *Nature, 536*(7615), 171–178.

Goldberg, A. E. (1995). *Constructions: A construction grammar approach to argument structure.* Chicago: University of Chicago Press.

Golfinopoulos, E., Tourville, J. A., Bohland, J. W., Ghosh, S. S., Nieto-Castanon, A., & Guenther, F. H. (2011). fMRI investigation of unexpected somatosensory feedback perturbation during speech. *NeuroImage, 55*(3), 1324–1338.

Golfinopoulos, E., Tourville, J. A., & Guenther, F. H. (2010). The integration of large-scale neural network modeling and functional brain imaging in speech motor control. *NeuroImage, 52*(3), 862–874.

Gonzalez, R., Rojas, M., & Ardila, A. (2020). Non-linguistic abilities in aphasia. *Journal of Neurolinguistics, 56*, 100916.

Goodale, M. A., & Milner, A. D. (1992). Separate visual pathways for perception and action. *Trends in Neurosciences, 15*(1), 20–25.

Goodale, M. A., & Milner, A. D. (2010). Two visual streams: Interconnections do not imply duplication of function. *Cognitive Neuroscience, 1*(1), 65–68.

Goodale, M. A., Milner, A. D., Jakobson, L. S., & Carey, D. P. (1991). A neurological dissociation between perceiving objects and grasping them. *Nature, 349*(6305), 154–156.

Goodglass, H. (1992). Diagnosis of conduction aphasia. In S. E. Kohn (Ed.), *Conduction aphasia* (pp. 39–49). Hillsdale, NJ: Lawrence Erlbaum Associates.

Goodglass, H. (1993). *Understanding aphasia.* San Diego: Academic Press.

Goodglass, H., Berko Gleason, J., Ackerman Bernholtz, N., & Hyde, M. R. (1972). Some linguistic structres in the speech of a Broca's aphasic. *Cortex, 8*(2), 191–212.

Goodglass, H., Christiansen, J. A., & Gallagher, R. (1993). Comparison of morphology and syntax in free narrative and structured tests: Fluent vs. nonfluent aphasics. *Cortex, 29*(3), 377–407.

Goodglass, H., & Mayer, J. (1958). Agrammatism in aphasia. *Journal of Speech and Hearing Disorders, 23*(1), 99–111.

Gordon, H. W. (1978). Left hemisphere dominance for rhythmic elements in dichotically-presented melodies. *Cortex, 14*(1), 58–70.

Gow, D. W., Jr. (2012). The cortical organization of lexical knowledge: A dual lexicon model of spoken language processing. *Brain and Language, 121*(3), 273–288.

Gracco, V. L., & Lofqvist, A. (1994). Speech motor coordination and control: Evidence from lip, jaw, and laryngeal movements. *Journal of Neuroscience, 14*(11 Pt 1), 6585–6597.

Graff-Radford, J., Jones, D. T., Strand, E. A., Rabinstein, A. A., Duffy, J. R., & Josephs, K. A. (2014). The neuroanatomy of pure apraxia of speech in stroke. *Brain and Language, 129,* 43–46.

Green, D. M., & Swets, J. A. (1966). *Signal detection theory and psychophysics.* New York: Wiley.

Greenberg, S. (2005). A multi-tier theoretical framework for understanding spoken language. In S. Greenberg & W. A. Ainsworth (Eds.), *Listening to speech: An auditory perspective* (pp. 411–433). Mahwah, NJ: Erlbaum.

Griffiths, T. D., & Warren, J. D. (2002). The planum temporale as a computational hub. *Trends Neurosciences, 25*(7), 348–353.

Grodzinsky, Y. (1986). Language deficits and the theory of syntax. *Brain and Language, 27*(1), 135–159.

Guenther, F. H. (2014). Auditory feedback control is involved at even sub-phonemic levels of speech production. *Language, Cognition and Neuroscience, 29*(1), 44–45.

Guenther, F. H. (2016). *Neural control of speech.* Cambridge, MA: MIT Press.

Guenther, F. H., Hampson, M., & Johnson, D. (1998). A theoretical investigation of reference frames for the planning of speech movements. *Psychological Review, 105*(4), 611–633.

Guenther, F. H., & Hickok, G. (2015). Role of the auditory system in speech production. *Handbook of Clinical Neurology, 129*, 161–175.

Gwilliams, L., Marantz, A., Poeppel, D., & King, J. R. (2024). Hierarchical dynamic coding coordinates speech comprehension in the brain. *bioRxiv*. doi: 10.1101/2024.04.19.590280.

Hackett, T. A., De La Mothe, L. A., Ulbert, I., Karmos, G., Smiley, J., & Schroeder, C. E. (2007a). Multisensory convergence in auditory cortex, II. Thalamocortical connections of the caudal superior temporal plane. *Journal of Comparative Neurology, 502*(6), 924–952.

Hackett, T. A., Smiley, J. F., Ulbert, I., Karmos, G., Lakatos, P., de la Mothe, L. A., & Schroeder, C. E. (2007b). Sources of somatosensory input to the caudal belt areas of auditory cortex. *Perception, 36*(10), 1419–1430.

Haegens, S., & Zion Golumbic, E. (2018). Rhythmic facilitation of sensory processing: A critical review. *Neuroscience & Biobehavioral Reviews, 86*, 150–165.

Hagoort, P. (2003). How the brain solves the binding problem for language: A neurocomputational model of syntactic processing. *NeuroImage, 20*(Suppl 1), S18–S29.

Hagoort, P. (2005). On Broca, brain, and binding: A new framework. *Trends in Cognitive Sciences, 9*(9), 416–423.

Hagoort, P. (2013). MUC (memory, unification, control) and beyond. *Frontiers in Psychology, 4*, 416.

Hamilton, L. S., Edwards, E., & Chang, E. F. (2018). A spatial map of onset and sustained responses to speech in the human superior temporal gyrus. *Current Biology, 28*(12), 1860–1871 e1864.

Hamilton, L. S., Oganian, Y., Hall, J., & Chang, E. F. (2021). Parallel and distributed encoding of speech across human auditory cortex. *Cell, 184*(18), 4626–4639 e4613.

Hannah, R., & Aron, A. R. (2021). Towards real-world generalizability of a circuit for action-stopping. *Nature Reviews Neuroscience, 22*(9), 538–552.

Harrington, A. (1985). Nineteenth-century ideas on hemispheric differences and "duality of mind." *Behavioral and Brain Sciences, 8*(4), 617–660.

Hart, J. J., & Gordon, B. (1990). Deliniation of single-word semantic comprehension deficits in aphasia, with anatomical correlation. *Annals of Neurology, 27*(3), 226–231.

Hartshorne, J. K., Tenenbaum, J. B., & Pinker, S. (2018). A critical period for second language acquisition: Evidence from 2/3 million English speakers. *Cognition, 177*, 263–277.

Harvey, D. Y., & Schnur, T. T. (2015). Distinct loci of lexical and semantic access deficits in aphasia: Evidence from voxel-based lesion-symptom mapping and diffusion tensor imaging. *Cortex, 67*, 37–58.

Haspelmath, M. (2023). Defining the word. *Word, 69*(3), 283–297.

Haxby, J. V., Gobbini, M. I., Furey, M. L., Ishai, A., Schouten, J. L., & Pietrini, P. (2001). Distributed and overlapping representations of faces and objects in ventral temporal cortex. *Science, 293*(5539), 2425–2430.

Heeschen, C., & Jürgens, R. (1977). Pragmatic-semantic and syntactic factors influencing ear differences in dichotic listening. *Cortex, 13*(1), 74–84.

Heeschen, C., & Kolk, H. (1988). Agrammatism and paragrammatism. *Aphasiology, 2*(3–4), 299–302.

Heffner, H. E., & Heffner, R. S. (1984). Temporal lobe lesions and perception of species-specific vocalizations by macaques. *Science, 226*(4670), 75–76.

Heider, F., & Simmel, M. (1944). An experimental study of apparent behavior. *American Journal of Psychology, 57*(2), 243–259.

Heilbrun, A. B., Jr. (1959). Lateralization of cerebral lesion and performance on spatial-temporal tasks. *Archives of Neurology, 1*(3), 282–287.

Heilman, K., & Scholes, R. J. (1976). The nature of comprehension errors in Broca's conduction and Wernicke's aphasics. *Cortex, 12*(3), 258–265.

Heilman, K. M., Scholes, R., & Watson, R. T. (1975). Auditory affective agnosia. Disturbed comprehension of affective speech. *Journal of Neurology, Neurosurgery and Psychiatry, 38*(1), 69–72.

Heilman, K. M., & Valenstein, E. (1985). *Clinical neuropsychology*. 2nd ed. Oxford: Oxford University Press.

Heilman, K. M., & Van Den Abell, T. (1980). Right hemisphere dominance for attention: The mechanism underlying hemispheric asymmetries of inattention (neglect). *Neurology, 30*(3), 327–330.

Heinks-Maldonado, T. H., Mathalon, D. H., Houde, J. F., Gray, M., Faustman, W. O., & Ford, J. M. (2007). Relationship of imprecise corollary discharge in schizophrenia to auditory hallucinations. *Archives of General Psychiatry, 64*(3), 286–296.

Henry, M. J., & Obleser, J. (2012). Frequency modulation entrains slow neural oscillations and optimizes human listening behavior. *Proceedings of the National Academy of Sciences, 109*(49), 20095–20100.

Hensch, T. K. (2004). Critical period regulation. *Annual Review of Neuroscience, 27*, 549–579.

Henschen, S. E. (1918). On the hearing sphere. *Acta Oto-Laryngologica, 1*(1), 423–485.

Henschen, S. E. (1926). On the function of the right hemisphere of the brain in relation to the left in speech, music and calculation. *Brain, 49*(1), 110–123.

Hermann, B. P., Wyler, A. R., & Somes, G. (1991). Language function following anterior temporal lobectomy. *Journal of Neurosurgery, 74*(4), 560–566.

Hickok, G. (2000). Speech perception, conduction aphasia, and the functional neuroanatomy of language. In Y. Grodzinsky, L. Shapiro, & D. Swinney (Eds.), *Language and the brain* (pp. 87–104). San Diego: Academic Press.

Hickok, G. (2009). Eight problems for the mirror neuron theory of action understanding in monkeys and humans. *Journal of Cognitive Neuroscience, 21*(7), 1229–1243.

Hickok, G. (2012a). Computational neuroanatomy of speech production. *Nature Reviews Neuroscience, 13*(2), 135–145.

Hickok, G. (2012b). The cortical organization of speech processing: feedback control and predictive coding the context of a dual-stream model. *Journal of Communication Disorders, 45*(6), 393–402.

Hickok, G. (2014a). The architecture of speech production and the role of the phoneme in speech processing. *Language Cognitive Process, 29*(1), 2–20.

Hickok, G. (2014b). *The myth of mirror neurons: the real neuroscience of communication and cognition*. New York: W.W. Norton & Company.

Hickok, G. (2014c). Toward an integrated psycholinguistic, neurolinguistic, sensorimotor framework for speech production. *Language Cognitive Process, 29*(1), 52–59.

Hickok, G. (2017). A cortical circuit for voluntary laryngeal control: Implications for the evolution of language. *Psychonomic Bulletin & Review, 24*(1), 56–63.

Hickok, G. (2018). The neural basis of word deafness: A frequency analysis. from http://www.talkingbrains.org/2018/08/the-neural-basis-of-word-deafness.html.

Hickok, G. (2022). The dual stream model of speech and language processing. In A. E. Hillis (Ed.), *Handbook of clinical neurology* (Vol. 185, pp. 57–69). Amsterdam: Elsevier.

Hickok, G. (2024). The "coordination conjecture" as an alternative to Patel's fortuitous enhancement hypothesis for the relation between vocal learning and beat-based dancing. *BMC Neuroscience, 25*(59), https://doi.org/10.1186/s12868-024-00868-x.

Hickok, G., & Avrutin, S. (1995). Representation, referentiality, and processing in agrammatic comprehension: Two case studies. *Brain and Language, 50*(1), 10–26.

Hickok, G., & Avrutin, S. (1996). Comprehension of wh-questions in two Broca's aphasics. *Brain and Language, 52*(2), 314–327.

Hickok, G., Bellugi, U., & Jones, W. (1995). Asymmetrical ability. *Science, 270*(5234), 219–220.

Hickok, G., Bellugi, U., & Klima, E. S. (1998). What's right about the neural organization of sign language? A perspective on recent neuroimaging results. *Trends in Cognitive Science, 2*(12), 465–468.

Hickok, G., Bellugi, U., & Klima, E. S. (2001). Sign language in the brain. *Scientific American, 284*(6), 58–65.

Hickok, G., Buchsbaum, B., Humphries, C., & Muftuler, T. (2003). Auditory-motor interaction revealed by fMRI: Speech, music, and working memory in area Spt. *Journal of Cognitive Neuroscience, 15*(5), 673–682.

Hickok, G., Costanzo, M., Capasso, R., & Miceli, G. (2011). The role of Broca's area in speech perception: Evidence from aphasia revisited. *Brain and Language, 119*(3), 214–220.

Hickok, G., Erhard, P., Kassubek, J., Helms-Tillery, A. K., Naeve-Velguth, S., Strupp, J. P., Strick, P. L., & Ugurbil, K. (2000). A functional magnetic resonance imaging study of the role of left posterior superior temporal gyrus in speech production: Implications for the explanation of conduction aphasia. *Neuroscience Letters, 287*(2), 156–160.

Hickok, G., Farahbod, H., & Saberi, K. (2015). The rhythm of perception: Entrainment to acoustic rhythms induces subsequent perceptual oscillation. *Psychological Science, 26*(7), 1006–1013.

Hickok, G., Houde, J., & Rong, F. (2011). Sensorimotor integration in speech processing: Computational basis and neural organization. *Neuron, 69*(3), 407–422.

Hickok, G., Klima, E. S., & Bellugi, U. (1996). The neurobiology of signed language and its implications for the neural basis of language. *Nature, 381*(6584), 699–702.

Hickok, G., Okada, K., Barr, W., Pa, J., Rogalsky, C., Donnelly, K., Barde, L., & Grant, A. (2008). Bilateral capacity for speech sound processing in auditory comprehension: Evidence from Wada procedures. *Brain and Language, 107*(3), 179–184.

Hickok, G., Okada, K., & Serences, J. T. (2009). Area Spt in the human planum temporale supports sensory-motor integration for speech processing. *Journal of Neurophysiology, 101*(5), 2725–2732.

Hickok, G., & Poeppel, D. (2000). Towards a functional neuroanatomy of speech perception. *Trends in Cognitive Sciences, 4*(4), 131–138.

Hickok, G., & Poeppel, D. (2004). Dorsal and ventral streams: A framework for understanding aspects of the functional anatomy of language. *Cognition, 92*(1-2), 67–99.

Hickok, G., & Poeppel, D. (2007). The cortical organization of speech processing. *Nature Reviews Neuroscience, 8*(5), 393–402.

Hickok, G., Rogalsky, C., Chen, R., Herskovits, E. H., Townsley, S., & Hillis, A. E. (2014). Partially overlapping sensorimotor networks underlie speech praxis and verbal short-term memory: Evidence from apraxia of speech following acute stroke. *Frontiers in Human Neuroscience, 8*, 649.

Hickok, G., Rogalsky, C., Matchin, W., Basilakos, A., Cai, J., Pillay, S., Ferrill, M., Mickelsen, S., Anderson, S. W., Love, T., . . . , & Fridriksson, J. (2018). Neural networks supporting audiovisual integration for speech: A large-scale lesion study. *Cortex, 103*, 360–371.

Hickok, G., & Saberi, K. (2012). Redefining the functional organization of the planum temporale region: Space, objects, and sensory-motor integration. In D. Poeppel, T. Overath, A. N. Popper, & R. R. Ray (Eds.), *The human auditory cortex* (Vol. 43, pp. 333–350). New York: Springer.

Hickok, G., Venezia, J., & Teghipco, A. (2023). Beyond Broca: Neural architecture and evolution of a dual motor speech coordination system. *Brain, 146*(5), 1775–1790.

Hickok, G., Wilson, M., Clark, K., Klima, E. S., Kritchevsky, M., & Bellugi, U. (1999). Discourse deficits following right hemisphere damage in deaf signers. *Brain and Language, 66*(2), 233–248.

Hickok, G., Zurif, E., & Canseco-Gonzalez, E. (1993). Structural description of agrammatic comprehension. *Brain and Language, 45*(3), 371–395.

Hillenbrand, J., Getty, L. A., Clark, M. J., & Wheeler, K. (1995). Acoustic characteristics of American English vowels. *Journal of the Acoustical Society of America, 97*(5 Pt 1), 3099–3111.

Hillis, A. E., Work, M., Barker, P. B., Jacobs, M. A., Breese, E. L., & Maurer, K. (2004). Re-examining the brain regions crucial for orchestrating speech articulation. *Brain, 127*(Pt 7), 1479–1487.

Hinke, R. M., Hu, X., Stillman, A. E., Kim, S.-G., Merkle, H., Salmi, R., & Ugurbil, K. (1993). Functional magnetic resonance imaging of Broca's area during internal speech. *Neuroreport, 4*(6), 675–678.

Hodges, J. R., & Patterson, K. (2007). Semantic dementia: A unique clinicopathological syndrome. *Lancet Neurol, 6*(11), 1004–1014.

Hodges, J. R., Patterson, K., Oxbury, S., & Funnell, E. (1992). Semantic dementia. Progressive fluent aphasia with temporal lobe atrophy. *Brain, 115*(Pt 6), 1783–1806.

Horwitz, B., Amunts, K., Bhattacharyya, R., Patkin, D., Jeffries, K., Zilles, K., & Braun, A. R. (2003). Activation of Broca's area during the production of spoken and signed language: A combined cytoarchitectonic mapping and PET analysis. *Neuropsychologia, 41*(14), 1868–1876.

Houde, J. F., Gill, J. S., Agnew, Z., Kothare, H., Hickok, G., Parrell, B., Ivry, R. B., & Nagarajan, S. S. (2019). Abnormally increased vocal responses to pitch feedback perturbations in patients with cerebellar degeneration. *Journal of the Acoustical Society of America, 145*(5), EL372.

Houde, J. F., & Jordan, M. I. (1998). Sensorimotor adaptation in speech production. *Science, 279*(5354), 1213–1216.

Houde, J. F., & Nagarajan, S. S. (2011). Speech production as state feedback control. *Frontiers in Human Neuroscience, 5*, 82.

Howard, M. F., & Poeppel, D. (2010). Discrimination of speech stimuli based on neuronal response phase patterns depends on acoustics but not comprehension. *Journal of Neurophysiology, 104*(5), 2500–2511.

Howell, P., Powell, D. J., & Khan, I. (1983). Amplitude contour of the delayed signal and interference in delayed auditory feedback tasks. *Journal of Experimental Psychology: Human Perception and Performance, 9*(5), 772–784.

Hughes, J. R. (1964). Responses from the visual cortex of unanesthetized monkeys. *International Review of Neurobiology, 6*, 99–152.

Hughlings Jackson, J. (1868/1915a). Notes on the physiology and pathology of the nervous system: Language and thought—the duality of mental processes. *Brain, 38*(1-2), 65–71.

Hughlings Jackson, J. (1868/1915b). On affections of speech from disease of the brain. *Brain, 38*(2), 107–129.

Hughlings Jackson, J. (1878). On affections of speech from disease of the brain. *Brain, 1*(3), 304–330.

Hullett, P. W., Hamilton, L. S., Mesgarani, N., Schreiner, C. E., & Chang, E. F. (2016). Human superior temporal gyrus organization of spectrotemporal modulation tuning derived from speech stimuli. *Journal of Neuroscience, 36*(6), 2014–2026.

Humphreys, G. F., & Lambon Ralph, M. A. (2015). Fusion and fission of cognitive functions in the human parietal cortex. *Cerebral Cortex, 25*(10), 3547–3560.

Humphries, C., Love, T., Swinney, D., & Hickok, G. (2005). Response of anterior temporal cortex to syntactic and prosodic manipulations during sentence processing. *Human Brain Mapping, 26*(2), 128–138.

Humphries, C., Willard, K., Buchsbaum, B., & Hickok, G. (2001). Role of anterior temporal cortex in auditory sentence comprehension: An fMRI study. *Neuroreport, 12*(8), 1749–1752.

Hung, C. P., Kreiman, G., Poggio, T., & DiCarlo, J. J. (2005). Fast readout of object identity from macaque inferior temporal cortex. *Science, 310*(5749), 863–866.

Hunt, J. (1869). On the localisation of the functions of the brain, with special reference to the faculty of language. *Anthropological Review, 7*(25), 201–214.

Iaccarino, L., Chieffi, S., & Iavarone, A. (2014). Utilization behavior: What is known and what has to be known? *Behavioural Neurology, 2014*(1), 297128.

Iacoboni, M., Woods, R. P., Brass, M., Bekkering, H., Mazziotta, J. C., & Rizzolatti, G. (1999). Cortical mechanisms of human imitation. *Science, 286*(5449), 2526–2528.

Idsardi, W., & Raimy, E. (2013). Three types of linearization and the temporal aspects of speech. In T. Biberauer & I. Roberts (Eds.), *Challenges to linearization* (pp. 31–56). De Gruyter Mouton.

Iijima, S., Ishimitsu, S., & Nakayama, M. (2019). Effects of gaze angle on vocalization. *ICIC Express Letters, Part B: Application, 10*(1), 63–69.

Indefrey, P. (2011). The spatial and temporal signatures of word production components: A critical update. *Fronters in Psychology, 2*, 255.

Indefrey, P., & Levelt, W. J. (2004). The spatial and temporal signatures of word production components. *Cognition, 92*(1–2), 101–144.

Ingle, D. (1973). Two visual systems in the frog. *Science, 181*(4104), 1053–1055.

Isenberg, A. L. (2012). *Neuroanatomical and functional investigations of sensory-motor integration in the speech domain.* University of California, Irvine.

Isenberg, A. L., Vaden, K. I., Jr., Saberi, K., Muftuler, L. T., & Hickok, G. (2012). Functionally distinct regions for spatial processing and sensory motor integration in the planum temporale. *Human Brain Mapping, 33*(10), 2453–2463.

Itabashi, R., Nishio, Y., Kataoka, Y., Yazawa, Y., Furui, E., Matsuda, M., & Mori, E. (2016). Damage to the left precentral gyrus is associated with apraxia of speech in acute stroke. *Stroke, 47*(1), 31–36.

Iwasaki, S. I., Yoshimura, K., Asami, T., & Erdogan, S. (2022). Comparative morphology and physiology of the vocal production apparatus and the brain in the extant primates. *Annals of Anatomy, 240*(7), 151887.

Jackendoff, R. (2002). *Foundations of language: Brain, meaning, grammar, evolution.* New York: Oxford University Press.

Jackendoff, R. (2007). A parallel architecture perspective on language processing. *Brain Research, 1146*, 2–22.

Jackendoff, R., & Audring, J. (2016). Morphological schemas: Theoretical and psycholinguistic issues. *Mental Lexicon, 11*(3), 467–493.

Jackendoff, R., & Audring, J. (2020). Relational morphology: A cousin of construction grammar. *Front Psychology, 11*, 2241.

Jacks, A., & Haley, K. L. (2021). Apraxia of speech. In J. S. Damico, A. Müller, & M. J. Ball (Eds.), *The handbook of language and speech disorders* (pp. 368–390). Online: Wiley.

Jackson, H. (1864). Clincal remarks on cases of defects of expression (by words, writing, signs, etc.) in diseases of the nervous system. *Lancet, 2*(2152), 604–605.

Jakobson, R. (1956). Two aspects of language and two types of aphasic disturbances. In R. Jakobson & M. Halle (Eds.), *Fundamentals of language*. The Hague: Mouton.

James, W. (1890). *The principles of psychology* (Vol. 2). New York: Henry Holt and Company.

Jansen, A. G., Mous, S. E., White, T., Posthuma, D., & Polderman, T. J. (2015). What twin studies tell us about the heritability of brain development, morphology, and function: A review. *Neuropsychology Review, 25*(1), 27–46.

January, D., Trueswell, J. C., & Thompson-Schill, S. L. (2009). Co-localization of stroop and syntactic ambiguity resolution in Broca's area: Implications for the neural basis of sentence processing. *Journal of Cognitive Neuroscience, 21*(12), 2434–2444.

Jarvis, E. D. (2019). Evolution of vocal learning and spoken language. *Science, 366*(6461), 50–54.

Jeannerod, M., Arbib, M. A., Rizzolatti, G., & Sakata, H. (1995). Grasping objects: The cortical mechanisms of visuomotor transformation. *Trends Neuroscience, 18*(7), 314–320.

Jefferies, E., & Lambon Ralph, M. A. (2006). Semantic impairment in stroke aphasia versus semantic dementia: A case-series comparison. *Brain, 129*(Pt 8), 2132–2147.

Jia, X., & Kohn, A. (2011). Gamma rhythms in the brain. *PLoS Biology, 9*(4), e1001045.

Johnson, J. S., & Newport, E. L. (1989). Critical period effects in second language learning: The influence of maturational state on the acquisition of English as a second language. *Cognitive Psychology, 21*(1), 60–99.

Jonides, J., Schumacher, E. H., Smith, E. E., Koeppe, R. A., Awh, E., Reuter-Lorenz, P. A., Marshuetz, C., & Willis, C. R. (1998). The role of parietal cortex in verbal working memory. *Journal of Neuroscience, 18*(13), 5026–5034.

Joshi, A. K., & Schabes, Y. (1997). Tree-adjoining grammars. In G. Rozenberg & A. Salomaa (Eds.), *Handbook of formal languages* (pp. 69–123). Berlin: Springer-Verlag.

Jürgens, U. (2002). Neural pathways underlying vocal control. *Neuroscience Biobehavioral Reviews, 26*(2), 235–258.

Jürgens, U. (2009). The neural control of vocalization in mammals: A review. *Journal of Voice, 23*(1), 1–10.

Just, M. A., Carpenter, P. A., Keller, T. A., Eddy, W. F., & Thulborn, K. R. (1996). Brain activation modulated by sentence comprehension. *Science, 274*(5284), 114–116.

Kaas, J. H. (2004). Evolution of somatosensory and motor cortex in primates. *Anatomical Record, Part A: Discoveries Molecular Cell Evolutionary Biology, 281*(1), 1148–1156.

Kahana, M. J., Sekuler, R., Caplan, J. B., Kirschen, M., & Madsen, J. R. (1999). Human theta oscillations exhibit task dependence during virtual maze navigation. *Nature, 399*(6738), 781–784.

Kang, H., Agus, T. R., & Pressnitzer, D. (2017). Auditory memory for random time patterns. *Journal of the Acoustical Society of America, 142*(4), 2219.

Kang, H., Auksztulewicz, R., An, H., Abi Chacra, N., Sutter, M. L., & Schnupp, J. W. H. (2021). Neural correlates of auditory pattern learning in the auditory cortex. *Frontiers in Neuroscience, 15*, 610978.

Kanwisher, N. (2000). Domain specificity in face perception. *Nature Neuroscience, 3*(8), 759–763.

Kaplan, R. M., & Bresnan, J. (1982). Lexical-functional grammar: A formal system for grammatical representation. In J. Bresnan (Ed.), *The mental representation of grammatical relations* (pp. 173–281). Cambridge, MA: MIT Press.

Kaspar, K., & Rübeling, H. (2011). Rhythmic versus phonemic interference in delayed auditory feedback. *Journal of Speech, Language, and Hearing Research, 54*(3), 932–943.

Kassubek, J., Hickok, G., & Erhard, P. (2004). Involvement of classical anterior and posterior language areas in sign language production, as investigated by 4 T functional magnetic resonance imaging. *Neuroscience Letters, 364*(3), 168–172.

Kazanina, N., Bowers, J. S., & Idsardi, W. (2018). Phonemes: Lexical access and beyond. *Psychonomic Bulletin Review, 25*(2), 560–585.

Keitel, C., Quigley, C., & Ruhnau, P. (2014). Stimulus-driven brain oscillations in the alpha range: Entrainment of intrinsic rhythms or frequency-following response? *Journal of Neuroscience, 34*(31), 10137–10140.

Kemmerer, D. (2015). *The cognitive neuroscience of langauge.* New York: Psychology Press.

Kertesz, A. (2012). *The western aphasia battery-revised.* San Antonio, TX: Pearson.

Kertesz, A., & McCabe, P. (1975). Intelligence and aphasia: Performance of aphasics on Raven's coloured progressive matrices (RCPM). *Brain and Language, 2*(4), 387–395.

Kertesz, A., Sheppar, A., & MacKenzie, R. (1982). Localization in transcortical sensory aphasia. *Archives of Neurology, 39*(8), 475–478.

Kimura, D. (1961). Some effects of temporal-lobe damage on auditory perception. *Canadian Journal of Psychology, 15*, 156–165.

Kimura, D. (1964). Left-right differences in the perception of melodies. *Quarterly Journal of Experimental Psychology, 14*(2), 355–358.

Kimura, D. (1967). Functional asymmetry of the brain in dichotic listening. *Cortex, 3*(2), 163–178.

Kirchner, H., Barbeau, E. J., Thorpe, S. J., Regis, J., & Liegeois-Chauvel, C. (2009). Ultra-rapid sensory responses in the human frontal eye field region. *Journal of Neuroscience, 29*(23), 7599–7606.

Klein, R., & Harper, J. (1956). The problem of agnosia in the light of a case of pure word deafness. *Journal of Mental Science, 102*(426), 112–120.

Kleist, K. (1914). Aphasie und geisteskrankheit. *Münchener Medizinische Wochenschrift, 6*(1), 8–12.

Knight, E. J., & Austin, S. F. (2020). The effect of head flexion/extension on acoustic measures of singing voice quality. *Journal of Voice, 34*(6), 964 e911–964 e921.

Koelsch, S. (2014). Brain correlates of music-evoked emotions. *Nature Reviews Neuroscience, 15*(3), 170–180.

Koelsch, S., Skouras, S., Fritz, T., Herrera, P., Bonhage, C., Kussner, M. B., & Jacobs, A. M. (2013). The roles of superficial amygdala and auditory cortex in music-evoked fear and joy. *NeuroImage, 81*, 49–60.

Kohn, S. E. (1984). The nature of the phonological disorder in conduction aphasia. *Brain and Language, 23*(1), 97–115.

Kolb, B., & Whishaw, I. Q. (1985). *Fundamental of human neuropsychology*. 2nd ed. New York: W. H. Freeman and Company.

Kosem, A., Bosker, H. R., Takashima, A., Meyer, A., Jensen, O., & Hagoort, P. (2018). Neural entrainment determines the words we hear. *Current Biology, 28*(18), 2867–2875 e2863.

Krauska, A., & Lau, E. (2023). Moving away from lexicalism in psycho- and neurolinguistics. *Frontiers in Language Science, 2,* 1125127.

Kravitz, D. J., Saleem, K. S., Baker, C. I., Ungerleider, L. G., & Mishkin, M. (2013). The ventral visual pathway: An expanded neural framework for the processing of object quality. *Trends in Cognitive Sciences, 17*(1), 26–49.

Krubitzer, L. (2009). In search of a unifying theory of complex brain evolution. *Annals of the New York Academy of Sciences, 1156*(1), 44–67.

Krubitzer, L., Clarey, J., Tweedale, R., Elston, G., & Calford, M. (1995). A redefinition of somatosensory areas in the lateral sulcus of macaque monkeys. *Journal of Neuroscience, 15*(5 Pt 2), 3821–3839.

Kuhl, P. K., & Miller, J. D. (1975). Speech perception by the chinchilla: Voiced-voiceless distinction in alveolar plosive consonants. *Science, 190*(4209), 69–72.

Kumar, V., Croxson, P. L., & Simonyan, K. (2016). Structural organization of the laryngeal motor cortical network and its implication for evolution of speech production. *Journal of Neuroscience, 36*(15), 4170–4181.

Kussmaul, A. (1877). Disturbances of speech. In H. Ziemssen (Ed.), *Cyclopaedia of the practice of medicine* (pp. 581–875). New York: William Wood.

Kutas, M., & Hillyard, S. A. (1980). Reading senseless sentences: Brain potentials reflect semantic incongruity. *Science, 207*(4427), 203–205.

Kuypers, H. G. (1958a). Corticobular connexions to the pons and lower brain-stem in man: An anatomical study. *Brain, 81*(3), 364–388.

Kuypers, H. G. (1958b). Some projections from the peri-central cortex to the pons and lower brain stem in monkey and chimpanzee. *Journal of Comparative Neurology, 110*(2), 221–255.

LaCroix, A. N., Diaz, A. F., & Rogalsky, C. (2015). The relationship between the neural computations for speech and music perception is context-dependent: An activation likelihood estimate study. *Frontiers in Psychology, 6,* 1138.

Ladd, D. R. (2008). *Intonational phonology* (2nd ed.). Cambridge: Cambridge University Press.

Ladd, D. R., Faulkner, D., Faulkner, H., & Schepman, A. (1999). Constant "segmental anchoring" of F0 movements under changes in speech rate. *Journal of the Acoustical Society of America, 106*(3 Pt 1), 1543–1554.

Lakatos, P., Karmos, G., Mehta, A. D., Ulbert, I., & Schroeder, C. E. (2008). Entrainment of neuronal oscillations as a mechanism of attentional selection. *Science, 320*(5872), 110–113.

Lakatos, P., Musacchia, G., O'Connel, M. N., Falchier, A. Y., Javitt, D. C., & Schroeder, C. E. (2013). The spectrotemporal filter mechanism of auditory selective attention. *Neuron, 77*(4), 750–761.

Lakatos, P., Shah, A. S., Knuth, K. H., Ulbert, I., Karmos, G., & Schroeder, C. E. (2005). An oscillatory hierarchy controlling neuronal excitability and stimulus processing in the auditory cortex. *Journal of Neurophysiology, 94*(3), 1904–1911.

Lam, N. H. L., Hulten, A., Hagoort, P., & Schoffelen, J.-M. (2018). Robust neuronal oscillatory entrainment to speech displays individual variation in lateralisation. *Language, Cognition and Neuroscience, 33*(8), 943–954.

Lanciego, J. L., Luquin, N., & Obeso, J. A. (2012). Functional neuroanatomy of the basal ganglia. *Cold Spring Harbor Perspectives in Medicine, 2*(12), a009621.

Landau, A. N., Schreyer, H. M., van Pelt, S., & Fries, P. (2015). Distributed attention is implemented through theta-rhythmic gamma modulation. *Current Biology, 25*(17), 2332–2337.

Landreth, A., & Richardson, R. C. (2014). Localization and the new phrenology: A review essay on William Uttal's *The New Phrenology*. *Philosophical Psychology, 17*(1), 107–123.

Language models and linguistic theories beyond words. (2023). *Nature Machine Intelligence, 5*, 677–678.

Larson, C. R., Altman, K. W., Liu, H., & Hain, T. C. (2008). Interactions between auditory and somatosensory feedback for voice F0 control. *Experimental Brain Research, 187*(4), 613–621.

Lashley, K. S. (1950). In search of the engram. In *Society of experimental biology symposium No. 4: Physiological mechanisms in animal behavior* (pp. 454–482). Cambridge: Cambridge University Press.

Lashley, K. S. (1951). The problem of serial order in behavior. In L. A. Jeffries (Ed.), *Cerebral mechanisms in behavior* (pp. 112–131). New York: Wiley.

Lasnik, H., & Lidz, J. (2017). The argument from the poverty of the stimulus. In I. Roberts (Ed.), *Oxford Handbook of Universal Grammar* (pp. 221–248). Oxford: Oxford University Press.

Lau, E. F., Phillips, C., & Poeppel, D. (2008). A cortical network for semantics: (De) constructing the N400. *Nature Reviews Neuroscience, 9*(12), 920–933.

Lee, B. S. (1950). Effects of delayed speech feedback. *Journal of the Acoustical Society of America, 22*(6), 824–826.

Lee, S., Potamianos, A., & Narayanan, S. (1999). Acoustics of children's speech: Developmental changes of temporal and spectral parameters. *Journal of the Acoustical Society of America, 105*(3), 1455–1468.

Leehey, S., Carey, S., Diamond, R., & Cahn, A. (1978). Upright and inverted faces: The right hemisphere knows the difference. *Cortex, 14*(3), 411–419.

Lenneberg, E. H. (1962). Understanding language without ability to speak: A case report. *Journal of Abnormal and Social Psychology, 65*(6), 419–425.

Leslie, K. R., Johnson-Frey, S. H., & Grafton, S. T. (2004). Functional imaging of face and hand imitation: Towards a motor theory of empathy. *NeuroImage, 21*(2), 601–607.

Levelt, W. J. (1983). Monitoring and self-repair in speech. *Cognition, 14*(1), 41–104.

Levelt, W. J. M. (1989). *Speaking: From intention to articulation*. Cambridge, MA: MIT Press.

Levelt, W. J. M., Praamstra, P., Meyer, A. S., Helenius, P., & Salmelin, R. (1998). An MEG study of picture naming. *Journal of Cognitive Neuroscience, 10*(5), 553–567.

Levelt, W. J. M., Roelofs, A., & Meyer, A. S. (1999). A theory of lexical access in speech production. *Behavioral and Brain Sciences, 22*(1), 1–38; discussion 38–75.

Levelt, W. J., & Wheeldon, L. (1994). Do speakers have access to a mental syllabary? *Cognition, 50*(1–3), 239–269.

Ley, R. G., & Bryden, M. P. (1979). Hemispheric differences in processing emotions and faces. *Brain and Language, 7*(1), 127–138.

Lhermitte, F. (1983). "Utilization behaviour" and its relation to lesions of the frontal lobes. *Brain, 106*(Pt 2), 237–255.

Li, W., Zhuang, J., Guo, Z., Jones, J. A., Xu, Z., & Liu, H. (2019). Cerebellar contribution to auditory feedback control of speech production: Evidence from patients with spinocerebellar ataxia. *Human Brain Mapping, 40*(16), 4748–4758.

Liberman, A. M. (1957). Some results of research on speech perception. *Journal of the Acoustical Society of America, 29*(1), 117–123.

Liberman, A. M., Cooper, F. S., Shankweiler, D. P., & Studdert-Kennedy, M. (1967). Perception of the speech code. *Psychological Review, 74*(6), 431–461.

Liberman, A. M., & Mattingly, I. G. (1985). The motor theory of speech perception revised. *Cognition, 21*(1), 1–36.

Lichtheim, L. (1885). On aphasia. *Brain, 7*(4), 433–484.

Lin, C. Y., Chen, P. H., Lin, H. H., & Huang, W. M. (2022). U(1) dynamics in neuronal activities. *Scientific Reports, 12*(1), 17629. doi:10.1038/s41598-022-22526-0.

Linebarger, M. C., Schwartz, M., & Saffran, E. (1983). Sensitivity to grammatical structure in so-called agrammatic aphasics. *Cognition, 13*(3), 361–393.

Liu, C., & Eddins, D. A. (2008). Effects of spectral modulation filtering on vowel identification. *Journal of the Acoustical Society of America, 124*(3), 1704–1715.

Loehr, D. P. (2012). Temporal, structural, and pragmatic synchrony between intonation and gesture. *Laboratory Phonology, 3*(1), 71–89.

Lorca-Puls, D. L., Gajardo-Vidal, A., Mandelli, M. L., Illan-Gala, I., Ezzes, Z., Wauters, L. D., Battistella, G., Bogley, R., Ratnasiri, B., Licata, A. E., . . . , & Gorno-Tempiri, M. L. (2024). Neural basis of speech and grammar symptoms in non-fluent variant primary progressive aphasia spectrum. *Brain, 147*(2), 607–626.

Lotto, A. J., Kluender, K. R., & Holt, L. L. (1997). Perceptual compensation for coarticulation by Japanese quail (*Coturnix japonica*). *Journal of the Acoustical Society of America, 102*(2 Pt 1), 1134–1140.

Love, T., & Oster, E. (2002). On the categorization of aphasic typologies: The S.O.A.P, a test of syntactic complexity. *Journal of Psycholinguistic Research, 31*(5), 503–529.

Lu, J., Zhao, Z., Zhang, J., Wu, B., Zhu, Y., Chang, E. F., Wu, J., Duffau, H., & Berger, M. S. (2021). Functional maps of direct electrical stimulation-induced speech arrest and anomia: A multicentre retrospective study. *Brain, 144*(8), 2541–2553.

Lukic, S., Borghesani, V., Weis, E., Welch, A., Bogley, R., Neuhaus, J., Deleon, J., Miller, Z. A., Kramer, J. H., . . . , & Gorno-Tempini, M. L. (2021). Dissociating nouns and verbs in temporal and perisylvian networks: Evidence from neurodegenerative diseases. *Cortex, 142*, 47–61.

Luo, H., Liu, Z., & Poeppel, D. (2010). Auditory cortex tracks both auditory and visual stimulus dynamics using low-frequency neuronal phase modulation. *PLoS Biology, 8*(8), e1000445.

Luo, H., & Poeppel, D. (2007). Phase patterns of neuronal responses reliably discriminate speech in human auditory cortex. *Neuron, 54*(6), 1001–1010.

Luria, A. R. (1970). *Traumatic aphasia*. The Hague: Mouton.

Luria, A. R., & Hutton, J. T. (1977). A modern assessment of the basic forms of aphasia. *Brain and Language, 4*(2), 129–151.

Lyn, H. (2012). Apes and the evolution of language: Taking stock of 40 years of research. In J. Vonk & T. K. Shackelford (Eds.), *The Oxford Handbook of Comparative Evolutionary Psychology* (pp. 356–378). Oxford: Oxford University Press.

MacDonald, M. C., & Seidenberg, M. S. (2006). Constraint satisfaction accounts of lexical and sentence comprehension. In M. J. Traxler & M. A. Gernsbacher (Eds.), *Handbook of Psycholinguistics* (pp. 581–611). San Diego: Academic Press.

MacNeilage, P. F. (1998). The frame/content theory of evolution of speech production. *Behavioral and Brain Sciences, 21*(4), 499–511; discussion 511–446.

Maess, B., Koelsch, S., Gunter, T. C., & Friederici, A. D. (2001). Musical syntax is processed in Broca's area: An MEG study. *Nature Neuroscience, 4*(5), 540–545.

Magnusdottir, S., Fillmore, P., den Ouden, D. B., Hjaltason, H., Rorden, C., Kjartansson, O., Boniha, L., & Fridriksson, J. (2013). Damage to left anterior temporal cortex predicts impairment of complex syntactic processing: A lesion-symptom mapping study. *Human Brain Mapping, 34*(10), 2715–2723.

Mailend, M. L., & Maas, E. (2020). To lump or to split? Possible subtypes of apraxia of speech. *Aphasiology, 35*(4), 592–613.

Majsak, M. J., Kaminski, T., Gentile, A. M., & Flanagan, J. R. (1998). The reaching movements of patients with Parkinson's disease under self-determined maximal speed and visually cued conditions. *Brain, 121*(Pt 4), 755–766.

Manns, J. R., & Eichenbaum, H. (2006). Evolution of declarative memory. *Hippocampus, 16*(9), 795–808.

Marantz, A. (2013). No escape from morphemes in morphological processing. *Language and Cognitive Processes, 28*(7), 905–916.

Marantz, A. (2016). Morphology. In G. Hickok & S. L. Small (Eds.), *Neurobiology of language* (pp. 153–163). New York: Elsevier.

Marslen-Wilson, W., & Tyler, L. K. (1980). The temporal structure of spoken language understanding. *Cognition, 8*(1), 1–71.

Martin, A. (2016). GRAPES–Grounding representations in action, perception, and emotion systems: How object properties and categories are represented in the human brain. *Psychonomic Bulletin & Review, 23*(4), 979–990.

Martin, N., & Saffran, E. M. (2002). The relationship of input and output phonological processing: An evaluation of models and evidence to support them. *Aphasiology, 16*(1–2), 107–150.

Martin, R. C. (2003). Language processing: Functional organization and neuroanatomical basis. *Annual Review of Psychology, 54*, 55–89.

Martins, P. T., & Boeckx, C. (2014). Attention mechanisms and the mosaic evolution of speech. *Front Psychology, 5*, 1463.

Massaro, D. W. (1972). Preperceptual images, processing time, and perceptual units in auditory perception. *Psychological Review, 79*(2), 124–145.

Massaro, D. W., & Chen, T. H. (2008). The motor theory of speech perception revisited. *Psychonomic Bulletin & Review, 15*(2), 453–457; discussion 458–462.

Matchin, W. G. (2018). A neuronal retuning hypothesis of sentence-specificity in Broca's area. *Psychonomic Bulletin & Review, 25*(5), 1682–1694.

Matchin, W., Basilakos, A., Stark, B. C., den Ouden, D. B., Fridriksson, J., & Hickok, G. (2020). Agrammatism and paragrammatism: A cortical double dissociation revealed by lesion-symptom mapping. *Neurobiology of Language, 1*(2), 208–225.

Matchin, W., den Ouden, D. B., Basilakos, A., Stark, B. C., Fridriksson, J., & Hickok, G. (2023). Grammatical parallelism in aphasia: A lesion-symptom mapping study. *Neurobiology of Language (Camb), 4*(4), 550–574.

Matchin, W., den Ouden, D. B., Hickok, G., Hillis, A. E., Bonilha, L., & Fridriksson, J. (2022). The Wernicke conundrum revisited: Evidence from connectome-based lesion-symptom mapping. *Brain, 145*(11), 3916–3930.

Matchin, W., Hammerly, C., & Lau, E. (2017). The role of the IFG and pSTS in syntactic prediction: Evidence from a parametric study of hierarchical structure in fMRI. *Cortex, 88*, 106–123.

Matchin, W., & Hickok, G. (2020). The cortical organization of syntax. *Cerebral Cortex, 30*(3), 1481–1498.

Matyas, F., Sreenivasan, V., Marbach, F., Wacongne, C., Barsy, B., Mateo, C., Aronoff, R., & Petersen, C. C. H. (2010). Motor control by sensory cortex. *Science, 330*(6008), 1240–1243.

Mazoyer, B. M., Tzourio, N., Frak, V., Syrota, A., Murayama, N., Levrier, O., et al. (1993). The cortical representation of speech. *Journal of Cognitive Neuroscience, 5*(4), 467–479.

McGurk, H., & MacDonald, J. (1976). Hearing lips and seeing voices. *Nature, 264*(5588), 746–748.

McIntosh, R. D., & Lashley, G. (2008). Matching boxes: Familiar size influences action programming. *Neuropsychologia, 46*(9), 2441–2444.

Meister, I. G., Wilson, S. M., Deblieck, C., Wu, A. D., & Iacoboni, M. (2007). The essential role of premotor cortex in speech perception. *Current Biology, 17*(19), 1692–1696.

Meltzer-Asscher, A., Mack, J. E., Barbieri, E., & Thompson, C. K. (2015). How the brain processes different dimensions of argument structure complexity: Evidence from fMRI. *Brain and Language, 142*, 65–75.

Meltzer-Asscher, A., Schuchard, J., den Ouden, D. B., & Thompson, C. K. (2013). The neural substrates of complex argument structure representations: Processing "alternating transitivity" verbs. *Language and Cognitive Processes, 28*(8), 1154–1168.

Meltzoff, A. N. (1988). The human infant as homo imitans. In Z. Zentall & B. Galef (Eds.), *Social learning: Psychological and biological perspectives* (pp. 319–342). Hillsdale, NJ: Lawrence Earlbaum Associates.

Meltzoff, A. N., & Decety, J. (2003). What imitation tells us about social cognition: A rapprochement between developmental psychology and cognitive neuroscience. *Philosophical Transactions of the Royal Society of London Series B: Biology Science, 358*(1431), 491–500.

Meltzoff, A. N., Kuhl, P. K., Movellan, J., & Sejnowski, T. J. (2009). Foundations for a new science of learning. *Science, 325*(5938), 284–288.

Mendez, M. F., & Rosenberg, S. (1991). Word deafness mistaken for Alzheimer's disease: Differential characteristics. *Journal of the American Geriatrics Society, 39*(2), 209–211.

Mesgarani, N., Cheung, C., Johnson, K., & Chang, E. F. (2014). Phonetic feature encoding in human superior temporal gyrus. *Science, 343*(6174), 1006–1010.

Mesulam, M. M. (1982). Slowly progressive aphasia without generalized dementia. *Annals of Neurology, 11*(6), 592–598.

Mesulam, M. M. (1990). Large-scale neurocognitive networks and distributed processing for attention, language, and memory. *Annals of Neurology, 28*(5), 597–613.

Mesulam, M. M. (1998). From sensation to cognition. *Brain, 121*(6), 1013–1052.

Mesulam, M. M., Rader, B. M., Sridhar, J., Nelson, M. J., Hyun, J., Rademaker, A., Geula, C., Bigio, E. H., Thompson, C. K., Gefen, T. D., . . . , & Rogalski, E. J. (2019). Word comprehension in temporal cortex and Wernicke area: A PPA perspective. *Neurology, 92*(3), e224–e233.

Mesulam, M. M., Thompson, C. K., Weintraub, S., & Rogalski, E. J. (2015). The Wernicke conundrum and the anatomy of language comprehension in primary progressive aphasia. *Brain, 138*(Pt 8), 2423–2437.

Meyer, D. E., & Schvaneveldt, R. W. (1971). Facilitation in recognizing pairs of words: Evidence of a dependence between retrieval operations. *Journal of Experimental Psychology, 90*(2), 227–234.

Mi, L., Tao, S., Wang, W., Dong, Q., Guan, J., & Liu, C. (2016). English vowel identification and vowel formant discrimination by native Mandarin Chinese- and native English-speaking listeners: The effect of vowel duration dependence. *Hearing Research, 333*, 58–65.

Miceli, G., Gainotti, G., Caltagirone, C., & Masullo, C. (1980). Some aspects of phonological impairment in aphasia. *Brain and Language, 11*(1), 159–169.

Miceli, G., Mazzucchi, A., Menn, L., & Goodglass, H. (1983). Contrasting cases of Italian agrammatic aphasia without comprehension disorder. *Brain and Language, 19*(1), 65–97.

Millett, D. (2001). Hans Berger: From psychic energy to the EEG. *Perspectives in Biology, 44*(4), 522–542.

Milner, A. D., & Goodale, M. A. (1995). *The visual brain in action*. Oxford: Oxford University Press.

Milner, A. D., & Goodale, M. A. (2008). Two visual systems re-viewed. *Neuropsychologia, 46*(3), 774–785.

Milner, B. (1958). Psychological defects produced by temporal lobe excision. *Research Publications of the Association for Research in Nervous & Mental Disease, 36*, 244–257.

Mirman, D., Chen, Q., Zhang, Y., Wang, Z., Faseyitan, O. K., Coslett, H. B., & Schwartz, M. F. (2015). Neural organization of spoken language revealed by lesion-symptom mapping. *Nature Communications, 6*, 6762.

Mirman, D., Landrigan, J. F., & Britt, A. E. (2017). Taxonomic and thematic semantic systems. *Psychological Bulletin, 143*(5), 499–520.

Mitchell, K. J. (2020). *Innate: How the wiring of our brains shapes who we are*. Princeton, NJ: Princeton University Press.

Mohr, J. P., Pessin, M. S., Finkelstein, S., Funkenstein, H. H., Duncan, G. W., & Davis, K. R. (1978). Broca's aphasia: Pathologic and clinical. *Neurology, 28*(4), 311–324.

Molinaro, N., Lizarazu, M., Lallier, M., Bourguignon, M., & Carreiras, M. (2016). Out-of-synchrony speech entrainment in developmental dyslexia. *Human Brain Mapping, 37*(8), 2767–2783.

Moses, D. A., Mesgarani, N., Leonard, M. K., & Chang, E. F. (2016). Neural speech recognition: Continuous phoneme decoding using spatiotemporal representations of human cortical activity. *Journal of Neural Engineering, 13*(5), 056004.

Motley, M. T., Camden, C. T., & Baars, B. J. (1982). Covert formulation and editing of anomalies in speech production: Evidence from experimentally elicited slips of the tongue. *Journal of Verbal Learning and Verbal Behavior, 21*(5), 578–594.

Mott, F. W. (1907). Bilateral lesion of the auditory cortical centre: Complete deafness and aphasia. *British Medical Journal, 2*(2432), 310–315.

Mummery, C. J., Patterson, K., Price, C. J., Ashburner, J., Frackowiak, R. S. J., & Hodges, J. R. (2000). A voxel-based morphometry study of the semantic dementia: Relationship between temporal lobe atrophy and semantic memory. *Annals of Neurology, 47*(1), 36–45.

Mummery, C. J., Patterson, K., Wise, R. J. S., Vandenbergh, R., Price, C. J., & Hodges, J. R. (1999). Disrupted temporal lobe connections in semantic dementia. *Brain, 122*(1), 61–73.

Mundy-Castle, A. C. (1951). Theta and beta rhythm in the electroencephalograms of normal adults. *Electroencephalography and Clinical Neurophysiology, 3*(4), 477–486.

Munhall, K. G., Jones, J. A., Callan, D. E., Kuratate, T., & Vatikiotis-Bateson, E. (2004). Visual prosody and speech intelligibility: Head movement improves auditory speech perception. *Psychological Science, 15*(2), 133–137.

Murphy, E. (2020). *The oscillatory nature of langauge*. Cambridge, UK: Cambridge University Press.

Nahum, A. S., & Liegeois, F. J. (2020). Language after childhood hemispherectomy: A systematic review. *Neurology, 95*(23), 1043–1056.

Narain, C., Scott, S. K., Wise, R. J., Rosen, S., Leff, A., Iversen, S. D., & Matthews, P. M. (2003). Defining a left-lateralized response specific to intelligible speech using fMRI. *Cerebral Cortex, 13*(12), 1362–1368.

Narayan, S. S., Temchin, A. N., Recio, A., & Ruggero, M. A. (1998). Frequency tuning of basilar membrane and auditory nerve fibers in the same cochleae. *Science, 282*(5395), 1882–1884.

Nath, A. R., & Beauchamp, M. S. (2012). A neural basis for interindividual differences in the McGurk effect, a multisensory speech illusion. *NeuroImage, 59*(1), 781–787.

Neef, N. E., Primassin, A., von Gudenberg, A. W., Dechent, P., Riedel, H. C., Paulus, W., & Sommer, M. (2021). Two cortical representations of voice control are differentially involved in speech fluency. *Brain Communication, 3*(2), fcaa232.

Neumann, O. (1990). Direct parameter specification and the concept of perception. *Psychological Research, 52*(2–3), 207–215.

Neville, H., Bavelier, D., Corina, D., Rauschecker, J., Karni, A., Lalwani, A., Braun, A., Clark, V., Jezzard, P., &, Turner, R. (1998). Cerebral organization for language in deaf and hearing subjects: Biological constraints and effects of experience. *Proceedings of the National Academy of Sciences, 95*(3), 922–929.

Neville, H., Nicol, J. L., Barss, A., Forster, K. I., & Garrett, M. F. (1991). Syntactically based sentence processing classes: Evidence from event-related brain potentials. *Journal of Cognitive Neuroscience, 3*(2), 151–165.

New, A. B., Robin, D. A., Parkinson, A. L., Duffy, J. R., McNeil, M. R., Piguet, O., Hornberger, M., Price, C. J., Eickhoff, S. B., & Ballard, K. J. (2015). Altered resting-state network connectivity in stroke patients with and without apraxia of speech. *NeuroImage: Clinical, 8*, 429–439.

Newport, E. L., Bavelier, D., & Neville, N. J. (2001). Critical thinking about critical periods: Perspectives on a critical period for language acquisition. In E. Dupous (Ed.), *Language, brain and cognitive development: Essays in honor of Jacques Mehler* (pp. 481–502). Cambridge, MA: MIT Press.

Niemi, M., Laaksonen, J. P., Aaltonen, O., & Happonen, R. P. (2004). Effects of transitory lingual nerve impairment on speech: An acoustic study of diphthong sounds. *Journal of Oral and Maxillofacial Surgery, 62*(1), 44–51.

Niemi, M., Laaksonen, J. P., Ojala, S., Aaltonen, O., & Happonen, R. P. (2006). Effects of transitory lingual nerve impairment on speech: an acoustic study of sibilant sound /s. *International Journal of Oral and Maxillofacial Surgery, 35*(10), 920–923.

Niemi, M., Laaksonen, J. P., Vahatalo, K., Tuomainen, J., Aaltonen, O., & Happonen, R. P. (2002). Effects of transitory lingual nerve impairment on speech: An acoustic study of vowel sounds. *Journal of Oral and Maxillofacial Surgery, 60*(6), 647–652; discussion 653.

Nieuwenhuys, R. (2013). The myeloarchitectonic studies on the human cerebral cortex of the Vogt-Vogt school, and their significance for the interpretation of functional neuroimaging data. *Brain Structure and Function, 218*(2), 303–352.

Niziolek, C. A., & Guenther, F. H. (2013). Vowel category boundaries enhance cortical and behavioral responses to speech feedback alterations. *Journal of Neuroscience, 33*(29), 12090–12098.

Nooteboom, S. G., & Quene, H. (2017). Self-monitoring for speech errors: Two-stage detection and repair with and without auditory feedback. *Journal of Memory and Language, 95*, 19–35.

Novick, J. M., Trueswell, J. C., & Thompson-Schill, S. L. (2005). Cognitive control and parsing: Reexamining the role of Broca's area in sentence comprehension. *Cognitive, Affective, & Behavioral Neuroscience, 5*(3), 263–281.

Nozari, N., Dell, G. S., & Schwartz, M. F. (2011). Is comprehension necessary for error detection? A conflict-based account of monitoring in speech production. *Cognitive Psychology, 63*(1), 1–33.

Nudo, R. J., & Masterton, R. B. (1990). Descending pathways to the spinal cord, III: Sites of origin of the corticospinal tract. *Journal of Comparative Neurology, 296*(4), 559–583.

Oberhuber, M., Hope, T. M. H., Seghier, M. L., Parker Jones, O., Prejawa, S., Green, D. W., & Price, C. J. (2016). Four functionally distinct regions in the left supramarginal gyrus support word processing. *Cerebral Cortex, 26*(11), 4212–4226.

Obleser, J., Herrmann, B., & Henry, M. J. (2012). Neural oscillations in speech: Don't be enslaved by the envelope. *Frontiers in Human Neuroscience, 6*, 250.

Ogar, J. M., Baldo, J. V., Wilson, S. M., Brambati, S. M., Miller, B. L., Dronkers, N. F., & Gorno-Tempini, M. L. (2011). Semantic dementia and persisting Wernicke's aphasia: Linguistic and anatomical profiles. *Brain and Language, 117*(1), 28–33.

Okada, K., & Hickok, G. (2006). Identification of lexical-phonological networks in the superior temporal sulcus using fMRI. *Neuroreport, 17*, 1293–1296.

Okada, K., Matchin, W., & Hickok, G. (2018). Neural evidence for predictive coding in auditory cortex during speech production. *Psychonomic Bulletin & Review, 25*(1), 423–430.

Okada, K., Rong, F., Venezia, J., Matchin, W., Hsieh, I. H., Saberi, K., Serences, J. T., & Hickok, G. (2010). Hierarchical organization of human auditory cortex: Evidence from acoustic invariance in the response to intelligible speech. *Cerebral Cortex, 20*(10), 2486–2495.

Okada, K., Venezia, J. H., Matchin, W., Saberi, K., & Hickok, G. (2013). An fMRI study of audiovisual speech perception reveals multisensory interactions in auditory cortex. *PLoS One, 8*(6), e68959.

O'Keefe, J., & Recce, M. L. (1993). Phase relationship between hippocampal place units and the EEG theta rhythm. *Hippocampus, 3*(3), 317–330.

Oransky, I. (2005). Joseph Bogen. *Lancet, 365*(9475), 1922.

Ornstein, R. E. (1972). *The psychology of consciousness*. San Francisco: W. H. Freeman and Company.

Ortiz-Rios, M., Kusmierek, P., DeWitt, I., Archakov, D., Azevedo, F. A., Sams, M., Jääskeläinen, I. P., Keliris, G., & Rauschecker, J. P. (2015). Functional MRI of the vocalization-processing network in the macaque brain. *Frontiers in Neuroscience, 9*, 113.

Overath, T., McDermott, J. H., Zarate, J. M., & Poeppel, D. (2015). The cortical analysis of speech-specific temporal structure revealed by responses to sound quilts. *Nature Neuroscience, 18*(6), 903–911.

Ozker, M., Doyle, W., Devinsky, O., & Flinker, A. (2022). A cortical network processes auditory error signals during human speech production to maintain fluency. *PLoS Biology, 20*(2), e3001493.

Pa, J., & Hickok, G. (2008). A parietal-temporal sensory-motor integration area for the human vocal tract: Evidence from an fMRI study of skilled musicians. *Neuropsychologia, 46*(1), 362–368.

Pagni, C. A., Naddeo, M., & Mascari, C. (1988). History of evoked potential recording in humans. In B. L. Grundy & R. M. Villani (Eds.), *Evoked potentials* (pp. 17–44). Vienna: Springer.

Pallier, C., Devauchelle, A. D., & Dehaene, S. (2011). Cortical representation of the constituent structure of sentences. *Proceedings of the National Academy of Sciences of the United States of America, 108*(6), 2522–2527.

Papathanasiou, I., & Coppens, P. (2017). *Aphasia and related neurogenic communication disorders* (2nd ed.). Burlington, MA: Jones & Bartlett Learning.

Parise, C. V., Knorre, K., & Ernst, M. O. (2014). Natural auditory scene statistics shapes human spatial hearing. *Proceedings of the National Academy of Sciences of the United States of America, 111*(16), 6104–6108.

Parisi, D., & Pizzamiglio, L. (1970). Syntactic comprehension in aphasia. *Cortex, 6*(2), 204–215.

Pasley, B. N., David, S. V., Mesgarani, N., Flinker, A., Shamma, S. A., Crone, N. E., Knight, R. T., & Chang, E. F. (2012). Reconstructing speech from human auditory cortex. *PLoS Biology, 10*(1), e1001251.

Patel, A. D. (2006). Musical rhythm, linguistic rhythm, and human evolution. *Music Perception, 24*(1), 99–104.

Patel, A. D. (2009). Music and the brain: Three links to language. In S. Hallam, I. Cross, & M. Thaut (Eds.), *Oxford handbook of music psychology* (pp. 208–216). Oxford: Oxford University Press.

Patel, A. D. (2021). Vocal learning as a preadaptation for the evolution of human beat perception and synchronization. *Philosophical Transactions of the Royal Society of London Series B: Biological Sciences, 376*(1835), 20200326.

Patel, A. D. (2024). Beat-based dancing to music has evolutionary foundations in advanced vocal learning. *BMC Neuroscience. 25*(65), https://doi.org/10.1186/s12868-024-00843-6.

Patel, A. D., Iversen, J. R., Bregman, M. R., & Schulz, I. (2009). Experimental evidence for synchronization to a musical beat in a nonhuman animal. *Current Biology, 19*(10), 827–830.

Patel, A. D., Iversen, J. R., Chen, Y., & Repp, B. H. (2005). The influence of metricality and modality on synchronization with a beat. *Experimental Brain Research, 163*(2), 226–238.

Patira, R., Ciniglia, L., Calvert, T., & Altschuler, E. L. (2017). Pure apraxia of speech due to infarct in premotor cortex. *Neurologia Neurochirurgia Polska, 51*(6), 519–524.

Patterson, K., Nestor, P. J., & Rogers, T. T. (2007). Where do you know what you know? The representation of semantic knowledge in the human brain. *Nature Reviews Neuroscience, 8*(12), 976–987.

Paus, T., Perry, D. W., Zatorre, R. J., Worsley, K. J., & Evans, A. C. (1996). Modulation of cerebral blood flow in the human auditory cortex during speech: Role of motor-to-sensory discharges. *European Journal of Neuroscience, 8*(11), 2236–2246.

Pause, M., Kunesch, E., Binkofski, F., & Freund, H. J. (1989). Sensorimotor disturbances in patients with lesions of the parietal cortex. *Brain, 112*(Pt 6), 1599–1625.

Peelle, J. E., & Davis, M. H. (2012). Neural oscillations carry speech rhythm through to comprehension. *Frontiers in Psychology, 3*, 320.

Peelle, J. E., Gross, J., & Davis, M. H. (2013). Phase-locked responses to speech in human auditory cortex are enhanced during comprehension. *Cerebral Cortex, 23*(6), 1378–1387.

Pell, M. D., & Baum, S. R. (1997). The ability to perceive and comprehend intonation in linguistic and affective contexts by brain-damaged adults. *Brain and Language, 57*(1), 80–99.

Persichetti, A. S., Denning, J. M., Gotts, S. J., & Martin, A. (2021). A data-driven functional mapping of the anterior temporal lobes. *Journal of Neuroscience, 41*(28), 6038-6049.

Petersen, S. E., Fox, P. T., Posner, M. I., Mintun, M., & Raichle, M. E. (1988). Positron emission tomographic studies of the cortical anatomy of single-word processing. *Nature, 331*(6157), 585–589.

Petersen, S. E., Fox, P. T., Posner, M. I., Mintun, M., & Raichle, M. E. (1989). Positron emission tomographic studies of the processing of single words. *Journal of Cognitive Neuroscience, 1*(2), 153–170.

Petersen, S. E., Seitzman, B. A., Nelson, S. M., Wig, G. S., & Gordon, E. M. (2024). Principles of cortical areas and their implications for neuroimaging. *Neuron, 112*(17), 2837-2853.

Petitto, L. A., Zatorre, R. J., Gauna, K., Nikelski, E. J., Dostie, D., & Evans, A. C. (2000). Speech-like cerebral activity in profoundly deaf people processing signed languages: Implications for the neural basis of human language. *Proceedings of the National Academy of Sciences of the United States of America, 97*(25), 13961–13966.

Piai, V., & Eikelboom, D. (2023). Brain areas critical for picture naming: A systematic review and meta-analysis of lesion-symptom mapping studies. *Neurobiology of Language (Camb), 4*(2), 280–296.

Piccolino, M. (1998). Animal electricity and the birth of electrophysiology: The legacy of Luigi Galvani. *Brain Research Bulletin, 46*(5), 381–407.

Piccolino, M. (2006). Luigi Galvani's path to animal electricity. *Comptes Rendus Biologies, 329*(5–6), 303–318.

Pillay, S. B., Binder, J. R., Humphries, C., Gross, W. L., & Book, D. S. (2017). Lesion localization of speech comprehension deficits in chronic aphasia. *Neurology, 88*(10), 970–975.

Pinker, S. (1991). Rules of language. *Science, 253*(5019), 530–535.

Pinker, S. (1999). *Words and rules: The ingredients of language.* New York: Basic Books.

Pinker, S., & Prince, A. (1988). On language and connectionism: Analysis of a parallel distributed processing model of language acquisition. *Cognition, 28*(1–2), 73–193.

Pinker, S., & Ullman, M. T. (2002). The past and future of the past tense. *Trends in Cognitive Sciences, 6*(11), 456–463.

Poe, E. A. (2000). The poetic principle. In *The works of Edgar Allan Poe* (Vol. 5). The Gutenberg Project. https://www.gutenberg.org/cache/epub/2151/pg2151-images.html#chap5.11.

Poeck, K., & Luzzatti, C. (1988). Slowly progressive aphasia in three patients. The problem of accompanying neuropsychological deficit. *Brain, 111*(Pt 1), 151–168.

Poeppel, D. (1996). A critical review of PET studies of phonological processing. *Brain and Language, 55*(3), 317–351; discussion 352–385.

Poeppel, D. (2001). Pure word deafness and the bilateral processing of the speech code. *Cognitive Science, 25*(5), 679–693.

Poeppel, D. (2003). The analysis of speech in different temporal integration windows: Cerebral lateralization as "asymmetric sampling in time." *Speech Communication, 41*(1), 245–255.

Poeppel, D., & Assaneo, M. F. (2020). Speech rhythms and their neural foundations. *Nature Reviews Neuroscience, 21*(6), 322–334.

Poeppel, D., & Hickok, G. (2004). Towards a new functional anatomy of language. *Cognition, 92*(1–2), 1–12.

Poizner, H., Klima, E. S., & Bellugi, U. (1987). *What the hands reveal about the brain.* Cambridge, MA: MIT Press.

Poljak, S. (1926). The connections of the acoustic nerve. *Journal of Anatomy, 60*(4), 465–469.

Prather, P., Shapiro, L., Zurif, E., & Swinney, D. (1991). Real-time examinations of lexical processing in aphasics. *Journal of Psycholinguistic Research, 20*(3), 271–281.

Pratt, C. C. (1930). The spatial character of high and low tones. *Journal of Experimental Psychology, 13*(3), 278–285.

Price, C. J., Wise, R. J. S., Warburton, E. A., Moore, C. J., Howard, D., Patterson, K., Frackowiak, R. S., & Friston, K, J. (1996). Hearing and saying: The functional neuroanatomy of auditory word processing. *Brain, 119*(3), 919–931.

Purcell, D. W., & Munhall, K. G. (2006). Adaptive control of vowel formant frequency: Evidence from real-time formant manipulation. *Journal of the Acoustical Society of America, 120*(2), 966–977.

Purves, D., Paydarfar, J. A., & Andrews, T. J. (1996). The wagon wheel illusion in movies and reality. *Proceedings of the National Academy of Sciences of the United States of America, 93*(8), 3693–3697.

Pylkkanen, L., Stringfellow, A., & Marantz, A. (2002). Neuromagnetic evidence for the timing of lexical activation: An MEG component sensitive to phonotactic probability but not to neighborhood density. *Brain and Language, 81*(1–3), 666–678.

Quigg, M., & Fountain, N. B. (1999). Conduction aphasia elicited by stimulation of the left posterior superior temporal gyrus. *Journal of Neurology, Neurosurgery, and Psychiatry, 66*(3), 393–396.

Rapp, B., Buchwald, A., & Goldrick, M. (2014). Integrating accounts of speech production: The devil is in the representational details. *Language, Cognition and Neuroscience, 29*(1), 24–27.

Ratajska, A., Brown, M. I., & Chabris, C. F. (2020). Attributing social meaning to animated shapes: A new experimental study of apparent behavior. *American Journal of Psychology, 133*(3), 295–312.

Rauschecker, J. P. (1998). Cortical processing of complex sounds. *Current Opinion in Neurobiology, 8*(4), 516–521.

Rauschecker, J. P., & Scott, S. K. (2009). Maps and streams in the auditory cortex: Nonhuman primates illuminate human speech processing. *Nature Neuroscience, 12*(6), 718–724.

Rauschecker, J. P., & Tian, B. (2000). Mechanisms and streams for processing of "what" and "where" in auditory cortex. *Proceedings of the National Academy of Sciences of the United States of America, 97*(22), 11800–11806.

Rauschecker, J. P., & Tian, B. (2004). Processing of band-passed noise in the lateral auditory belt cortex of the rhesus monkey. *Journal of Neurophysiology, 91*(6), 2578–2589.

Rauschecker, J. P., Tian, B., & Hauser, M. (1995). Processing of complex sounds in the macaque nonprimary auditory cortex. *Science, 268*(5207), 111–114.

Reed, C. L., & Caselli, R. J. (1994). The nature of tactile agnosia: A case study. *Neuropsychologia, 32*(5), 527–539.

Reed, C. L., Caselli, R. J., & Farah, M. J. (1996). Tactile agnosia. Underlying impairment and implications for normal tactile object recognition. *Brain, 119*(Pt 3), 875–888.

Reilly, J., & Martin, N. (2018). Semantic processing in transcortical sensory aphasia. In A. M. Raymer & J. G. Rothi (Eds.), *The Oxford handbook of aphasia and langauge disorders* (pp. 97–120). Oxford: Oxford University Press.

Richardson, D. H., & Knights, R. M. (1970). A bibliography on dichotic listening. *Cortex, 6*(2), 236–240.

Richardson, J. D., Fillmore, P., Rorden, C., Lapointe, L. L., & Fridriksson, J. (2012). Re-establishing Broca's initial findings. *Brain and Language, 123*(2), 125–130.

Rimmele, J. M., Morillon, B., Poeppel, D., & Arnal, L. H. (2018). Proactive sensing of periodic and aperiodic auditory patterns. *Trends in Cognitive Sciences, 22*(10), 870–882.

Rine, R. M., & Skavenski, A. A. (1997). Extraretinal eye position signals determine perceived target location when they conflict with visual cues. *Vision Research, 37*(6), 775–787.

Risser, A. H. (2018). Katharine McBride, 1935, and "Aphasia." In W. B. Barr & L. A. Bieliauskas (Eds.), *The Oxford handbook of history of clinical neuropsychology* (pp. 48–66). Oxford: Oxford University Press.

Rizzolatti, G., & Craighero, L. (2004). The mirror-neuron system. *Annual Review of Neuroscience, 27*, 169–192.

Rizzolatti, G., Fogassi, L., & Gallese, V. (1997). Parietal cortex: From sight to action. *Current Opinion in Neurobiology, 7*(4), 562–567.

Robson, H., Keidel, J. L., Ralph, M. A., & Sage, K. (2012). Revealing and quantifying the impaired phonological analysis underpinning impaired comprehension in Wernicke's aphasia. *Neuropsychologia, 50*(2), 276–288.

Roelofs, A., & Ferreira, V. S. (2019). The architecture of speaking. In P. Hagoort (Ed.), *Human language: From genes to brains to behavior* (pp. 35–50). Cambridge, MA: MIT Press.

Roelofs, A., Meyer, A. S., & Levelt, W. J. (1998). A case for the lemma/lexeme distinction in models of speaking: Comment on Caramazza and Miozzo (1997). *Cognition, 69*(2), 219–230.

Rogalsky, C., Basilakos, A., Rorden, C., Pillay, S., LaCroix, A. N., Keator, L., Mickelsen, S., Anderson, S. W., Love, T., Fridriksson, J., . . . , & Hickok, G. (2022). The neuroanatomy of speech processing: A large-scale lesion study. *Journal of Cognitive Neuroscience, 34*(8), 1355–1375.

Rogalsky, C., & Hickok, G. (2009). Selective attention to semantic and syntactic features modulates sentence processing networks in anterior temporal cortex. *Cerebral Cortex, 19*(4), 786–796.

Rogalsky, C., LaCroix, A. N., Chen, K. H., Anderson, S. W., Damasio, H., Love, T., & Hickok, G. (2018). The neurobiology of agrammatic sentence comprehension: A lesion study. *Journal of Cognitive Neuroscience, 30*(2), 234–255.

Rogalsky, C., Love, T., Driscoll, D., Anderson, S. W., & Hickok, G. (2011). Are mirror neurons the basis of speech perception? Evidence from five cases with damage to the purported human mirror system. *Neurocase, 17*(2), 178–187.

Rogalsky, C., Matchin, W., & Hickok, G. (2008). Broca's area, sentence comprehension, and working memory: An fMRI study. *Frontiers in Human Neuroscience, 2*, 14.

Rogalsky, C., Pitz, E., Hillis, A. E., & Hickok, G. (2008). Auditory word comprehension impairment in acute stroke: Relative contribution of phonemic versus semantic factors. *Brain and Language, 107*(2), 167–169.

Rogalsky, C., Poppa, T., Chen, K. H., Anderson, S. W., Damasio, H., Love, T., & Hickok, G. (2015). Speech repetition as a window on the neurobiology of auditory-motor integration for speech: A voxel-based lesion symptom mapping study. *Neuropsychologia, 71*, 18–27.

Rogalsky, C., Rong, F., Saberi, K., & Hickok, G. (2011). Functional anatomy of language and music perception: Temporal and structural factors investigated using functional magnetic resonance imaging. *Journal of Neuroscience, 31*(10), 3843–3852.

Rogers, S. L., & Friedman, R. B. (2008). The underlying mechanisms of semantic memory loss in Alzheimer's disease and semantic dementia. *Neuropsychologia, 46*(1), 12–21.

Romanski, L. M., Tian, B., Fritz, J., Mishkin, M., Goldman-Rakic, P. S., & Rauschecker, J. P. (1999). Dual streams of auditory afferents target multiple domains in the primate prefrontal cortex. *Nature Neuroscience, 2*(12), 1131–1136.

Ronan, L., & Fletcher, P. C. (2015). From genes to folds: A review of cortical gyrification theory. *Brain Structure and Function, 220*(5), 2475–2483.

Rong, F., Isenberg, A. L., Sun, E., & Hickok, G. (2018). The neuroanatomy of speech sequencing at the syllable level. *PLoS One, 13*(10), e0196381.

Ross, E. D. (1981). The aprosodias: Functional-anatomic organization of the affective components of language in the right hemisphere. *Archives of Neurology, 38*(9), 561–569.

Ross, E. D., & Mesulam, M. M. (1979). Dominant language functions of the right hemisphere? *Archives of Neurology, 36*(3), 144–148.

Ross, E. D., & Monnot, M. (2008). Neurology of affective prosody and its functional-anatomic organization in right hemisphere. *Brain and Language, 104*(1), 51–74.

Ross, L. A., Saint-Amour, D., Leavitt, V. M., Javitt, D. C., & Foxe, J. J. (2007). Do you see what I am saying? Exploring visual enhancement of speech comprehension in noisy environments. *Cerebral Cortex, 17*(5), 1147–1153.

Roubeau, B., Chevrie-Muller, C., & Lacau Saint Guily, J. (1997). Electromyographic activity of strap and cricothyroid muscles in pitch change. *Acta Oto-Laryngologica, 117*(3), 459–464.

Roux, F. E., Miskin, K., Durand, J. B., Sacko, O., Rehault, E., Tanova, R., & Demonet, J. F. (2015). Electrostimulation mapping of comprehension of auditory and visual words. *Cortex, 71*, 398–408.

Rumelhart, D. E., & McClelland, J. L. (1986a). On learning the past tenses of English verbs. In J. L. McClelland & D. E. Rumelhart (Eds.), *Parallel distributed processing: Exploring the microstructure of cognition* (Vol. 2, pp. 216–271). Cambridge, MA: MIT Press.

Rumelhart, D. E., & McClelland, J. L. (Eds.). (1986b). *Parallel distributed processing: Explorations in the microstructure of cognition*. Cambridge, MA: MIT Press.

Rutten, G.-J. (2017). *The Broca-Wernicke doctrine: A historical and clinical perspective on localization of language functions*. Berlin: Springer.

Saberi, K., & Hickok, G. (2023). Forward entrainment: Psychophysics, neural correlates, and function. *Psychonomic Bulletin & Review, 30*(3), 803–821.

Sabin, A. T., Eddins, D. A., & Wright, B. A. (2012). Perceptual learning evidence for tuning to spectrotemporal modulation in the human auditory system. *Journal of Neuroscience, 32*(19), 6542–6549.

Sacks, O. (1973). *Awakenings*. London: Duckworth.

Sacks, O. (2008). *Musicophilia: Tales of music and the brain*. New York: Knopf.

Saffran, E. M., Marin, O. S., & Yeni-Komshian, G. H. (1976). An analysis of speech perception in word deafness. *Brain and Language, 3*(2), 209–228.

Santoro, R., Moerel, M., De Martino, F., Goebel, R., Ugurbil, K., Yacoub, E., & Formisano, E. (2014). Encoding of natural sounds at multiple spectral and temporal resolutions in the human auditory cortex. *PLOS Computational Biology, 10*(1), e1003412.

Santoro, R., Moerel, M., De Martino, F., Valente, G., Ugurbil, K., Yacoub, E., & Formisano, E. (2017). Reconstructing the spectrotemporal modulations of real-life sounds from fMRI response patterns. *Proceedings of the National Academy of Sciences of the United States of America, 114*(18), 4799–4804.

Sathian, K. (2016). Analysis of haptic information in the cerebral cortex. *Journal of Neurophysiology, 116*(4), 1795–1806.

Sathian, K., Lacey, S., Stilla, R., Gibson, G. O., Deshpande, G., Hu, X., Laconte, S., & Glielmi, C. (2011). Dual pathways for haptic and visual perception of spatial and texture information. *NeuroImage, 57*(2), 462–475.

Savini, N., Babiloni, C., Brunetti, M., Caulo, M., Del Gratta, C., Perrucci, M. G., Rossini, P. M., Romani, G. L., & Ferretti, A. (2010). Passive tactile recognition of geometrical shape in humans: An fMRI study. *Brain Research Bulletin, 83*(5), 223–231.

Schell, M., Zaccarella, E., & Friederici, A. D. (2017). Differential cortical contribution of syntax and semantics: An fMRI study on two-word phrasal processing. *Cortex, 96*, 105–120.

Schenk, T. (2010). Visuomotor robustness is based on integration not segregation. *Vision Research, 50*(24), 2627–2632.

Schenk, T., & McIntosh, R. D. (2010). Do we have independent visual streams for perception and action? *Cognitive Neuroscience, 1*(1), 52–62.

Schlosser, M. J., Aoyagi, N., Fulbright, R. K., Gore, J. C., & McCarthy, G. (1998). Functional MRI studies of auditory comprehension. *Human Brain Mapping, 6*(1), 1–13.

Schneider, G. E. (1969). Two visual systems. *Science, 163*(3870), 895–902.

Schneider, R. J., Friedman, D. P., & Mishkin, M. (1993). A modality-specific somatosensory area within the insula of the rhesus monkey. *Brain Research, 621*(1), 116–120.

Schomers, M. R., Kirilina, E., Weigand, A., Bajbouj, M., & Pulvermuller, F. (2014). Causal influence of articulatory motor cortex on comprehending single spoken words: TMS evidence. *Cerebral Cortex, 25*(10), 3894–3902.

Schonwiesner, M., Rubsamen, R., & von Cramon, D. Y. (2005). Hemispheric asymmetry for spectral and temporal processing in the human antero-lateral auditory belt cortex. *European Journal of Neuroscience, 22*(6), 1521–1528.

Schroeder, C. E., & Lakatos, P. (2009). Low-frequency neuronal oscillations as instruments of sensory selection. *Trends in Neuroscience, 32*(1), 9–18.

Schroeder, C. E., Smiley, J., Fu, K. G., McGinnis, T., O'Connell, M. N., & Hackett, T. A. (2003). Anatomical mechanisms and functional implications of multisensory convergence in early cortical processing. *International Journal of Psychophysiology, 50*(1–2), 5–17.

Schroll, H., & Hamker, F. H. (2013). Computational models of basal-ganglia pathway functions: Focus on functional neuroanatomy. *Frontiers in Systems Neuroscience, 7*, 122.

Schubotz, R. I., & von Cramon, D. Y. (2004). Sequences of abstract nonbiological stimuli share ventral premotor cortex with action observation and imagery. *Journal of Neuroscience, 24*(24), 5467–5474.

Schwartz, M. F. (2014). Theoretical analysis of word production deficits in adult aphasia. *Philosophical Transactions of the Royal Society of London Series B: Biological Sciences, 369*(1634), 20120390.

Schwartz, M. F., Faseyitan, O., Kim, J., & Coslett, H. B. (2012). The dorsal stream contribution to phonological retrieval in object naming. *Brain, 135*(Pt 12), 3799–3814.

Schwartz, M. F., Kimberg, D. Y., Walker, G. M., Brecher, A., Faseyitan, O. K., Dell, G. S., Mirman, D., & Coslett, H. B. (2011). Neuroanatomical dissociation for taxonomic and thematic knowledge in the human brain. *Proceedings of the National Academy of Sciences of the United States of America, 108*(20), 8520–8524.

Schwartz, M. F., Kimberg, D. Y., Walker, G. M., Faseyitan, O., Brecher, A., Dell, G. S., & Coslett, H. B. (2009). Anterior temporal involvement in semantic word retrieval: Voxel-based lesion-symptom mapping evidence from aphasia. *Brain, 132*(Pt 12), 3411–3427.

Scott, S. K., Blank, C. C., Rosen, S., & Wise, R. J. S. (2000). Identification of a pathway for intelligible speech in the left temporal lobe. *Brain, 123*(12), 2400–2406.

Semon, R. W. (1921). *The mneme*. New York: The MacMillan Company.

Senghas, A., Kita, S., & Ozyurek, A. (2004). Children creating core properties of language: Evidence from an emerging sign language in Nicaragua. *Science, 305*(5691), 1779–1782.

Sengpiel, F. (2007). The critical period. *Current Biology, 17*(17), R742–R743.

Sergent, J. (1982). Theoretical and methodological consequences of variations in exposure duration in visual laterality studies. *Perception & Psychophysics, 31*(5), 451–461.

Shamma, S. (2001). On the role of space and time in auditory processing. *Trends in Cognitive Sciences, 5*(8), 340–348.

Shankweiler, D. (1966). Effects of temporal-lobe damage of perception of dichotically presented melodies. *Journal of Comparative Physiology, 62*(1), 115–119.

Shankweiler, D., Strange, W., & Verbrugge, R. (1977). Speech perception and the problem of perceptual constance. In R. Shaw & J. Bransford (Eds.), *Perceiving, acting, and knowing: Toward an ecological psychology* (pp. 315–346). New York: Lawrence Earlbaum Associates.

Shankweiler, D., & Studdert-Kennedy, M. (1967). Identification of consonants and vowels presented to left and right ears. *Quarterly Journal of Experimental Psychology, 19*(1), 59–63.

Shattuck-Hufnagel, S. (2015). Prosodic frames in speech production. In M. A. Redford (Ed.), *The handbook of speech production* (pp. 419–444). Oxford, UK: John Wiley & Sons.

Shattuck-Hufnagel, S., & Turk, A. E. (1996). A prosody tutorial for investigators of auditory sentence processing. *Journal of Psycholinguistic Research, 25*(2), 193–247.

Shimotake, A., Matsumoto, R., Ueno, T., Kunieda, T., Saito, S., Hoffman, P., Kikuchi, T., Fukuyama, H., Miyamoto, S., Takahashi, R., . . . , & Lambon Ralph, M. A. (2015). Direct exploration of the role of the ventral anterior temporal lobe in semantic memory: Cortical stimulation and local field potential evidence from subdural grid electrodes. *Cerebral Cortex, 25*(10), 3802–3817.

Silveri, M. C., Incordino, F., Lo Monaco, R., Bizzarro, A., Masullo, C., Piludu, F., & Colosimo, C. (2017). Neural substrates of the "low-level" system for speech articulation: Evidence from primary opercular syndrome. *Journal of Neuropsychology, 11*(3), 450–457.

Simonyan, K. (2014). The laryngeal motor cortex: Its organization and connectivity. *Current Opinion in Neurobiology, 28,* 15–21.

Simonyan, K., Ackermann, H., Chang, E. F., & Greenlee, J. D. (2016). New developments in understanding the complexity of human speech production. *Journal of Neuroscience, 36*(45), 11440–11448.

Simonyan, K., & Horwitz, B. (2011). Laryngeal motor cortex and control of speech in humans. *Neuroscientist, 17*(2), 197–208.

Simonyan, K., & Jürgens, U. (2005). Afferent cortical connections of the motor cortical larynx area in the rhesus monkey. *Neuroscience, 130*(1), 133–149.

Simpson, G. B., Petersen, R. R., Casteel, M. A., & Bergess, C. (1989). Lexical and sentence context effects on word recognition. *Journal of Experimental Psychology: Learning, Memory, and Cognition, 15*(1), 88–97.

Singer, W. (1993). Synchronization of cortical activity and its putative role in information processing and learning. *Annual Review of Physiology, 55,* 349–374.

Skinner, B. F. (1957). *Verbal behavior.* New York: Appleton-Century-Crofts.

Smith, E. E., & Jonides, J. (1997). Working memory: A view from neuroimaging. *Cognitive Psychology, 33*(1), 5–42.

Smith, E. E., Jonides, J., & Koeppe, R. A. (1996). Dissociating verbal and spatial working memory using PET. *Cerebral Cortex, 6*(1), 11–20.

Smith, K. R., Hsieh, I. H., Saberi, K., & Hickok, G. (2010). Auditory spatial and object processing in the human planum temporale: No evidence for selectivity. *Journal of Cognitive Neuroscience, 22*(4), 632–639.

Smith, K. R., Okada, K., Saberi, K., & Hickok, G. (2004). Human cortical motion areas are not motion selective. *Neuroreport, 15*(9), 1523–1526.

Smith, K. R., Saberi, K., & Hickok, G. (2007). An event-related fMRI study of auditory motion perception: No evidence for a specialized cortical system. *Brain Research, 1150,* 94–99.

Snow, J. C., Goodale, M. A., & Culham, J. C. (2015). Preserved haptic shape processing after bilateral LOC lesions. *Journal of Neuroscience, 35*(40), 13745–13760.

Snowden, J. S., Goulding, P. J., & Neary, D. (1989). Semantic dementia: A form of circumscribed cerebral atrophy. *Behavioral Neurology, 2*(3), 167–182.

Sohoglu, E., Peelle, J. E., Carlyon, R. P., & Davis, M. H. (2012). Predictive top-down integration of prior knowledge during speech perception. *Journal of Neuroscience, 32*(25), 8443–8453.

Spencer, K. A., & Slocomb, D. L. (2007). The neural basis of ataxic dysarthria. *Cerebellum, 6*(1), 58–65.

Sperry, R. W. (1961). Cerebral organization and behavior: The split brain behaves in many respects like two separate brains, providing new research possibilities. *Science, 133*(3466), 1749–1757.

Spinelli, E., Grimault, N., Meunier, F., & Welby, P. (2010). An intonational cue to word segmentation in phonemically identical sequences. *Attention Perception Psychophysics, 72*(3), 775–787.

Springer, S. P., & Deutsch, G. (1981). *Left brain, right brain.* San Francisco: W. H. Freeman.

Sprouse, J., & Almeida, D. (2013). The empirical status of data in syntax: A reply to Gibson and Fedorenko. *Language and Cognitive Processes, 28*(3), 222–228.

Steinthal, H. (1871). *Einleitung in die Psychologie und Sprachwissenschaft.* Berlin: F. Dummler.

Stengel, E. (1947). A clinical and psychological study of echo-reactions. *Journal of Mental Science, 93*(392), 598–612.

Stevens, K. N. (1980). Acoustic correlates of some phonetic categories. *Journal of the Acoustical Society of America, 68*(3), 836–842.

Stockall, L., & Marantz, A. (2006). A single route, full decomposition model of morphological complexity. *Mental Lexicon, 1*(1), 85–123.

Stokes, R. C., Venezia, J. H., & Hickok, G. (2019). The motor system's [modest] contribution to speech perception. *Psychonomic Bulletin & Review, 26*(4), 1354–1366.

Stowe, L. A., Paans, A. M., Wijers, A. A., Zwarts, F., Mulder, G., & Vaalburg, W. (1999). Sentence comprehension and word repetition: A positron emission tomography investigation. *Psychophysiology, 36*(6), 786–801.

Stromswold, K., Caplan, D., Alpert, N., & Rauch, S. (1996). Localization of syntactic comprehension by positron emission tomography. *Brain and Language, 52*(3), 452–473.

Strong, O. S., & Elwyn, A. (1948). *Human neuroanatomy.* 2nd ed. Baltimore: Williams and Wilkins.

Stuart, A., Kalinowski, J., Rastatter, M. P., & Lynch, K. (2002). Effect of delayed auditory feedback on normal speakers at two speech rates. *Journal of the Acoustical Society of America, 111*(5 Pt 1), 2237–2241.

Sumby, W. H., & Pollack, I. (1954). Visual contributions to speech intelligibility in noise. *Journal of the Acoustical Society of America, 26*(2), 212–215.

Summers, W. V., Pisoni, D. B., Bernacki, R. H., Pedlow, R. I., & Stokes, M. A. (1988). Effects of noise on speech production: acoustic and perceptual analyses. *Journal of the Acoustical Society of America, 84*(3), 917–928.

Suzuki, T., Itoh, S., Arai, N., Kouno, M., Noguchi, M., Takatsu, M., & Takeda, K. (2012). Ambient echolalia in a patient with germinoma around the bilateral ventriculus lateralis: A case report. *Neurocase, 18*(4), 330–335.

Suzuki, W. A., & Amaral, D. G. (1994). Perirhinal and parahippocampal cortices of the macaque monkey: cortical afferents. *Journal of Comparative Neurology, 350*(4), 497–533.

Swinney, D. (1979). Lexical access during sentence comprehension: (Re)consideration of context effects. *Journal of Verbal Learning and Verbal Behavior, 18*(6), 645–660.

Swinney, D., Zurif, E., & Nicol, J. (1989). The effects of focal brain damage on sentence processing: An examination of the neurological organization of a mental module. *Journal of Cognitive Neuroscience, 1*(1), 25–37.

Tal, N., & Amedi, A. (2009). Multisensory visual-tactile object related network in humans: Insights gained using a novel crossmodal adaptation approach. *Experimental Brain Research, 198*(2–3), 165–182.

Tallal, P., Miller, S., & Fitch, R. H. (1993). Neurobiological basis of speech: A case for the preeminence of temporal processing. *Annals of the New York Academy of Sciences, 682*, 27–47.

Taylor, J. R. (Ed.). (2015). *The Oxford handbook of the word.* Oxford: Oxford University Press.

Teghipco, A., Okada, K., Murphy, E., & Hickok, G. (2023). Predictive coding and internal error correction in speech production. *Neurobiology of Language (Camb), 4*(1), 81–119.

Ten Oever, S., Schroeder, C. E., Poeppel, D., van Atteveldt, N., Mehta, A. D., Megevand, P., et al. (2017). Low-frequency cortical oscillations entrain to subthreshold rhythmic auditory stimuli. *Journal of Neuroscience, 37*(19), 4903–4912.

Ten Oever, S., Schroeder, C. E., Poeppel, D., van Atteveldt, N., & Zion-Golumbic, E. (2014). Rhythmicity and cross-modal temporal cues facilitate detection. *Neuropsychologia, 63*, 43–50.

Tesak, J., & Code, C. (2008). *Milestones in the history of aphasia: Theories and protagonists.* London: Psychology Press.

Thothathiri, M., Kimberg, D. Y., & Schwartz, M. F. (2012). The neural basis of reversible sentence comprehension: Evidence from voxel-based lesion symptom mapping in aphasia. *Journal of Cognitive Neuroscience, 24*(1), 212–222.

Titze, I. R., Luschei, E. S., & Hirano, M. (1989). Role of the thyroarytenoid muscle in regulation of fundamental frequency. *Journal of Voice, 3*(3), 213–224.

Topalidou, M., Neftci, E., & Hickok, G. (2018). *A computational model of speech production with internal error detection and correction mechanism.* Paper presented at COSYNE, Denver.

Tourville, J. A., & Guenther, F. H. (2011). The DIVA model: A neural theory of speech acquisition and production. *Language and Cognitive Processes, 26*(7), 952–981.

Tourville, J. A., Reilly, K. J., & Guenther, F. H. (2008). Neural mechanisms underlying auditory feedback control of speech. *NeuroImage, 39*(3), 1429–1443.

Tremblay, P., & Dick, A. S. (2016). Broca and Wernicke are dead, or moving past the classic model of language neurobiology. *Brain and Language, 162,* 60–71.

Tremblay, S., Shiller, D. M., & Ostry, D. J. (2003). Somatosensory basis of speech production. *Nature, 423*(6942), 866–869.

Trousseau, A. (1865). *Clinique medicale de l'Hotel-Dieu de Paris.* Paris: Bailliere et Fils.

Tsuboi, N., Francis, W. S., & Jameson, J. T. (2021). How word comprehension exposures facilitate later spoken production: Implications for lexical processing and repetition priming. *Memory, 29*(1), 39–58.

Turkeltaub, P. E., & Coslett, H. B. (2010). Localization of sublexical speech perception components. *Brain and Language, 114*(1), 1–15.

Turker, S., Kuhnke, P., Eickhoff, S. B., Caspers, S., & Hartwigsen, G. (2023). Cortical, subcortical, and cerebellar contributions to language processing: A meta-analytic review of 403 neuroimaging experiments. *Psychological Bulletin, 149*(11–12), 699–723.

Tyler, L. K., & Marslen-Wilson, W. (2008). Fronto-temporal brain systems supporting spoken language comprehension. *Philosophical Transactions of the Royal Society of London Series B: Biological Sciences, 363*(1493), 1037–1054.

Uddin, L. Q., Nomi, J. S., Hebert-Seropian, B., Ghaziri, J., & Boucher, O. (2017). Structure and function of the human insula. *Journal of Clinical Neurophysiology, 34*(4), 300–306.

Ullman, M. T., Corkin, S., Coppola, M., Hickok, G., Growdon, J. H., Koroshetz, W. J., & Pinker, S. (1997). A neural dissociation within language: Evidence that the mental dictionary is part of declarative memory, and that grammatical rules are processed by the procedure system. *Journal of Cognitive Neuroscience, 9*(2), 266–276.

Ungerleider, L. G., & Mishkin, M. (1982). Two cortical visual systems. In D. J. Ingle, M. A. Goodale, & R. J. W. Mansfield (Eds.), *Analysis of visual behavior* (pp. 549–586). Cambridge, MA: MIT Press.

Utianski, R. L., Duffy, J. R., Clark, H. M., Strand, E. A., Botha, H., Schwarz, C. G., Machulda, M., Senjem, M. L., Spychalla, A. J., Jack, C. R. Jr., . . . , & Josephs, K. A.

(2018). Prosodic and phonetic subtypes of primary progressive apraxia of speech. *Brain and Language, 184*, 54–65.

Uttal, W. (2001). *The new phrenology: The limits of localizing cognitive processes in the brain.* Cambridge, MA: MIT Press.

Valenza, N., Ptak, R., Zimine, I., Badan, M., Lazeyras, F., & Schnider, A. (2001). Dissociated active and passive tactile shape recognition: A case study of pure tactile apraxia. *Brain, 124*(Pt 11), 2287–2298.

Vallar, G., & Perani, D. (1986). The anatomy of unilateral neglect after right-hemisphere stroke lesions. A clinical/CT-scan correlation study in man. *Neuropsychologia, 24*(5), 609–622.

van Dijk, H., Schoffelen, J. M., Oostenveld, R., & Jensen, O. (2008). Prestimulus oscillatory activity in the alpha band predicts visual discrimination ability. *Journal of Neuroscience, 28*(8), 1816–1823.

VanRullen, R. (2016). Perceptual cycles. *Trends in Cognitive Science, 20*(10), 723–735.

Vanthornhout, J., Decruy, L., Wouters, J., Simon, J. Z., & Francart, T. (2018). Speech intelligibility predicted from neural entrainment of the speech envelope. *Journal of the Association for Research in Otolaryngology, 19*(2), 181–191.

Van Wagenen, W. P., & Herren, R. Y. (1940). Surgical division of the commissural pathways in the corpus callosum: Relation to spread of an epliptic attack. *Archives of Neurology & Psychiatry, 44*(4), 740–759.

van Wyth, J. (2002). The authority of human nature: The schadellehre of Franz Joseph Gall. *British Journal of the History of Science, 35*(1), 17–42.

Vargha-Khadem, F., Gadian, D. G., Copp, A., & Mishkin, M. (2005). FOXP2 and the neuroanatomy of speech and language. *Nature Reviews Neuroscience, 6*(2), 131–138.

Venezia, J. H., Herrera, C., Whittle, N., Leek, M. R., Barnes, S., Holshouser, B., & Yi, A. (2022). Task-modulated sensitivity to vocal pitch in the dorsal premotor cortex during multitalker speech recognition. *Journal of Cognitive Neuroscience, 34*(11), 2189–2214.

Venezia, J. H., Hickok, G., & Richards, V. M. (2016). Auditory "bubbles": Efficient classification of the spectrotemporal modulations essential for speech intelligibility. *Journal of Acoustical Society of America, 140*(2), 1072.

Venezia, J. H., Richards, V. M., & Hickok, G. (2021). Speech-driven spectrotemporal receptive fields beyond the auditory cortex. *Hearing Research, 408*, 108307.

Venezia, J. H., Thurman, S. M., Richards, V. M., & Hickok, G. (2019). Hierarchy of speech-driven spectrotemporal receptive fields in human auditory cortex. *NeuroImage, 186*, 647–666.

Venezia, J. H., Vaden, K. I., Jr., Rong, F., Maddox, D., Saberi, K., & Hickok, G. (2017). Auditory, visual and audiovisual speech processing streams in superior temporal sulcus. *Frontiers in Human Neuroscience, 11*, 174.

Ventura, M. I., Nagarajan, S. S., & Houde, J. F. (2009). Speech target modulates speaking induced suppression in auditory cortex. *BMC Neuroscience, 10*, 58.

Vigliocco, G., Antonini, T., & Garrett, M. F. (1998). Grammatical gender is on the tip of Italian tongues. *Psychological Science, 8*(4), 314–317.

Viinikainen, M., Katsyri, J., & Sams, M. (2012). Representation of perceived sound valence in the human brain. *Human Brain Mapping, 33*(10), 2295–2305.

Visser, M., Jefferies, E., Embleton, K. V., & Lambon Ralph, M. A. (2012). Both the middle temporal gyrus and the ventral anterior temporal area are crucial for multimodal semantic processing: Distortion-corrected fMRI evidence for a double gradient of information convergence in the temporal lobes. *Journal of Cognitive Neuroscience, 24*(8), 1766–1778.

Vitevitch, M. S., & Luce, P. A. (1999). Probabilistic phonotactics and neighborhood activation in spoken word recognition. *Journal of Memory and Language, 40*(3), 374–408.

Vitevitch, M. S., Luce, P. A., Pisoni, D. B., & Auer, E. T. (1999). Phonotactics, neighborhood activation, and lexical access for spoken words. *Brain and Language, 68*(1–2), 306–311.

Vogt, O. (1910). Die myeloarchitektonische Felderung des Menschlichen Stirnhirns. *Journal für Psychologie und Neurologie, 15*(4–5), 221–232.

von Békésy, G. (1960). *Experiments in hearing.* Oxford, UK: McGraw Hill.

von Stein, A., & Sarnthein, J. (2000). Different frequencies for different scales of cortical integration: from local gamma to long range alpha/theta synchronization. *International Journal of Psychophysiology, 38*(3), 301–313.

Wada, J., & Rasmussen, T. (1960). Intracarotid injection of sodium amytal for the lateralization of cerbral speech dominance. *Journal of Neurosurgery, 17*(2), 266–282.

Wagner, M., & Watson, D. G. (2010). Experimental and theoretical advances in prosody: A review. *Language and Cognitive Processes, 25*(7–9), 905–945.

Walker, G. M., Fridriksson, J., Hillis, A. E., den Ouden, D. B., Bonilha, L., & Hickok, G. (2022). The severity-calibrated aphasia naming test. *American Journal of Speech-Language Pathology, 31*(6), 2722–2740.

Walker, G. M., & Hickok, G. (2016). Bridging computational approaches to speech production: The semantic-lexical-auditory-motor model (SLAM). *Psychonomic Bulletin & Review, 23*(2), 339–352.

Walker, G. M., Hickok, G., & Fridriksson, J. (2018). A cognitive psychometric model for assessment of picture naming abilities in aphasia. *Psychol Assess, 30*(6), 809–826.

Walker, G. M., Rollo, P. S., Tandon, N., & Hickok, G. (2021). Effect of bilateral opercular syndrome on speech perception. *Neurobiology of Language, 2*(3), 335–353.

Walker, G. M., Schwartz, M. F., Kimberg, D. Y., Faseyitan, O., Brecher, A., Dell, G. S., & Coslett, H. B. (2011). Support for anterior temporal involvement in semantic error production in aphasia: New evidence from VLSM. *Brain and Language, 117*(3), 110–122.

Walker, P., Bremner, J. G., Lunghi, M., Dolscheid, S., B, D. B., & Simion, F. (2018). Newborns are sensitive to the correspondence between auditory pitch and visuospatial elevation. *Developmental Psychobiology, 60*(2), 216–223.

Warren, J. D., & Griffiths, T. D. (2003). Distinct mechanisms for processing spatial sequences and pitch sequences in the human auditory brain. *Journal of Neuroscience, 23*(13), 5799–5804.

Warren, J. D., Zielinski, B. A., Green, G. G., Rauschecker, J. P., & Griffiths, T. D. (2002). Perception of sound-source motion by the human brain. *Neuron, 34*(1), 139–148.

Warren, R. M. (1970). Perceptual restoration of missing speech sounds. *Science, 167*(3917), 392–393.

Warrington, E. K. (1975). The selective impairment of semantic memory. *Quarterly Journal of Experimental Psychology, 27*(4), 187–199.

Warrington, E. K., & Crutch, S. J. (2004). A circumscribed refractory access disorder: A verbal semantic impairment sparing visual semantics. *Cognitive Neuropsychology, 21*(2), 299–315.

Watkins, K. E., Strafella, A. P., & Paus, T. (2003). Seeing and hearing speech excites the motor system involved in speech production. *Neuropsychologia, 41*(8), 989–994.

Weigl, E., & Bierwisch, M. (1970). Neuropsychology and linguistics: Topics of common research. *Foundations of Language, 6*(1), 1–18.

Weisenburg, T., & McBride, K. E. (1935). *Aphasia: A clinical and psychological study*. New York: Commonwealth Fund.

Wernicke, C. (1874/1969). The symptom complex of aphasia: A psychological study on an anatomical basis. In R. S. Cohen & M. W. Wartofsky (Eds.), *Boston studies in the philosophy of science* (pp. 34–97). Dordrecht, Netherlands: D. Reidel Publishing Company.

Wernicke, C. (1874/1977). Der aphasische symptomencomplex: Eine psychologische studie auf anatomischer basis. In G. H. Eggert (Ed.), *Wernicke's works on aphasia: A sourcebook and review* (pp. 91–145). The Hague: Mouton.

Westerlund, M., & Pylkkanen, L. (2014). The role of the left anterior temporal lobe in semantic composition vs. semantic memory. *Neuropsychologia, 57*, 59–70.

Whitaker, H. (1976). A case of the isolation of the language function. In H. Whitaker & H. A. Whitaker (Eds.), *Studies in neurolinguistics* (Vol. 2, pp. 1–58). San Diego: Academic Press.

Wigan, A. L. (1844). *A new view of insanity: The duality of the mind.* London: Longman, Brown, Green & Longmans.

Williams, A. (2015). *Agruments in syntax and semantics. Key topics in syntax.* Cambridge: Cambridge University Press.

Wilshire, C. E. (2015). Conduction aphasia: Impaired phonological processing. In A. M. Raymer & L. J. Gonzalez-Rothi (Eds.), *The Oxford handbook of aphasia and language disorders* (pp. 75–96). Oxford: Oxford University Press.

Wilshire, C. E., & McCarthy, R. A. (1996). Experimental investigations of an impairment in phonological encoding. *Cognitive Neuropsychology, 13*(7), 1059–1098.

Wilson, M. (2001). The case for sensorimotor coding in working memory. *Psychonomic Bulletin & Review, 8*(1), 44–57.

Wilson, S. M. (2009). Speech perception when the motor system is compromised. *Trends in Cognitive Sciences, 13*(8), 329–330.

Wilson, S. M., Bautista, A., & McCarron, A. (2018). Convergence of spoken and written language processing in the superior temporal sulcus. *NeuroImage, 171*, 62–74.

Wilson, S. M., Galantucci, S., Tartaglia, M. C., & Gorno-Tempini, M. L. (2012). The neural basis of syntactic deficits in primary progressive aphasia. *Brain and Language, 122*(3), 190–198.

Wilson, S. M., Henry, M. L., Besbris, M., Ogar, J. M., Dronkers, N. F., Jarrold, W., Miller, B. L., & Gorno-Tempini, M. L. (2010). Connected speech production in three variants of primary progressive aphasia. *Brain, 133*(Pt 7), 2069–2088.

Wilson, S. M., Saygin, A. P., Sereno, M. I., & Iacoboni, M. (2004). Listening to speech activates motor areas involved in speech production. *Nature Neuroscience, 7*(7), 701–702.

Winner, E., Brownell, H., Happe, F., Blum, A., & Pincus, D. (1998). Distinguishing lies from jokes: Theory of mind deficits and discourse interpretation in right hemisphere brain-damaged patients. *Brain and Language, 62*(1), 89–106.

Winslow, F. (1911, June 25). "Murderers I have met," by Dr. Forbes L. Winslow. *New York Times*, 61.

Winson, J. (1978). Loss of hippocampal theta rhythm results in spatial memory deficit in the rat. *Science, 201*(4351), 160–163.

Wise, R., Chollet, F., Hadar, U., Friston, K., Hoffner, E., & Frackowiak, R. (1991). Distribution of cortical neural networks involved in word comprehension and word retrieval. *Brain, 114*(Pt 4)(5), 1803–1817.

Wise, R. J. S., Scott, S. K., Blank, S. C., Mummery, C. J., Murphy, K., & Warburton, E. A. (2001). Separate neural sub-systems within "Wernicke's area." *Brain, 124*(1), 83–95.

Wolpert, D. M., Miall, R. C., & Kawato, M. (1998). Internal models in the cerebellum. *Trends in Cognitive Sciences, 2*(9), 338–347.

Woolf, V. (1984). In L. DeSalvo & M. Leaska (Eds.), *The letters of Vita Sackville-West and Virginia Woolf.* San Francisco: Cleis Press.

Wray, A. (2015). Why are we so sure we know what a word is? In J. R. Taylor (Ed.), *Oxford handbook of the word* (pp. 725–750). Oxford: Oxford University Press.

Xu, Y. (2002). *Articulatory constraints and tonal alignment.* Paper presented at the 1st International Conference on Speech Prosody, Aix-en-Provence, France.

Xu, Y., & Sun, X. (2002). Maximum speed of pitch change and how it may relate to speech. *Journal of the Acoustical Society of America, 111*(3), 1399–1413.

Yates, A. J. (1963). Delayed auditory feedback. *Psychological Bulletin, 60*(3), 213–251.

Yau, J. M., Kim, S. S., Thakur, P. H., & Bensmaia, S. J. (2016). Feeling form: The neural basis of haptic shape perception. *Journal of Neurophysiology, 115*(2), 631–642.

Yeo, B. T., Krienen, F. M., Sepulcre, J., Sabuncu, M. R., Lashkari, D., Hollinshead, M., Roffman, J. L., Smoller, J. W., Zöllei, L., Polmeini, J. R., . . . , & Buckner, R. L. (2011). The organization of the human cerebral cortex estimated by intrinsic functional connectivity. *Journal of Neurophysiology, 106*(3), 1125–1165.

Zaccarella, E., & Friederici, A. D. (2017). The neurobiological nature of syntactic hierarchies. *Neuroscience & Biobehavioral Reviews, 81*(Pt B), 205–212.

Zaidel, E. (1985). Language in the right hemisphere. In D. F. Benson & E. Zaidel (Eds.), *The dual brain: Hemispheric specialization in humans* (pp. 205–231). New York: Guilford Press.

Zarco, W., Merchant, H., Prado, L., & Mendez, J. C. (2009). Subsecond timing in primates: Comparison of interval production between human subjects and rhesus monkeys. *Journal Neurophysiology, 102*(6), 3191–3202.

Zatorre, R. J. (1988). Pitch perception of complex tones and human temopral-lobe function. *Journal of the Acoustical Society of America, 84,* 566–572.

Zatorre, R. J. (2015). Musical pleasure and reward: Mechanisms and dysfunction. *Annals of the New York Academy of Sciences, 1337*(1), 202–211.

Zatorre, R. J., Belin, P., & Penhune, V. B. (2002). Structure and function of auditory cortex: Music and speech. *Trends in Cognitive Sciences, 6*(1), 37–46.

Zatorre, R. J., Bouffard, M., Ahad, P., & Belin, P. (2002). Where is "where" in the human auditory cortex? *Nature Neuroscience, 5*(9), 905–909.

Zatorre, R. J., Evans, A. C., Meyer, E., & Gjedde, A. (1992). Lateralization of phonetic and pitch discrimination in speech processing. *Science, 256*(5058), 846–849.

Zhao, L., Silva, A. B., Kurteff, G. L., & Chang, E. F. (2023). Inhibitory control of speech production in the 1 human premotor frontal cortex. *BioRxiv*. 2023.03.01.530698; doi: https://doi.org/10.1101/2023.03.01.530698.

Zhuang, J., Randall, B., Stamatakis, E. A., Marslen-Wilson, W. D., & Tyler, L. K. (2011). The interaction of lexical semantics and cohort competition in spoken word recognition: An fMRI study. *Journal of Cognitive Neuroscience, 23*(12), 3778–3790.

Zion Golumbic, E. M., Poeppel, D., & Schroeder, C. E. (2012). Temporal context in speech processing and attentional stream selection: a behavioral and neural perspective. *Brain and Language, 122*(3), 151–161.

Zurif, E. B., Caramazza, A., & Myerson, R. (1972). Grammatical judgments of agrammatic aphasics. *Neuropsychologia, 10*(4), 405–417.

Zurif, E. B., Green, E., Caramazza, A., & Goodenough, C. (1976). Grammatical intuitions of aphasic patients: sensitivity to functions. *Cortex, 12*(2), 183–186.

Zurif, E. B., & Sait, P. E. (1970). The role of syntax in dichotic listening. *Neuropsychologia, 8*(2), 239–244.

Index

Publisher contact:
The MIT Press
Massachusetts Institute of Technology
77 Massachusetts Avenue, Cambridge, MA 02139
mitpress.mit.edu

EU Authorised Representative:
Easy Access System Europe, Mustamäe tee 50,
10621 Tallinn, Estonia
gpsr.requests@easproject.com

Printed by Integrated Books International,
United States of America